Lecture Notes Electrical Engineering

Volume 20

S.C. Mukhopadhyay · G.S. Gupta

Smart Sensors and Sensing Technology

Springer

Subhas Chandra Mukhopadhyay
School of Engineering and
Advanced Technology (SEAT)
Massey University (Turitea Campus)
Palmerston North
New Zealand
S.C.Mukhopadhyay@massey.ac.nz

Gourab Sen Gupta
School of Engineering
and Advanced Technology (SEAT)
Massey University (Turitea Campus)
Palmerston North
New Zealand
G.SenGupta@massey.ac.nz

ISBN: 978-3-642-09852-9 e-ISBN: 978-3-540-79590-2

Cover design: eStudio Calamar S.L.

Printed on acid-free paper

9 8 7 6 5 4 3 2 1

springer.com

Contents

List of Contributors

Derek Abbott
Centre for Biomedical Engineering (CBME), School of Electrical and Electronic Engineering, University of Adelaide, Australia,
e-mail: dabbott@eleceng.adelaide.edu.au

Moawia O. Ahmed
School of Materials Science and Engineering, Nanyang Technological University
50 Nanyang Avenue, Singapore 639798

Said F. Al-Sarawi
Centre for High Performance Integrated Technologies and Systems (CHiPTec), School of Electrical and Electronic Engineering, University of Adelaide, Australia,
e-mail: alsarawi@eleceng.adelaide.edu.au

C. Ambrosino
Engineering Department, University of Sannio, C.so Garibaldi 107 82100 Benevento, Italy, e-mail: carmen.ambrosino@unisannio.it

M. Andersson
Department of Physics, Chemistry and Biology, Linköping University, Sweden

Kunihiro Asada
Department of Electronic Engineering, VLSI Design and Education Center (VDEC), The University of Tokyo, e-mail: asada@silicon.t.u-tokyo.ac.jp

Syed Mahfuzul Aziz
School of Electrical and Information Engineering, University of South Australia, Mawson Lakes, SA 5095, Australia, e-mail: mahfuz.aziz@unisa.edu.au

Takehito Azuma
Department of Electrical and Electronics Engineering, Utsunomiya University, 7-1-2 Yoto, Utsunomiya 321-8585, Japan, e-mail: tazuma@cc.utsunomiya-u.ac.jp

J. Belmonte
Department of Materials, Imperial College London, UK

Francis Bony
Université de Toulouse, LOSE, INPT, ENSEEIHT, Toulouse, France

Thierry Bosch
Université de Toulouse, LOSE, INPT, ENSEEIHT, Toulouse, France,
e-mail: thierry.bosch@enseeiht.fr

G. Breglio
University "Federico II" Department of Electronic and Telecommunication Engineering via Claudio 21, Naples, Italy

Graham M. Brooker
Australian Centre for Field Robotics, University of Sydney, Sydney, Australia,
e-mail: gbrooker@acfr.usyd.edu.au

K. Buchholt
Department of Physics, Chemistry and Biology, Linköping University, Linköping, Sweden, e-mail: kribu@ifm.liu.se

X. Chen
Department of Pure and Applied Chemistry, Faculty of Science and Technology, Tokyo University of Science, Yamazaki 2641, Noda, Chiba,
e-mail: xqchen39@hotmail.com

Moi-Tin Chew
School of Engineering and Technology, Massey University, New Zealand,
e-mail: M.Chew@massey.ac.nz

N. Cioffi
Dipartimento di Chimica Università degli Studi di Bari, Bari, Italy

L. Colaianni
Dipartimento di Chimica Università degli Studi di Bari, Bari, Italy

A. Cusano
Engineering Department, University of Sannio, C.so Garibaldi 107 82100 Benevento, Italy, e-mail: carmen.ambrosino@unisannio.it

A. Cutolo
Engineering Department, University of Sannio, C.so Garibaldi 107 82100 Benevento, Italy, e-mail: carmen.ambrosino@unisannio.it

A. Dawson
School of Chemical & Physical Sciences, Victoria University of Wellington, P.O. Box 600, Wellington, New Zealand, e-mail: dawsonand@hotmail.com

Serge Demidenko
School of Engineering and Technology, Massey University, New Zealand; School of Engineering, Monash University, Australia-Malaysia

R. Dev Das
IC Design and Fabrication Center, Department of Electronics and Telecommunication Engineering, Jadavpur University, Kolkata-700032, e-mail: sahahiranmay@yahoo.com

G. Diodati
C.I.R.A. (Italian Aerospace Research Center), via Maiorise 81043 Capua CE, Italy

Don W. Dissanayake
Centre for High Performance Integrated Technologies and Systems (CHiPTec), School of Electrical and Electronic Engineering, University of Adelaide, Australia, e-mail: don@eleceng.adelaide.edu.au

Winncy Y. Du
San Jose State University, San Jose, USA, e-mail: winncy.du@sjsu.edu

S.K. Dutta
Department of Physics, City College, e-mail: swap9885@dataone.in

Anton Fuchs
Institute of Electrical Measurement and Measurement Signal Processing, Graz University of Technology, Graz, Austria, e-mail: anton.fuchs@tugraz.at

Anuroop Gaddam
School of Engineering and Advanced Technology, Massey University, Palmerston North, New Zealand, e-mail: anuroop24@gmail.com

M. Giordano
National Research Council, Institute for Composite and Biomedical Materials, Naples, Italy

I. Gracia
National Centre for Microelectronics (CNM), Bellaterra, Spain

Milos Grcic
School of Electrical and Information Engineering, University of South Australia, Mawson Lakes, SA 5095, Australia

G. Gouws
School of Chemical & Physical Sciences, Victoria University of Wellington, P.O. Box 600, Wellington, New Zealand, e-mail: Gideon.Gouws@vuw.ac.nz

Gourab Sen Gupta
School of Engineering and Advanced Technology, Massey University, Palmerston North, New Zealand

P. Harris
Industrial Research Ltd., P.O. Box 31-310, Lower Hutt, New Zealand, e-mail: P.Harris@irl.cri.nz

Michael Healy
Optical Fibre Sensors Research Centre, Department of Electronic and Computer Engineering, University of Limerick, Limerick, Ireland, e-mail: michael.healy@ul.ie

Steven Hinckley
Physics Research Group, Edith Cowan University, Joondalup, Western Australia

Gert Holler
Institute of Electrical Measurement and Measurement Signal Processing, Graz University of Technology, Graz, Austria, e-mail: gert.holler@tugraz.at

E. Ieva
Dipartimento di Chimica Università degli Studi di Bari, Bari, Italy

Makoto Ikeda
Department of Electronic Engineering, VLSI Design and Education Center (VDEC), The University of Tokyo, e-mail: ikeda@silicon.t.u-tokyo.ac.jp

Aimé Lay-Ekuakille
D.I.I.– University of Salento (Italy), e-mail: aime.lay.ekuakille@unile.it

Muhammad Saqib Ikram
Department of Electrical and Computer Systems Engineering, Monash University, Bldg. 72, P.O. Box 35, Clayton, VIC Australia 3800

David G. Johnson
Australian Centre for Field Robotics, University of Sydney, Sydney, Australia, e-mail: d.johnson@acfr.usyd.edu.au

P.O. Käll
Department of Physics, Chemistry and Biology, Linköping University, Linköping, Sweden

J. Kanungo
IC Design and Fabrication Center, Department of Electronics and Telecommunication Engineering, Jadavpur University, Kolkata-700032, e-mail: sahahiranmay@yahoo.com

Nemai Chandra Karmakar
Department of Electrical and Computer Systems Engineering, Monash University, Bldg. 72, P.O. Box 35, Clayton, VIC Australia 3800, e-mail: nemai.karmakar@eng.monash.edu.au

A. Laudati
Optosmart s.r.l., via Pontano 61, 80121 Napoli, Italy, e-mail: info@optosmart.com

Elfed Lewis
Optical Fibre Sensors Research Centre, Department of Electronic and Computer Engineering, University of Limerick, Limerick, Ireland, e-mail: elfed.lewis@ul.ie

A. Lloyd Spetz
Department of Physics, Chemistry and Biology, Linköping University, Linköping, Sweden

Craig Lobsey
Australian Centre for Field Robotics, University of Sydney, Sydney, Australia

S. Majhi
Indian Association of Cultivation of Science, Kolkata-700032

Kate McWilliams
University of Cape Town, South Africa

Tom Moir
School of Engineering and Advanced Technology, Massey University at Auckland, Auckland, New Zealand

Shabbir Moochhala
DSO National Laboratories, 27 Medical Drive, Singapore 117510

Gareth Monkman
University of Applied Sciences Regensburg, Germany,
e-mail: gareth.monkman@e-technik.fh-regensburg.de

S. Motojima
Gifu University, Gifu 501-1193, Japan, e-mail: motojima@apchem.gifu-u.ac.jp

Subhas Chandra Mukhopadhyay
School of Engineering and Advanced Technology, Massey University, Palmerston North, New Zealand

Thomas Newe
Optical Fibre Sensors Research Centre, Department of Electronic and Computer Engineering, University of Limerick, Limerick, Ireland, e-mail: thomas.newe@ul.ie

Yoke-Fei Ng
School of Engineering, Monash University, Australia-Malaysia

Al. Palaniappan
Department of Mechanical Engineering, National University of Singapore, Singapore, e-mail: alps@nus.edu.sg

R.E. Pearce
Department of Chemistry, Imperial College London, UK; Department of Physics, Chemistry and Biology, Linköping University, Sweden,
e-mail: ruth@ifm.liu.se

Somrak Petchartee
Ferderal Arms Force University Munich, Germany,
e-mail: somrak.petchartee@gmail.com

Nicky C.L. Phua
Department of Mechanical Engineering, National University of Singapore, Singapore

Ziming Qi
Applied Technology Institute, Unitec New Zealand, Auckland, New Zealand,
e-mail: tqi@unitec.ac.nz

Sushim Mukul Roy
Department of Electrical and Computer Systems Engineering, Monash University, Bldg. 72, P.O. Box 35, Clayton, VIC Australia 3800

H. Saha
IC Design and Fabrication Center, Department of Electronics and Telecommunication Engineering, Jadavpur University, Kolkata-700032,
e-mail: sahahiranmay@yahoo.com

Shinsaku Saijo
Department of Electrical and Electronics Engineering, Utsunomiya University, 7-1-2 Yoto, Utsunomiya 321-8585, Japan

N. Sawada
Gifu University, Gifu 501-1193, Japan, e-mail: motojima@apchem.gifu-u.ac.jp

M.S.P. Shaffer
Department of Chemistry, Imperial College London, UK

J. Shaw
Tyco Safety Products, UK

F. Söderlind
Department of Physics, Chemistry and Biology, Linköping University, Linköping, Sweden

M. M. Stevens
Department of Materials, Institute of Biomedical Engineering, Imperial College London, UK

Iwao Sugimoto
School of Bionics, Tokyo University of Technology, Tokyo, Japan,
e-mail: sugimoto@cc.teu.ac.jp

Anan Suebsomran
King Mongkut's University of Technology, North Bangkok, Thailand,
e-mail: asr@kmutnb.ac.th

C. RoyChaudhuri
Department of Electronics and Telecommunication Engineering, Bengal Engineering and Science University Shibpur, Howrah-711103,
e-mail: chirosreepram@yahoo.com

S. RoyChaudhuri
Consultant Physician, Anandalok Hospitals, Kolkata,
e-mail: soumyocardium@yahoo.com

Kazuhiko Takahashi
Department of Information Systems Design, Doshisha University, Kyoto, Japan,
e-mail: katakaha@mail.doshisha.ac.jp

Francis E.H. Tay
Department of Mechanical Engineering, National University of Singapore, Singapore

Amerigo Trotta
D.E.E.-Polytechnic of Bari, Bari, Italy

L. Torsi
Dipartimento di Chimica Università degli Studi di Bari, Bari, Italy

Tharshan Vaithianathan
Mawson Institute, University of South Australia, Mawson Lakes, SA 5095, Australia

Giuseppe Vendramin
D.I.I.- University of Salento, Italy

Graham Wild
Physics Research Group, Edith Cowan University, Joondalup, Western Australia, e-mail: g.wild@ecu.edu.au

Yusuke Yachide
Department of Electronic Engineering, The University of Tokyo, e-mail: yachide@silicon.t.u-tokyo.ac.jp

S. Yang
Gifu University, Gifu 501-1193, Japan, e-mail: smyang@apchem.gifu-u.ac.jp

Scott W. Yelich
San Jose State University, San Jose, USA, e-mail: swyelich@sbcglobal.net

Usman Zabit
Université de Toulouse, LOSE, INPT, ENSEEIHT, Toulouse, France

Hubert Zangl
Institute of Electrical Measurement and Measurement Signal Processing, Graz University of Technology, Graz, Austria, e-mail: hubert.zangl@tugraz.at

Guest Editorial

This special issue titled "Smart Sensors and Sensing Technology" in the book series of "Lecture Notes in Electrical Engineering" contains the extended version of the papers selected from those presented at the 2nd International Conference on Sensing Technology (ICST 2007) which was held on November 26–28, 2007 at Massey University, Palmerston North, New Zealand. A total of 116 papers were presented at ICST 2007, of which 25 papers have been selected for this special issue.

The first three papers are in the category of Sensors and Sensor Network. In the first paper M. Healy et al. have compared the performance of securing data on a sensor network by using hardware platform against using software implementations of some of the most popular cipher algorithms suitable for WSN. In their paper T. Azuma et al., have reported the development of an autonomous vehicle system and demonstrated the use of congestion controllers for networks based on state predictive control. In the networked autonomous vehicle, images from a camera are sent to a computer via computer networks and are processed to recognize circumstances around the vehicle. A cost-effective power line communication system, based on the use of the PowerBus technology, has been reported by Moi-Tin Chew et al. The system employs a dedicated control networking originally intended by the manufacturer for mere electrical control purposes. In addition to the controlling functions, the system provides textual and data transmission capabilities.

In the next category of Gas/Bio sensors we have five papers. Kazuhiko Takahashi and Iwao Sugimoto have proposed a smart gas sensing system to achieve emotion recognition using breath gas information. A breath gas sensing system has been designed by using a quartz crystal resonator with a plasma-polymer film as a sensor. For emotion recognition, machine learning-based approaches, such as neural network (NN) and support vector machine (SVM), have been investigated. The utilisation of electrochemically synthesised gold and palladium nanoparticles as catalytically active gate material on gas sensitive field effect sensor devices have been reported by K. Buchholt et al. The synthesised nanoparticles have been characterised by TEM and XPS, and the morphology of the thermally treated nano-structured sensing layers has been investigated using SEM and XPS. Al. Palaniappan et al. have tested Cobalt Phthalocyanine (CoPc) functionalized silica matrix as receptors for gaseous bio-markers detection. Bio-marker such as Nitric Oxide (NO) is identified and quantified by tetra amino phthalocyanine (Co-TAPc) functionalized silica matrix. CoPc is anchored chemically on to a meso-porous silica network which is deposited on a Quartz Crystal Microbalance (QCM). Time of Flight Secondary Ion Mass Spectroscopy (Tof-SIMS) is used to characterize the CoPc/silica hybrids. Multi walled carbon nanotubes (MWCNTs) are known to respond well to a range of gases and vapours. Metallic additives are often introduced to improve the sensitivity and selectivity to some gases. The difference in response between MWCNT sensors, with and without a metal additive, has been discussed for a range of gases by R.E. Pearce and her group. C. RoyChaudhuri et al., have reported the electrical sensing of biochemical solutions using macroporous silicon formed by anodic etching of silicon with DMF and HF solutions. The fabricated porous silicon layers are around 30 μm thick with pore diameter around 1 μm and have been electrically characterised over a wide frequency range.

In the next category we have three papers on SAW sensors. G. Brooker and his group have outlined the design and implementation of a combined IR and acoustic beacon tracker and its evaluation as the guidance sensor on a radio controlled car. The implementation is unique insofar as it combines the broad acquisition angle of IR sensors (nearly 180°) with the precision angle measurement capability of an acoustic monopulse tracker. Additionally it uses the time difference of arrival of the IR and acoustic signals to measure the range to the beacon. A control loop was designed using the sensor to allow the car to follow the beacon at a fixed distance. A novel integration of low power operated Surface Acoustic Wave (SAW) devices with the electrostatic micro-actuators for micro-fluidic applications has been presented by Don W. Dissanayake et al. Passive, low power and small area devices can be interrogated wirelessly using SAW devices, which can respond to a uniquely coded signal for a secure and reliable operation. The approach relies on converting the interrogating coded signal to surface acoustic wave that is then correlated with an embedded code. Phase-Locked Loops (PLLs) are critical components of most modern communication and radar systems, commonly using Direct Digital Synthesis to form the reference signal. David G Johnson and Graham M Brooker have employed an alternative linearization method using a Delay Line Discriminator within a PLL. Experimental results show better than 10 cm range resolution between a pair of closely-located targets over a linearized bandwidth of ~4 GHz.

The next two papers are on fibre bragg grating sensors. C. Ambrosino and his group have presented an active vibration control hybrid system using Fiber Bragg Grating (FBG) sensors and piezoelectric (PZT) actuators for vibration suppression. An aluminium test-structure has been provided with FBG sensors bonded below the same number of PZT actuators in a co-located configuration. A novel bonding procedure was used to realize the co-located sensors/actuators cells combining epoxy based glue filled with glass micro-balloons matching the size of optical fibres. Closed loop operation was implemented by using a Proportional-Derivative controller and power amplifiers to drive the actuators based on the FBGs response. In the next paper G. Wild and S. Hinckley have presented a Transmit Reflect Detection System (TRDS) for a FBG acoustic emissions and transmissions sensor.

The TRDS utilises both the transmitted and reflected signals from the FBG. A narrow line width laser source is tuned to one of the 3dB points of the FBG. The transmitted and reflected signals are detected via two photoreceivers. The difference between the transmitted and reflected signals is then taken, doubling the total received signal. The TRDS improves the sensitivity and efficiency of the FBG acoustic emission and transmission sensor. The TRDS and FBG sensor was used to successfully receive actively generated ultrasonic signals, including acoustic communications signals, and to detect acoustic emissions.

In the next paper S.M. Aziz and his group have presented the design and implementation of a tracking system using multiple magnetic sensors for visualizing the location of a magnetically marked diagnostic capsule in real time. The intended application is the gastrointestinal (GI) tract which is regarded by gastroenterologists as a black box. An endoscopic capsule capable of releasing diagnostic biomarker probes to areas of the gastrointestinal tract, that are inaccessible via conventional means, will enable gastroenterologists to accurately determine the changes in functionality of the GI tract due to diseases. This requires a tracking system that can show the location of the capsule in real-time as it travels down the digestive tract.

A. Dawson and his group have investigated the influence of high frequency (30–70 MHz) ultrasonic wave propagation through microstructure and presented an ultrasound system, wave propagation simulation results and experimental results for highly structured porous aluminium samples.

In the next paper A. Fuchs et al. have presented the applications of capacitive techniques for moisture sensing of bulk solids. Based on experimental measurement data for three test materials, theoretical considerations on moisture models in bulk solids are undertaken and the reliability of these models is discussed. The sensitivity of the capacitive setup on the moisture distribution within the particles is analyzed by means of a multilayer model and Finite Element Analysis methods.

X. Chen et al. have produced biomimetic proximity/tactile sensors by mimicking Meissner's corpuscles, carbon microcoils (CMCs) embedded into an elastic resin. The CMC sensors were found to have high elasticity, high sensitivity, high discrimination ability, and a high performance, as well as being easily made in a micron size. These sensors have potential applications in robotic surgery, medical treatment, and diagnosis, etc.

In the next paper Winncy Y. Du and Scott W. Yelich have reviewed the conventional and emerging technologies used in pipeline leak detection. These technologies are discussed in terms of the phenomena employed, principles, system applicability, and positive and negative virtues of each when applied to various pipeline defect detection.

Y. Yachide et al. have shown a camera and projector calibration scheme based on the light section method using spheres for multiple rangefinders. The proposed calibration method fits the scanning data, based on the light-section method, to the shapes of each sphere. The procedure of the proposed calibration consists of two steps in camera and projector calibration.

In the next paper Z. Qi and T. Moir have presented a real-time Wiener filter for noise reduction, improving the signal to noise ratio by significant amounts in a car environment. Somrak Petchartee et al. in the next paper have proposed a surface recognition algorithm capable of determining contact surfaces types by means of tactile sensor fusion. It is a recognition process for 3-dimensional deformations in a 2-dimensional parametric domain. Tactile information is extracted by physical contact with a grasped object through a sensing medium. Information is obtained directly at the interface between the object and the sensing device and relates to three-dimensional position and orientation of the object in the presence of noise. The technique called "eigenvalue trajectory analysis" is introduced and adopted for specifying the margin of classification and classification thresholds.

In the next paper N.C. Karmakar and his group have reported the state of the art of the development of a smart antenna for RFID readers. The smart antenna enhances the performance of automatic identification systems, asset tracking in real time and inventory control in warehouses.

A. Lay-Ekuakille et al. have focussed on a proposal of a vehicle acoustic imaging that allows the driver to recognize obstacles and other cars around him in a safe way. The proposed front end electronic and sensing systems are included in a specific architecture supervised by a microcontroller. The algorithm used for gathering data belongs to LCMV and LCMP beam forming methods.

A novel idea of home monitoring, especially of elder people, using wireless sensors and network has been reported by A. Gaddam et al. Smart sensors are employed to monitor the use of electrical appliances at defined time intervals and based on their use warning signals can be generated and sent to the caregiver.

In the last paper Usman Zabit and his group have reported the development of a displacement sensor with a resolution of 32 nm using a laser diode self-mixing sensor. In this paper, they have focussed on the signal processing required to reach such a resolution and notably on the Nelder-Mead simplex method used for the joint estimation of parameters like the linewidth enhancement factor and feedback level described by the coupling factor.

We hope that the readers will find this special issue interesting and useful in their research as well as in practical engineering work in the area of smart sensors and sensing technology. We are pleased to be able to offer the readers such a diverse special issue, both in terms of its topical coverage and geographic representation.

Finally, we would like to whole-heartedly thank all the authors for their contribution to this special issue.

Subhas Chandra Mukhopadhyay, Guest Editor
School of Engineering and Advanced Technology (SEAT),
Massey University (Turitea Campus)
Palmerston North, New Zealand
S.C.Mukhopadhyay@massey.ac.nz

Gourab Sen Gupta, Guest Editor
School of Engineering and Advanced Technology (SEAT),
Massey University (Turitea Campus)
Palmerston North, New Zealand
G.SenGupta@massey.ac.nz

About the Editors

Dr. Subhas Chandra Mukhopadhyay graduated from the Department of Electrical Engineering, Jadavpur University, Calcutta, India in 1987 with a Gold medal and received the Master of Electrical Engineering degree from Indian Institute of Science, Bangalore, India in 1989. He obtained the PhD (Eng.) degree from Jadavpur University, India in 1994 and Doctor of Engineering degree from Kanazawa University, Japan in 2000.

During 1989–90 he worked almost 2 years in the research and development department of Crompton Greaves Ltd., India. In 1990 he joined as a Lecturer in the Electrical Engineering department, Jadavpur University, India and was promoted to Senior Lecturer of the same department in 1995.

Obtaining Monbusho fellowship he went to Japan in 1995. He worked with Kanazawa University, Japan as Researcher and Assistant Professor until September 2000.

In September 2000 he joined as Senior Lecturer in the Institute of Information Sciences and Technology, Massey University, New Zealand where he is working currently as an Associate professor. His fields of interest include Sensors and Sensing Technology, Electromagnetics, control, electrical machines and numerical field calculation etc.

He has authored 185 papers in different international journals and conferences, co-authored a book and written a book chapter and edited eight conference proceedings. He has also edited two special issues of international journals (IEEE Sensors Journal and IJISTA) as guest editor and a book with Springer-Verlag.

He is a Fellow of IET (UK), a senior member of IEEE (USA), an associate editor of IEEE Sensors journal and IEEE Transactions on Instrumentation and Measurements. He is in the editorial board of e-Journal on Non-Destructive Testing, Sensors and Transducers, Transactions on Systems, Signals and Devices (TSSD), Journal on the Patents on Electrical Engineering, Journal of Sensors. He is in the technical programme committee of IEEE Sensors conference, IEEE IMTC conference and IEEE DELTA conference. He was the Technical Programme Chair of ICARA 2004 and ICARA 2006. He was the General chair of ICST 2005, ICST 2007. He is organizing the IEEE Sensors conference 2008 at Leece, Italy as General Co-chair and IEEE Sensors conference 2009 at Christchurch, New Zealand as General Chair.

Mr. Gourab Sen Gupta graduated with a Bachelor of Engineering (Electronics) from the University of Indore, India in 1982. In 1984 he received his Master of Electronics Engineering from the University of Eindhoven, The Netherlands. In 1984 he joined Philips India and worked as an Automation Engineer in the Consumer Electronics division until 1989. Thereafter he worked as a Senior Lecturer in the School of Electrical and Electronic Engineering at Singapore Polytechnic, Singapore. He has been with Massey University, Palmerston North, New Zealand, since September 2002 as a Senior Lecturer. He is currently carrying out studies for his PhD degree.

He has published over 70 papers in international journals and conference proceedings, co-authored two books on programming, edited four conference proceedings, and edited two special issues of international journals (IEEE Sensors Journal and IJISTA) as guest editor. He is a senior member of IEEE. His area of interest is robotics, autonomous systems, embedded controllers and vision processing for real-time applications. He was the General Chair of ICARA 2004 and ICARA 2006 conferences. He was the Technical Programme chair of ICST 2005 and ICST 2007 conferences.

Part I
Sensors and Sensors Network

Analysis of Hardware Encryption Versus Software Encryption on Wireless Sensor Network Motes

Michael Healy, Thomas Newe and Elfed Lewis

Abstract Due to the sensitive and often personal nature of sensor data that many wireless sensor networks collect, the security of this data must be guaranteed. This is fast becoming an important concern for sensor networks which are finding applications in the military and home health domains. The best and often the only way to secure this data is to encrypt it using a secure encryption algorithm before it is transmitted over the air ways. Due to the constrained nature of the resources, memory and clock speeds, available on sensor nodes however, the cost, both in terms of power consumption and speed of encryption, of a software based encryption procedure can often outweigh the risks of the transmission being intercepted. This paper presents a solution to reduce this cost of employing encryption by taking advantage of a resource already available on many sensor nodes, including the Crossbow MICAz and MoteIV's TmoteSKY; this resource being the AES encryption module available on the Chipcon CC2420 transceiver chip. The performance of using this method of securing data on a sensor network against using software implementations of some of the most popular cipher algorithms suitable for WSN is then analysed for both hardware platforms.

Keywords Sensor data security · wireless sensor networks · AES · CC2420

Michael Healy
Optical Fibre Sensors Research Centre, Department of Electronic and Computer Engineering, University of Limerick, Limerick, Ireland, e-mail: michael.healy@ul.ie

Thomas Newe
Optical Fibre Sensors Research Centre, Department of Electronic and Computer Engineering, University of Limerick, Limerick, Ireland, e-mail: thomas.newe@ul.ie

Elfed Lewis
Optical Fibre Sensors Research Centre, Department of Electronic and Computer Engineering, University of Limerick, Limerick, Ireland, e-mail: elfed.lewis@ul.ie

S.C. Mukhopadhyay, G.S. Gupta (eds.), *Smart Sensors and Sensing Technology*,

1 Introduction

Technological advancements in recent years have enabled the development of tiny, cheap disposable and self contained battery powered computers, known as sensor nodes or “motes”, which can accept input from an attached sensor, process this input and transmit the results wirelessly to some interested device(s). When a number of these nodes work together, conceivably up to hundreds of thousands, a Wireless Sensor Network (WSN) is formed.

Research in the area of wireless sensor networks has become increasingly widespread in recent years, partly due to their wide range of potential uses and also partly due to the fact that the technology enabling such networks is now widely available from many different suppliers, such as: Crossbow, MoteIV, Intel and SUN (java based motes).

These wireless sensor networks have the potential to allow a level of integration between computers and the physical world that, to date, has been virtually impossible. The uses for such networks is almost limitless and include such diverse applications as a counter sniper system for urban warfare [1] tracking the path of a forest fire [2], determining the structural stability of a building after an earthquake [3], or tracking people or objects inside a building [4], etc.

Advances in wireless communication have been a major factor in allowing the development of large networks of sensors. However, as stated in the IEEE 802.15.4 standard specification [5], the wireless connectivity of the sensors is not so much a feature of the sensors but rather an application enabler (unlike the majority of wireless applications currently available). This is the case because wired sensor networks on the required scale would be very costly to build and maintain and also very costly to install, making them impractical for many application areas, such as: environmental monitoring, home health and military to list but a few.

Despite making such sensor networks possible, the wireless nature of the sensors presents a number of problems to the developer. Chief among these is the problem of security. Many WSNs are designed to collect data which is sensitive in some way, such as information about a person’s health, confidential data about a company’s manufacturing process, troop locations on a battlefield, etc. Without adequate security this data can easily be read and/or modified by an attacker. Due to the wireless nature of the sensor networks detecting and preventing eavesdropping and modification of the data is greatly complicated. Also the constrained nature of resources on the wireless sensor nodes means that security architectures used for traditional wireless networks are not viable.

As power and RAM are normally the most constrained resources on a wireless node, a memory and power efficient cryptographic algorithm is a very important part of any proposed security architecture for a WSN. Traditionally RC5 [6] and Skipjack [7] are considered to be the most suitable algorithms for WSNs [8, 9] due to their speed and memory usage but as shown by Law et al. [10] AES [11] or MISTY1 [12] can be more suitable, depending on the situation.

A solution to reduce power and memory costs, as well as greatly increasing speed of encryption, is to use a hardware implementation of a cipher. As AES is currently

Table 1 Features of currently available Wireless Sensor Nodes [20]

Platform	MCU	RAM (KB)	Program memory (KB)	Non-volatile data mem (KB)	Radio chip
BTnode3	ATMega 128	64	128	180	CC1000/ ZV4002
Cricket	ATMega 128	4	128	512	CC1000
imote2	Intel PXA271	256	32,000	0	CC2420
mica2	ATMega 128	4	128	512	CC1000
mica2Dot	ATMega 128	4	128	512	CC1000
micaz	ATMega 128	4	128	512	CC2420
Shimmer	TI MSP430	10	48	Up to 2 GB	CC2420/ WML-C46
TelosA	TI MSP430	2	60	512	CC2420
TelosB	TI MSP430	10	48	1	CC2420
Tmote Sky	TI MSP430	10	48	1	CC2420
XYZ	ARM 7	32	256	256	CC2420

the most widely used and secure block cipher available it has a larger range of commercial implementations than for any other algorithm. However none of the currently available wireless sensor nodes has such a chip already included and adding one can prove difficult and very time consuming, especially for large deployments. Another solution is available however. As can be seen by Table 1 the majority of the currently available sensor nodes, both those commercially available and those created by research institutions, use the same IEEE 802.15.4 compliant transceiver, the Chipcon CC2420 [13]. The CC2420 features a number of security operations, including AES encryption [14]. We test the performance of using this hardware encryption on the CC2420 compared with a software implementation of the AES algorithm and the TinySec [8] implementations of RC5 and Skipjack. Two platforms are used for this testing; Crossbow's MICAz and MoteIV's TmoteSKY.

2 Background

In this section the target hardware and the operating system that was used is described. An explanation of why AES is considered an appropriate choice for use in wireless sensor network data protection is also provided as well as a description of the block ciphers traditionally used in wireless sensor networks.

2.1 Sensor Node Hardware

The sensor nodes that work together to form a WSN are composed of four subsystems; a computing sub-system, a communication sub-system, a power subsystem and a sensing sub-system.

The computing sub-system consists of a processor and memory, which includes program memory, RAM and possibly non-volatile data memory. An important aspect of processors in sensor nodes is different operational modes, usually Active, Idle and Sleep. This is important so as to preserve power as much as possible without impeding the operation of the processor when it is required.

The communication sub-system is required to enable the sensor nodes to communicate with each other and with a base station. Generally the communication sub-system is a short range radio but the use of infrared communication, ultrasound and inductive fields has also been explored. Most currently available sensor nodes use a radio chip which conforms to the IEEE 802.15.4 standard, but as can be seen from Table I some nodes use Bluetooth as an alternative.

The power sub-system consists of a battery which supplies power to the sensor node. Due to the long term unattended operation of nodes in a WSN the developer must ensure every aspect of the network such as communication algorithms, localization algorithms, sensing devices, etc., must be as efficient as possible in their power usage. A power generator may also be included to recharge the battery onsite. Photovoltaic, motion/vibration and thermoelectric energy conversion are all possible sources of power, depending on the location of the node [15].

Sensor transducers translate physical phenomena to electrical signals. Therefore the sensing sub-system of the node is its link to the outside world. The output of the sensors may be digital or analog signals. If the output is analog the node must also include an Analog to Digital Converter (ADC) in order to allow the processor to read the data. Some nodes have sensors built in, but many do not, instead providing suitable ports allowing a variety of sensors to be attached for more versatility.

As already mentioned the CC2420 transceiver chip from Chipcon is currently the most popular radio chip on wireless sensor nodes. The CC2420 works in the 2.4Ghz band, is IEEE 802.15.4 compliant, ZigBee ready, low cost, and designed for low-voltage and low-power wireless applications. These features, along with a host of other features make the CC2420 ideal for use on sensor nodes.

The CC2420 feature that is most interesting is its ability to perform hardware security operations. The CC2420 is capable of performing IEEE 802.15.4 MAC security operations, including counter (CTR) mode encryption and decryption, CBC-MAC authentication and CCM encryption plus authentication. The CC2420 also offers plain stand-alone encryption of 128 bit blocks. Each of these security functions are based on AES encryption using 128 bit keys.

One limitation of the CC2420 security wise is that it does not offer AES decryption. However the impact of this limitation is negated by the fact that for many applications any decryptions are performed by the base station which can be a much more powerful device and/or has less limitation on its power supply. Another option is to use a block cipher mode of operation which does not require a decryption function, such as Cipher Feed Back (CFB) mode, Output Feed Back (OFB) mode, or Counter (CTR) mode.

2.2 TinyOS

As can be seen from Table 1 [20] the specifications of the target hardware for the tested system are varied. The variety of processors, radio chips as well as the varying amount of RAM and program memory means the software written for each node would need to be significantly different. However a number of operating systems already exist to solve this problem. The TinyOS [16] operating system was chosen as the development platform, mainly because it currently supports the largest range of hardware and has good power management [17].

TinyOS was the first operating system specifically designed for WSNs and as a result it still has the largest user base and is the standard by which the other operating systems are judged. TinyOS was initially developed at the University of California, Berkeley but is now being developed by a consortium and is open source, making it easy for developers to customize it as required.

TinyOS is not an operating system in the traditional sense; rather it is a programming framework which contains a set of components that allows an application specific OS to be constructed for each particular need, consisting of selected system components and custom components. The principal goal of this component based architecture is to allow application designers to build components that can easily work together to produce a complete, concurrent system but still performs extensive compile time checks.

2.3 TinySec

TinySec [8] is a link layer security architecture for wireless sensor networks, implemented for the TinyOS operating system. In order to overcome the processor, memory and energy constraints of sensor nodes TinySec leverages the inherent sensor network limitations, such as low bandwidth and relatively short lifetime for which the messages need to remain secure, to choose the parameters of the cryptographic primitives used. Using these parameters TinySec adds less than 10% energy, latency and bandwidth overhead.

TinySec has two modes of operation: authenticated encryption (TinySec-AE) and (TinySec-Auth). With authenticated encryption, TinySec encrypts the data payload and authenticates the packet with a Message Authentication Code (MAC). The MAC is computed over the encrypted data and the packet header. In authentication only mode, TinySec authenticates the entire packet with a MAC, but the data payload is not encypted.

An important feature of TinySec is its ease of use, as many application developers will either implement the security features incorrectly or leave out any security entirely if the security API is difficult to use. TinySec solves this problem by integrating into TinyOS at a low level. To enable TinySec-Auth mode, only one line needs to be added to the application's Makefile, the application code itself does not need to be touched. TinySec-Auth mode is the default mode of operation for

TinySec. To enable TinySec-AE mode the TinySecC configuration file needs to be wired into the application and one function call needs to be given.

All of TinySec's security primitives are built on top of the employed block cipher, and currently users can choose between RC5 of Skipjack.

2.4 Block Ciphers

The block cipher employed is probably the most important cryptographic building block to get right for securing a wireless sensor network. A balancing act between security, memory requirements and power consumption is required. Obviously the cipher needs to be secure as if it is easily broken there is no point in using it, the data might as well be sent unencrypted. However, if the cipher used takes up so much program memory, RAM or energy as to seriously impede the sensing, data processing and data transmission functions or the effective lifetime of the sensor node, the use of the cipher becomes impractical.

An important point to consider when determining the cipher to be used is that the length of time the data needs to be kept secure in a wireless sensor network is often much shorter than in a traditional wireless network. As a result, a less secure, but more memory and energy efficient cipher can often be more appropriate.

2.4.1 AES/Rijndael

Rijndael [11] was chosen as the Advanced Encryption Standard (AES) by the National Institute of Standards and Technology of the United States in 2001 to replace the existing Data Encryption Standard (DES). It is also one of the ciphers recommended for use by Japans CRYPTREC and by the New European Schemes for Signature, Integrity and Encryption (NESSIE) consortium.

Rijndael is designed to work with three different key lengths of 128 bits, 192 bits or 256 bits depending on security requirements. A key length of 128 bits is considered more than sufficient for securing wireless sensor networks because the length of time any collected data needs to remain confidential will have long since expired before an exhaustive key search on this length key can be performed using currently available technology. However for increased security, at a cost of memory and speed, one of the longer key lengths can also easily be employed.

One of the criteria used to choose AES was the cipher's efficiency and performance on a variety of platforms, from 8-bit smart cards to high end processors. For this reason Rijndael should be well suited for use on wireless sensor nodes, despite their limited processing power.

Rijndael has come under intense scrutiny both during the review process before it was chosen as AES and also since then. As of January 2008 no feasible attacks against the algorithm have been published and so it is still considered secure.

2.4.2 RC5

RC5 [6] is a block cipher designed by Ronald Rivest in 1994 and is noted for its simplicity. RC5 is also noted for its flexibility, with a variable block size of 32, 64 or 128 bits, variable key size of 0 to 2040 bits and a variable number of rounds of 0 to 255.

TinySec uses a block size of 64 bits, key size of 64 bits and 12 rounds. For traditional networks these parameters would no longer be considered secure because 12 round RC5 using 64 bit blocks is susceptible to a differential attack [19]. However for this attack 244 chosen plaintexts are required so it is highly unlikely to be an effective attack during the lifetime of any particular sensor network.

The key length of 64 bits is the biggest weakness of the TinySec implementation of RC5. Limiting the key to 64 bits means it would be possible to complete an exhaustive key search attack within the lifetime of some networks.

Despite being very fast and still sufficiently secure RC5 is not widely used due to the fact that it is a patented algorithm. Patent free alternatives with comparable properties are widely available.

2.4.3 Skipjack

Skipjack [7] was initially a classified algorithm developed by the National Security Agency in the USA. Since its declassification in 1998 Skipjack has come under intense scrutiny and is still considered secure.

Skipjack uses 64 bit blocks, an 80 bit key and has 32 rounds. While an 80 bit key is significantly more secure than the 64 bit key used in the TinySec's RC5 implementation, with current technology trends it is quite likely that this length key will be rendered insecure to an exhaustive key search attack within the next 5 years.

3 Implementation

The secure application was implemented under TinyOS which employs the CC2420 hardware encryption features in order to encrypt 16 bytes of data at a time and then send this data to a listening base station. This application was then run on two of the most popular sensor node platforms which use the CC2420 radio, Crossbow's MICAz and MoteIV's Tmote SKY.

Using the stand alone encryption of the CC2420 is relatively straight forward. First off the two Security Control registers, SECCTRL0 and SECCTRL1, which control the security operations of the CC2420, need to be set. As we are using key 0 as our stand-alone key we can set both of these registers to 0. We then write the 128 bit key and the plaintext to the appropriate RAM locations on the CC2420; from location 0×100 and 0×120 respectively. The SAES command strobe is then sent in order to begin the encryption operation. We can tell when the encryption operation is completed by reading the ENC_BUSY status bit, which tells us whether the encryption module is busy or not. We read the CC2420's status bits by sending the SNOP command strobe. The final step is to read back the cipher text. As the encryption module overwrites the plaintext with the cipher text this is simply a case of reading 16 bytes from the location the plaintext was originally written to, i.e. from location 0×120. It is worth noting that the CC2420 RAM write operation also outputs the data currently at the location being written to, so that a new plaintext can be written at the same time as reading out the previous cipher text in order to streamline encrypting data longer than 128 bits.

It is also worth noting that when using the other security operations supported by the CC2420, i.e. the IEEE 802.15.4 MAC security modes, the encryption happens after the message is buffered in the TXFIFO buffer, and the decryption happens before the message is read out of the RXFIFO buffer. This means that the encrypted message does not have to be read from the CC2420 at any stage, which would only have to be sent back to the transceiver for another operation anyway, so using these modes is even simpler and more transparent for the application developer than using the stand-alone encryption mode.

The only difference between the versions of the application on each platform is that for the MICAz version the function to read from the RAM on the CC2420 chip had to be implemented and integrated whereas for the Tmote SKY this function was already made available by TinyOS.

For comparison purposes a number of software implementations of the AES encryption algorithm were also created. The first implementation is used for reference and is as simple as possible based on the Rijndael AES proposal [11]. The second software based AES version has some simple optimisations. These optimisations are primarily using pre-computed lookup tables and, as we are using a fixed block size and key size of 128 bits each (to match the parameters of the stand-alone encryption on the CC2420), we were able to unroll some loops.

The default TinyOS operation is to load the lookup tables into RAM at boot up. As this consumes a significant amount of RAM we created a second implementation which prevents TinyOS from doing this.

As already mentioned RC5 and Skipjack are the conventional algorithms used for wireless sensor networks. In order to compare the performance of the AES encryption on the CC2420 to the performance of these algorithms we implemented a TinyOS application which encrypts 16 bytes of data using each of these algorithms, these implementations being based on the highly optimised TinySec versions. Currently TinySec is only available for the MICA2 platform and TinySec's block ciphers are optimised for use on this platform, using inline assembly code to speed up the most common operations. However, as can be seen in Table 1, the MICA2 and MICAz nodes use the same microprocessor, the ATMega 128L, so this optimised code runs unmodified on the MICAz. The Tmote SKY's processor is a Texas Instruments MSP430, and as a result our RC5 and Skipjack implementations do not run on this platform.

As already mentioned the TinySec implementations of the RC5 and Skipjack encryption algorithms are highly optimised and as a result our versions of these

Table 2 Comparison of implemented block ciphers

	Block size	Key size	No. rounds
AES	128 bits	128 bits	10
RC5	64 bits	64 bits	12
Skipjack	64 bits	80 bits	32

algorithms do not offer the same level of security as our AES applications. This is the case because, as can be seen in Table 2, a 64 bit key is used for RC5 and a 80 bit key is used for Skipjack, whereas a 128 bit key is used for AES, both the hardware and software versions.

4 Results

Table 3 shows the amount of program memory and RAM each implementation uses on each platform [18]. As can be clearly seen the AES implementation that uses the CC2420 encryption module uses almost 1 KB less ROM than the nearest competitor on both platforms and also comes out top in terms of RAM usage. The version of Skipjack used minimized RAM usage, as is the case with the TinySec implementation, where the lookup tables that are required were placed in program memory, however, as will be seen this comes at a cost of performance.

The AES software reference version generates the reference tables required at runtime in order to save program memory. This technique can also be used to save RAM. However as we saved these generated tables in static global arrays, they only needed to be generated once so we did not reap the RAM saving benefits. The time it takes to generate these tables is shown in Table 4. In order to create a software implementation of AES which optimizes both RAM and program memory usage these tables would need to be generated every time they were required.

Tables 5 and 6 show the amount of time it takes to execute various parts of the code for each implementation. Table 5 shows the amount of time it takes each

Table 3 Program memory and RAM usage (bytes)

	MICAz		Tmote SKY	
	ROM	RAM	ROM	RAM
CC2420 AES	10058	437	11872	407
AES software Ref.	11980	2844	13206	2811
AES software	12720	1915	13980	1833
AES software min RAM	12748	625	14200	887
RC5	11050	533	–	–
Skipjack	11696	451	–	–

Table 4 Time to generate lookup tables

MICAz	Tmote SKY
7.88137 ms	13.29879 ms

Table 5 Time to run up setup procedure

	MICAz	Tmote SKY
CC2420 AES	294.003 µs	2.06034 ms
AES software Ref.	652.15 µs	940.83 µs
AES software	565.37 µs	752.27 µs
AES software min RAM	574.85 µs	934.83 µs
RC5	1.68591 ms	–
Skipjack	687.674 µs	–

application to run the setup procedure before encryption can commence. For the CC2420 AES version this involves setting the setting the security registers and loading the key and plaintext into the CC2420's RAM and the time taken to read the resulting cipher text back is also included. For the other five versions the setup procedure involves performing any key expansion required. Overall the AES implementations came out top in this category with the hardware version being by far the quickest and the three software versions being significantly quicker than the RC5 or Skipjack implementations. For the five software based versions the setup procedure only needs to be run each time the encryption key changes but for the hardware version some of the tasks we have included in the time taken for the setup procedure need to be run for each 128 bit block being encrypted, i.e. writing the plaintext to the CC2420's RAM and reading back the cipher text. However as mentioned in Sect. 3 these two operations can be performed at the same time so the impact of this is not significant.

Table 6 shows the amount of time it takes to actually perform the encryption routine on 16 bytes of data. It is in this category that the CC2420 AES implementation really shows its worth, being over 15 times faster than the RC5 algorithm, over 24

Table 6 Time taken to encrypt 16 bytes

	MICAz	Tmote SKY
CC2420 AES	29.8310 µs	449.203 µs
AES software Ref.	1.49578 ms	1.94108 ms
AES software	1.45753 ms	1.87064 ms
AES software min RAM	1.54772 ms	1.94374 ms
RC5	465.685 µs	–
Skipjack	721.446 µs	–

times faster than the Skipjack version, and nearly 49 times faster than either of the AES software based implementations on the MICAz.

On the Tmote SKY the hardware based AES version is only about 4 times faster than the software based AES versions. Also the Tmote SKY is nearly 15 times slower than the MICAz when using the same hardware to perform the encryption. Also, as seen in table 5, the CC2420 AES version's setup procedure on the Tmote SKY is very slow compared to that of the MICAz's. These timing differences cannot be adequately explained by the processor and clock differences between the two platforms and so we suspect that the difference is due to TinyOS's platform specific routines that read from and write to the RAM on the CC2420.

As seen in Tables 5 and 6, the RAM optimised software implementation of AES is slower than the un-optimised versions. This increase in processing time is due to the extra overhead of reading from the lookup tables placed in program memory over that of reading the same tables in RAM. This is a cost the Skipjack implementation also pays, which, as a result, could be modified to have slightly improved execution times at a cost of more using more RAM if required.

AES was generally not considered viable for use on wireless sensor nodes because of its execution speed and RAM usage. However our RAM minimized software implementation of AES, which still has a lot of scope for more optimisation, only uses 92 bytes more RAM than an optimised RC5 implementation and is executed only about 2 times slower than an optimised Skipjack implementation. Therefore, as using the CC2420 to speed up encryption results in using at least 96 less bytes of RAM and executes at a minimum of 15 times faster than the alternatives discussed here, using this hardware aided method to secure data results in a significant boost in performance.

5 Conclusions

It has been shown that using hardware encryption instead of software based encryption is beneficial in terms of speed and memory usage for wireless sensor network motes. In particular we have shown that using a resource, the Chipcon CC2420 transceiver chip, already existing on two of today's most popular wireless sensor motes (Crossbow's MICAz and MoteIV's TmoteSKY) one can significantly reduce the cost of securing data on a sensor network with regards to power consumption and speed of encryption. Indeed it can be seen that the MICAz offers significant improvement over the TmoteSKY mote for hardware based encryption. The authors suspect that this is due to TinyOS's platform specific routines that read from and write to the RAM on the CC2420. This work is ongoing.

Acknowledgements The authors wish to thank the following for their financial support:

- SFI Research Frontiers Programme grant number 05/RFP/CMS0071
- The Embark Initiative and Intel, who fund this research through the Irish Research Council for Science, Engineering and Technology (IRCSET) postgraduate Research Scholarship Scheme.

References

1. Á. Lédeczi, A. Nádas, P. Völgyesi, G. Balogh, B. Kusy, J. Sallai, G. Pap, S. Dóra, K. Molnár, M. Maróti, and G. Simon (2005) Counter sniper System for Urban Warfare. In ACM Transactions on Sensor Networks (TOSN), vol. 1, pp. 157–177.
2. C.-L. Fok, G.-C. Roman, and C Lu (2005) Mobile Agent Middleware for Sensor Networks: An Application Case Study. In Proc. 4th International Conference on Information Processing in Sensor Networks (IPSN'05), Los Angeles, California.
3. T. Schmid, H. Dubois-Ferrière, and M. Vetterli (2005) SensorScope: Experiences with a Wireless Building Monitoring Sensor Network. In Proc. Workshop on Real-World Wireless Sensor Networks (REALWSN'05), Stockholm, Sweden.
4. R. Want, A. Hopper, V. Falcão, and J. Gibbons (1992) The Active Badge Location System. In ACM Transactions on Information Systems (TOIS), vol. 10, pp. 91–102.
5. 802.15.4: Wireless Access Control (MAC) and Physical Layer (PHY) Specifications for Low-Rate Wireless Personal Area Networks (LR-WPANs). (2003) New York: IEEE Standards Association.
6. R. Rivest (1994) The RC5 Encryption Algorithm. In Proc. 1994 Leuven Workshop on Fast Software Encryption, Leuven, Belgium.
7. NIST (1998) Skipjack and KEA Algorithm Specifications Version 2.0. NIST.
8. C. Karlof, N. Sastry, and D. Wagner (2004) TinySec: A Link Layer Security Architecture for Wireless Sensor Networks. In Proc. 2nd International Conference on Embedded Networked Sensor Systems, Baltimore, MD, USA.
9. A. Vitaletti and G. Palombizio (2006) Rijndael for Sensor Networks: Is Speed the Main Issue?. In Proc. 2nd Workshop on Cryptography for Ah Hoc Networks, Venice, Italy.
10. Y.W. Law, J. Doumen, and P. Hartel (2006) Survey and Benchmark of Block Ciphers for Wireless Sensor Networks. In ACM Transactions on Sensor Networks (TOSN), vol. 2, pp. 65–93.
11. J Daemen and V. Rijmen (1999) AES Proposal: Rigndael.
12. M. Matsui (1997) New Block Encryption Algorithm MISTY. In Proc. 4th International Workshop on Fast Software Encryption, Haifa, Israel.
13. Chipcon (2004) CC2420 Datasheet [online], available: http://www.chipcon.com/files/CC2420_Data_Sheet_1_3.pdf [accessed 9 Feb 2007].
14. M. Healy, T. Newe and E. Lewis (2007) Resource Implications for Data Security in Wireless Sensor Network nodes. In Proc. International Conference on Sensor Technologies and Applications, Valencia, Spain.
15. S. Roundy, P.K. Wright, and J.M. Rabaey (2003) Energy Scavenging for Wireless Sensor Networks: With Special Focus on Vibrations, 1st ed: Springer.
16. J. Hill, R. Szewczyk, A. Woo, S. Hollar, D.E. Culler, and K.S.J. Pister (2000) System Architecture Directions for Networked Sensors. In Proc. 9th International Conference on Architectural Support for Programming Languages and Operating Systems, Cambridge, Massachusetts, USA.
17. M. Healy, T. Newe, and E. Lewis (2007) Power Management in Operating Systems for Wireless Sensor Nodes. In Proc. IEEE Sensors Applications Symposium (SAS 2007), San Diego, California, USA.
18. K. Sun, P. Ning, and C. Wang (2006) TinySeRSync: Secure and Resilient Time Synchronization in Wireless Sensor Networks. In Proc. 13th ACM Conference on Computer and Communications Security, Alexandria, Virginia, USA.
19. A. Biryukov, E. Kushilevitz (1998) Improved Cryptanalysis of RC5. In Proc. Advances in Cryptology—EUROCRYPT '98, International Conference on the Theory and Application of Cryptographic Techniques. LNCS, Springer-Verlag, vol. 1403, pp. 85–99.
20. M. Healy, T. Newe, and E. Lewis (2007) Securing Sensor Data on a Wireless Sensor Network. In Proc. 2nd International Conference on Sensing Technology, Palmerston North, New Zealand. pp. 176–181, ISBN 978-0-473-12432-8.

Development and Experiment of Networked Control Systems with Congestion Control

Takehito Azuma and Shinsaku Saijo

Abstract The purpose of this paper is to develop an autonomous vehicle system including communication networks and to demonstrate congestion controllers for networks based on state predictive control via the developed system. In the networked autonomous vehicle, the vehicle has one camera as sensor. Images from camera are sent to a computer via computer networks and are processed to recognize circumstances around the vehicle. Using the processed camera image, the vehicle is controlled over computer networks. By considering that this system uses computer networks, congestion problems are important because congestion causes instability of the whole system. For these problems, congestion controllers are designed based on the systems & control theory. In this paper, new congestion controllers are designed by using state predictive control to assure stability of computer networks. Effectiveness of congestion controllers is verified in experiments using the developed system.

Keywords Networked control systems · congestion control · camera sensor · autonomous vehicle · state predictive control

1 Introduction

Recently communication networks are developed widely [1] and it uses to control many machines from a distant place. For example, a surgery is done between USA and Euro using surgery manipulators. This surgery is called as laparoscope-assisted surgery. In the control engineering research area, this kind of researches are called as NCSs (Networked Control Systems) and many researches have been done [3, 4].

Takehito Azuma
Department of Electrical and Electronics Engineering, Utsunomiya University, 7-1-2 Yoto, Utsunomiya 321-8585, Japan, e-mail: tazuma@cc.utsunomiya-u.ac.jp

Shinsaku Saijo
Department of Electrical and Electronics Engineering, Utsunomiya University, 7-1-2 Yoto, Utsunomiya 321-8585, Japan

S.C. Mukhopadhyay, G.S. Gupta (eds.), *Smart Sensors and Sensing Technology*,

On the other hand, one of communication networks is computer network and Internet or LAN (Local Area Network) is famous as a computer network [1, 2]. A main problem in the computer network is to transfer data smoothly in the network. If data transfer is not smooth, the network is busy and this status is called as congestion. It is possible to avoid congestion in the computer network by using systems & control theory. Some researches focus on this problem and some methods are proposed [5, 6].

In this paper, NCSs and congestion problems in the computer network are both considered, a NCSs is developed actually and a congestion controller is designed based on state predictive control. Moreover some experiments are done in the developed NCSs and the efficacy of congestion controllers is verified in the real networked system. It seems original to show the efficacy of congestion controllers in the developed NCSs.

Firstly an autonomous vehicle system including computer networks is developed In the system, the vehicle can be controlled via radio transmission and a camera is mounted on the vehicle as sensor to recognize circumstances. Visual images from camera are sent to the developed vehicle control system and the image data is processed at a receiver computer through a simple computer network. Based on the image processing, a reference trajectory is detected and the vehicle is controlled from a distant place.

Secondly considering dynamics of computer networks, congestion controllers are designed. The dynamics of computer networks is nonlinear but it is possible to obtain a linearized model with an information delay, which causes from the round trip time of data transfer. The information delay is called as time-delay in the systems & control theory. For this linear time-delay system, a congestion controller is designed to avoid occurring congestion in the computer network.

Finally applying the designed congestion controller to the developed autonomous vehicle system, efficacy of the designed congestion controller is verified by having experiments without a congestion controller and with the designed congestion controller.

2 Development of Networked Control Systems

2.1 System Configuration

A networked autonomous vehicle system has been developed and the system is shown in Fig. 1. In this figure, a wire-less vehicle is tracking the black cycle line that denotes a reference trajectory for the vehicle. The wire-less vehicle is controlled via visual feedback control. Visual feedback control is a control method by using camera as sensor. The vehicle control system consists of data sending and receiving parts via wire-less radio transmission. The vehicle control system is connected through a RS-232C cable. Sender is a computer that is connected to computer networks. Here LAN (Local Area Network) is considered as the computer network.

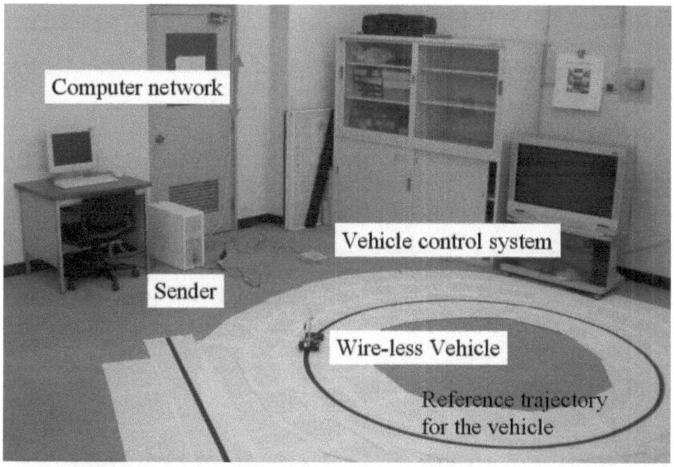

Fig. 1 The developed autonomous vehicle system

A router and a receiver are shown in Fig. 2. The LAN cable from the computer network in Fig. 1 goes to the router in Fig. 2. The receiver is connected to the router via a LAN cable. The room in Fig. 1 is different from the room shown in Fig. 2. The room in Fig. 1 is vacant during some experiments and the wire-less vehicle is controlled from another room through the computer network.

Figure 3 shows whole system configuration of our developed autonomous vehicle systems combined with computer networks. In this figure, one more sender is added and the sender is used to cause congestion in the computer network. The purpose of this paper is to verify effectiveness of the designed congestion controller that is implemented in the router. The design problem of congestion controllers is discussed in the next section. Congestion controllers stabilize queue sizes of routers in computer networks.

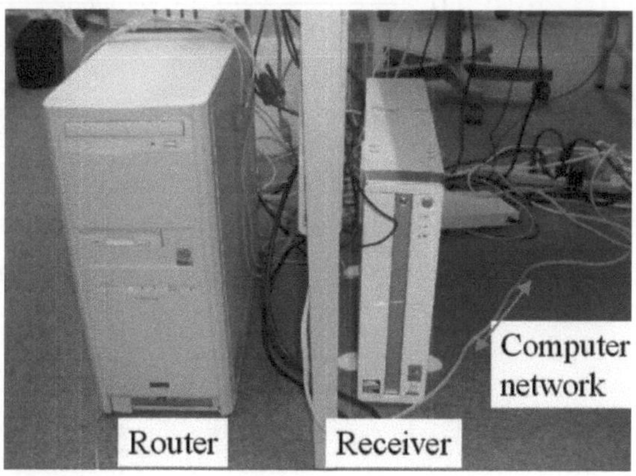

Fig. 2 System configuration for computer networks

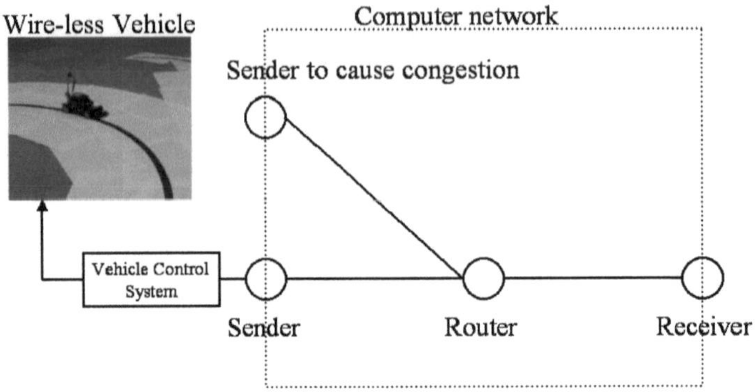

Fig. 3 Network topology and autonomous vehicle system

2.2 Dynamics of the Wire-Less Vehicle and Automatic Control

In the Fig. 4, the wire-less vehicle is shown and it can be seen that a wire-less camera is mounted on the vehicle. The dynamical model of this vehicle can be derived from a two-wheel model that can be obtained as an approximated model of a four-wheel model (the Fig. 4). The dynamics of the vehicle is given as follows,

$$\begin{aligned} \dot{x} &= Ax + B\delta, \\ y &= Cx, \end{aligned} \tag{1}$$

where

$$A = \begin{bmatrix} 0 & 0 & 1 & 0 \\ 0 & 0 & 0 & 1 \\ 0 & \dfrac{2(K_f+K_r)}{m} & -\dfrac{2(K_f+K_r)}{mV} & -\dfrac{2(l_fK_f+l_rK_r)}{mV} \\ 0 & \dfrac{2(l_fK_f-l_rK_r)}{I} & -\dfrac{2(l_fK_f-l_rK_r)}{IV} & -\dfrac{2(l_f^2K_f-l_r^2K_r)}{IV} \end{bmatrix},$$

Fig. 4 Wire-less vehicle with a wire-less camera

$$B = \begin{bmatrix} 0 \\ 0 \\ \dfrac{2K_f}{m} \\ \dfrac{2l_f K_f}{I} \end{bmatrix}, \quad C = [\,1 \quad l_s \quad 0 \quad 0\,].$$

In the state equation (1), the state variable $x(t) \in R^4$ denotes errors for vehicle center of gravity in Fig. 4 and it can be observed by using the camera image. The input vector $\delta(t) \in R$ denotes the steering angle of the vehicle and it can be applied to the vehicle as a voltage of an electrical motor. The other parameters are constant and it can be determined by measuring the vehicle.

For the vehicle dynamics (1), a LQG (Linear Quadratic Gaussian) controller can be designed. The LQG controller is general and useful as a basic method to construct automatic control systems based on systems and control theory. The LQG controller design is given as follows,

1. Solve the matrix $P > 0$ which satisfies the following matrix equation (Riccati equation).
$$PA + A'P - PBR^{-1}B'P + C'QP = 0 \tag{2}$$
where matrices R, Q are weighting matrix for LQG control.
2. Solve the matrix $M > 0$ which satisfies the following matrix equation.
$$MA' + AM - MC'CM + BGG'B' = 0 \tag{3}$$
where G denotes a system matrix for the state equation (1) in case of considering disturbances.
3. By using the above matrices $P > 0, M > 0$, the LQG controller is given as follows,
$$\begin{aligned} \dot{x}_K &= A_K x_K + B_K y, \\ \delta &= C_K x_K, \end{aligned} \tag{4}$$
where
$$\begin{aligned} A_K &= A - HC - BF, \\ B_K &= H, C_K = -F, \\ F &= R^{-1}B'P, H = CM. \end{aligned}$$

Here note that LQG control minimizes the following cost function.

$$J = \int_0^{\infty} \{x'(t)QPx(t) + u'(t)Ru(t)\}dt$$

Moreover this designed controller is implemented as a program in the receiver shown in Fig. 2.

3 Congestion Controller Design Using the State Predictive Control

3.1 Dynamics of Computer Networks

In this paper, the network topology is given in Fig. 3 as a computer network. This kind of network topology can be considered as The network consists of n nodes (Sender) , 1 node (Receiver) and 1 bottleneck router which sends packets form the Sender to the Receiver. This network topology denotes 1 server machine to multiple client machines in a computer network. Large scale networks can be simplified as this network topology in case of designing congestion controllers if only one router is bottleneck in the large-scale computer network.

From the paper [5], the following nonlinear models describe behaviors of the average window size in senders and the queue size in the bottleneck router in the network.

$$\begin{aligned} \dot{w}_s(t) &= \frac{1}{R(t)} - \frac{w_s(t)}{2}\frac{w_s(t-R_r(t))}{R(t-R(t))}p(t-R(t)), \\ \dot{q}(t) &= N(t)\frac{w_s(t)}{R(t)} - C(t), \end{aligned} \tag{5}$$

where $w_s \in [0, w_{s\max}]$ is the average TCP windows size, $w_{s\max}$ is the maximum window size, $q \in [0, q_{\max}]$ is the queue length at the bottleneck router, $q_{\max}$ is the maximum queue size, N is a number of TCP sessions, $p \in [0,1]$ is probability of packets loss at the bottleneck router, C is a link capacity from the router to the receiver and R is an information delay. The information delay R is called as the round trip time and can be described as

$$R = \frac{q}{C} + T_p,$$

where T_p is a propagation delay on links of the computer networks. The nonlinear model with a time delay has a property that the time delay R is linearly dependent on the queue size which is one of the state variables of the nonlinear model given. This property makes it difficult to design congestion controllers. Introducing an equilibrium point, the nonlinear model is linearized.

To obtain the linearized model of the nonlinear model, it is assumed that the number of TCP sessions N and the propagation delay T_p are constant. The equilibrium point is defined as (w_{so}, q_o, p_o) and new variables are defined as follows,

$$\delta w_s(t) = w_s(t) - w_{so}, \quad \delta q(t) = q(t) - q_o, \quad \delta p(t) = p(t) - p_o.$$

Then the linearized model is given as follows,

$$\begin{aligned} \dot{x}_p(t) &= A_p x_p + B_p u(t - R_0), \\ y_p(t) &= C_p x_p(t), \end{aligned} \tag{6}$$

where

$$A_p = \begin{bmatrix} -\frac{2N}{R_0^2 C} & 0 \\ \frac{N}{R_0} & -\frac{1}{R_0} \end{bmatrix}, \quad B_p = \begin{bmatrix} -\frac{R_0 C^2}{2N^2} \\ 0 \end{bmatrix}, \quad C_p = \begin{bmatrix} 0 & 1 \end{bmatrix},$$

and $x_p(t) = \left[\, \delta w_s(t)\ \delta q\,(t) \right]^T$, $y_p(t) = \delta q(t)$ and $u(t) = \delta p(t)$.

The designed congestion controller, which is an AQM mechanism, is given as a state feedback controller whose input signal is $x_p(t) = \left[\, \delta w_s(t)\ \delta q\,(t) \right]^T$ and whose output signal is $u(t) = \delta p(t)$. The controller is implemented as a program in the bottleneck router of the Fig. 2. Now define the object of the design as stabilizing the queue size at the equilibrium point q_0 for the TCP/AQM network consisting of a TCP network and the designed AQM mechanism. In this paper, the equilibrium point q_0 is given as half of the maximum value of the queue size to achieve no congestion and effective utility of the queue size. Moreover the time delay $R(t)$ become constant if the queue size is stable near the equilibrium point $q_0 = 0.5q_{\max}$.

3.2 Congestion Controller Design

In this section, congestion controllers are designed by using state predictive control to satisfy the control object defined in the end of the previous section. Figure 5 shows the block diagram of the closed loop system designed finally in this paper though it is needed that this block diagram is same one for the closed loop system of the error system.

The procedure of the controller design is as follows,

1. Designing an observer for the input time delay system given by equation (6).
2. Constructing an augmented system from the designed observer and an integrator. The integrator is introducing to improve the steady state characteristic.

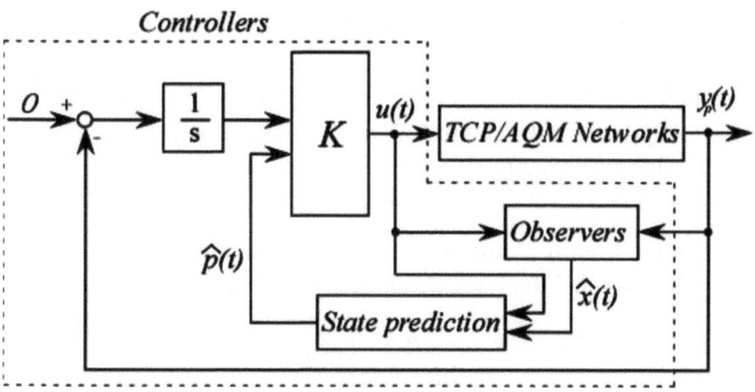

Fig. 5 The block diagram of the closed loop system using state predictive control

3. Transforming the augmented system into a linear time invariant system by using state predictive control because the augmented system is expressed as an input time delay system.
4. Designed a LQI (Linear Quadratic with Integrators) controller for the derived linear time invariant system.

The dotted line area is the designed congestion controller from the above procedure and the controller is embedded as a program in the bottleneck router that is equivalent to an AQM mechanism. Here note that this is a controller for the error system and it is of no meaning for the controller to be implemented directly as Fig. 2.

Design of the observer gain L_p: It is impossible to obtain the state of equations (6) directly because the controller is embedded in the bottleneck router and the TCP window size is not available on-line. The state $x_p(t)$ is estimated by using the following observer.

$$\dot{\hat{x}}_p(t) = A_p\hat{x}_p(t) + B_p u(t-R_0) - L_p(y_p(t) - C_p\hat{x}_p(t)),$$

where L_p is an observer gain which is designed as $A_p + L_pC_p$ becomes stable. Considering an estimated error of the state $e_p(t) = x_p(t) - \hat{x}_p(t)$, the error system is given as follows,

$$\begin{aligned}
\dot{e}_p(t) &= \dot{x}_p(t) - \dot{\hat{x}}_p(t), \\
&= A_p(x_p(t) - \hat{x}_p(t)) + L_pC_p(x_p(t) - \hat{x}_p(t)), \\
&= (A_p + L_pC_p)(x_p(t) - \hat{x}_p(t)), \\
&= (A_p + C_pL_p)e_p(t).
\end{aligned}$$

Thus it is possible to design observers if (C_p, A_p) is observable.

Now it is assumed that (C_p, A_p) is observable and the observer gain L_p is designed as a synthesis problems of a Kalman filter in case of considering that the small change of the input time delay is a white noise. The general description of linear systems with white noises is given as follows,

$$\begin{aligned}
\dot{x}(t) &= Ax(t) + B_1u(t) + B_2\xi(t), \\
y(t) &= Cx(t) + D_1u(t) + D_2\xi(t) + \eta(t),
\end{aligned}$$

where $A = A_p$, $B_1 = B_p$, $B_2 = \left[\, w_{s0}{}^{*}2 \;\; q_0/2 \,\right]^T$, $C = C_p$, $D_1 = 0$, $D_2 = 0$, $E\left[\xi(t)\xi(t)^T\right] = 1$ and $E\left[\eta(t)\eta(t)^T\right] = 1$. B_2 denotes the small change of the input time delay. Finally the observer gain L_p is easily designed as a synthesis problem of a Kalman filter for this system by using MATLAB if (C_p, A_p) is observable.

Design of the state feedback gain *K* based on state predictive control : The system consisting of the observer and the error dynamics is given as follows,

$$\begin{aligned}
\dot{\hat{x}}_p(t) &= A_p\hat{x}_p(t) + B_pu(t-R_0) - L_pC_pe_p(t), \\
\dot{e}_p(t) &= (A_p + C_pL_p)e_p(t).
\end{aligned}$$

Considering the integrator

$$\dot{x}_i(t) = r(t) - y_p(t) = -y_p(t),$$

the augmented system is obtained such as

$$\begin{aligned}
\dot{\hat{x}}(t) &= A\hat{x}(t) + Bu(t - R_0) - \begin{bmatrix} L_p C_p \\ 0 \end{bmatrix} e_p(t), \\
\dot{e}_p(t) &= (A_p + C_p L_p) e_p(t), \\
\hat{x}(t) &= \begin{bmatrix} \hat{x}_p(t) \\ x_i(t) \end{bmatrix}, \; A = \begin{bmatrix} A_p & 0 \\ -C_p & 0 \end{bmatrix}, \; B = \begin{bmatrix} B_p \\ 0 \end{bmatrix}.
\end{aligned}$$

Here note that the integrator is used to improve the steady state characteristic because steady state errors appear and the control object cannot be achieved in case of no integrators on SIMULINK simulations and *ns-2* simulations. Moreover the dynamics of the augmented system is divided into the above two equations to reduce the dimension of the augmented system and to make it easy to apply state predictive control.

Now introduce the following auxiliary variable.

$$\hat{p}(t) = \hat{x}(t) + \int_{-R_0}^{0} e^{-A(\beta + R_0)} Bu(t + \beta) d\beta.$$

Differentiating the auxiliary variable $\hat{p}(t)$ and considering the input time delay system, the following linear time invariant system is derived.

$$\dot{\hat{p}}(t) = A\hat{p}(t) + e^{-AR_0} Bu(t) - \begin{bmatrix} L_p C_p \\ 0 \end{bmatrix} e_p(t).$$

The linear system with no time delays can be described as the next equation by considering a state feedback controller $u(t) = K\hat{p}(t)$.

$$\dot{\hat{p}}(t) = (A + e^{-AR_0} BK)\hat{p}(t) - \begin{bmatrix} L_p C_p \\ 0 \end{bmatrix} e_p(t).$$

If $(A, e^{-AR_0} B)$ is controllable, the state feedback gain K exists and it can be given as a solution of linear quadratic control problems. For example, defining Q_{aug}, R_{aug} as weighting matrices, the state feedback gain K is decided as follows,

$$K = -R_{aug}^{-1} B^T e^{-A^T R_0} P,$$

where the matrix P is a solution of the following Riccati equation.

$$PA + A^T P - PBe^{-AR_0} R_{aug}^{-1} B^T e^{-A^T R_0} P + Q_{aug} = 0.$$

Thus the next equation is the control input which stabilizes the input time delay system with the state feedback gain.

$$\begin{aligned} u(t) &= K\hat{p}(t) \\ &= K\left\{\hat{x}(t) + \int_{-R_0}^{0} e^{-A(\beta+R_0)}Bu(t+\beta)d\beta\right\}. \end{aligned}$$

4 Experimental Results

In this section, a networked autonomous vehicle system is developed and effectiveness of congestion controllers is verified by using the developed vehicle system.

4.1 Results without Congestion Control

The networked autonomous vehicle system is actually constructed such as Figs. 1 and 2. The designed LQG controller is embedded in the receiver of Fig. 2. Here note that parameters are omitted to show in this paper.

The image from camera is shown in Fig. 5. The receiver processes camera images and provides a line for LQG control (See the Fig. 6(c)). The LQG controller can make a voltage for the electrical motor for the wire-less vehicle to track the detected line. In this experiment, the reference trajectory is given as a circle.

In the developed system, it is possible to occur congestion in the computer network by using another sender in Fig. 3. To show the efficacy of congestion

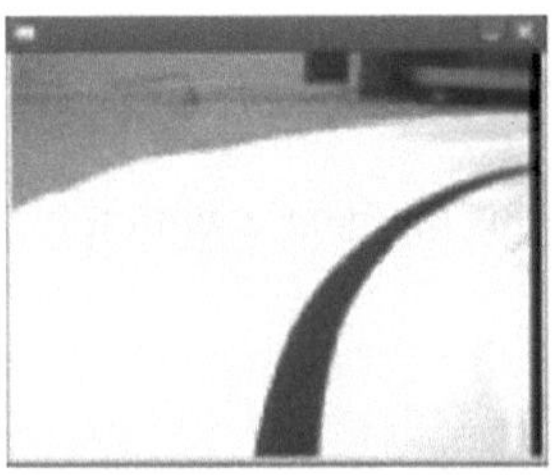

(a) Original image

(b) 2-bit image after first image processing

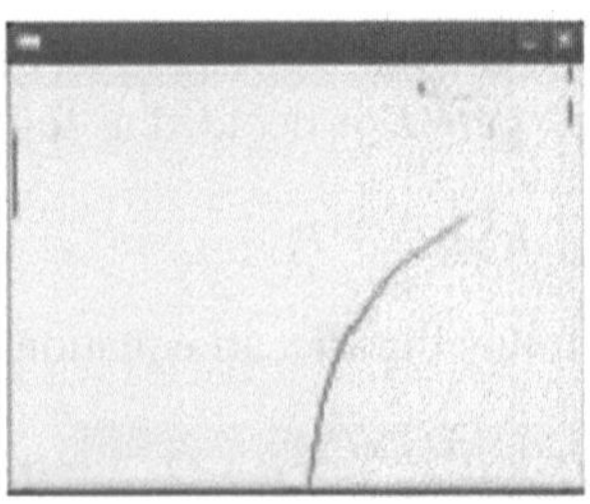

(c) Detecting a reference trajectory for autonomous vehicle after second image processing

Fig. 6 Image processing based on visual information from camera

Fig. 7 Queue size in the router (Congestion case)

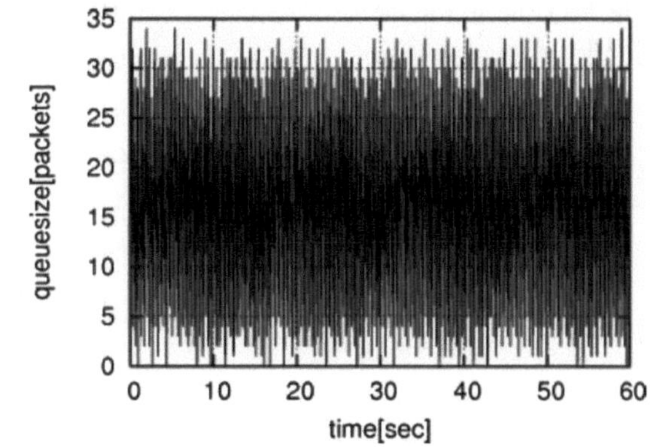

controllers, we make congestion occur in the computer network. The Fig. 7 shows the queue size in the router of the Fig. 3. It is known that the queue size is largely changing and many congestions occurs in the computer network. In this case, the autonomous vehicle system becomes unstable and the wire-less vehicle is out of control. The Fig. 8 shows the error between the detected reference trajectory and vehicle center of gravity. After 15[s], the vehicle is out of control.

4.2 Results with Congestion Control

The designed congestion controller is applied to the router and congestion is avoided in the computer network.

Figure 9 shows the queue size in the router. It is easy seen that congestion does not appear in the computer network and the amplitude of oscillation is made small in comparison from Fig. 7. It also known that the designed congestion controller works well and can avoid to occur congestion in the computer network.

Figure 10 shows the error between the detected reference trajectory and vehicle center of gravity. At 9 [s] and 15 [s], the error become large but the designed LQG

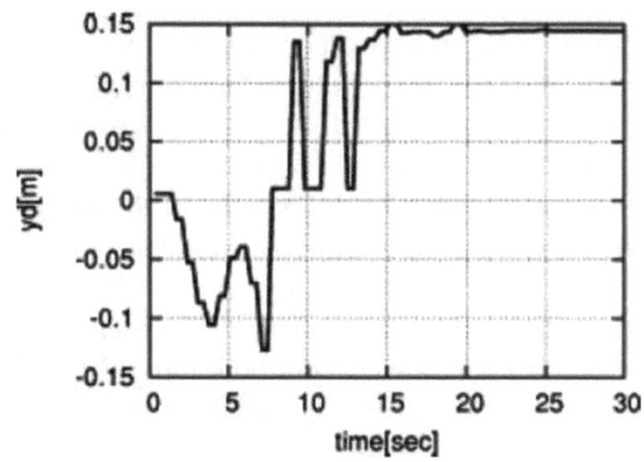

Fig. 8 Error from a reference trajectory (Congestion case)

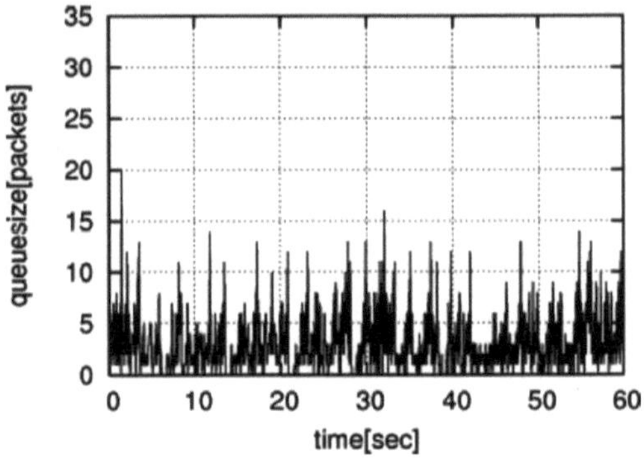

Fig. 9 Queue size in the router

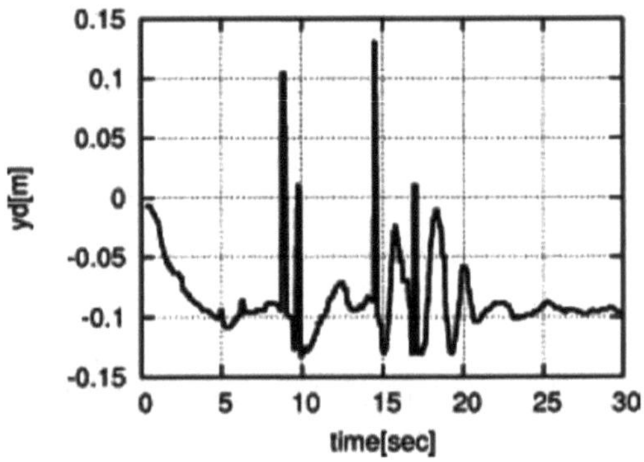

Fig. 10 Error from a reference trajectory

controller makes the system stable. The wire-less vehicle can drive automatically on the near reference trajectory. Thus the networked autonomous vehicle system can be developed. It is also seen that the designed congestion controller have important role in the networked autonomous vehicle system.

5 Conclusion

In this paper, a networked autonomous vehicle system has been developed. Especially dynamical models of the computer network are considered and congestion problems have been discussed. To avoid congestion in the computer network, a congestion controller has been designed based on state predictive control. Some experiments have been done in the developed vehicle system and effectiveness of the designed congestion controller has been verified from a experiment for tracking control of a circle trajectory. The result of this paper is located between researches about NCSs (Networked Control Systems) and Control of Networks. There are few researches to discuss about both topics of NCSs and control of networks. Moreover almost no researches are available to develop NCSs actually and design congestion controllers for computer networks.

References

1. S. Low, F. Paganini and J. Doyle, "Internet Congestion Control: An Analytical Perspective", *IEEE Control Systems Magazine*, Vol. 22, No. 1, pp. 28–43 (2002).
2. T. Azuma and M. Fujita, "Compensation of Time-Delays on Computer Networks", *Journal of SICE*, Vol. 45, No. 10, pp. 893–898 (2006).
3. J. Nilsson, B. Bernhardsson and B. Wittenmark, "Stochastic Analysis and Control of Real-time Systems with Random Time Delays", *Automatica*, Vol. 34, No. 1, pp. 57–64 (1998).
4. W. Zhang, M. S. Branicky and S. M. Philips, "Stability of Networked Control Systems", *IEEE Control Systems Magazine*, Vol. 21, No. 1, pp. 84–99 (2001).
5. C. V. Hollot, V. Misra, D. Towsley and W. Gong, "Analysis and Design of Controllers for AQM Routers Supporting TCP Flows", *IEEE Transactions on Automatic Control,* Vol. 47, No. 6, pp. 945–959 (2002).
6. T. Azuma, M. Uchida and Y. Kawai, "Gainscheduled congestion control for TCP/AQM networks using state predictive observers", *Proceedings of the 3rd International Conference on Autonomous Robots and Agents*, pp. 653–658 (2006).

Data Messaging and Control in Sensor-Actuator Networks using PowerBus Technology

Moi-Tin Chew, Serge Demidenko and Yoke-Fei Ng

Abstract The paper presents a cost-effective power line communication system based on the use of the PowerBus technology. The system employs a dedicated control networking originally intended by the manufacturer for merely electrical control purposes. In addition to the controlling functions the system provides textual and data transmission capabilities. The tailor-made software (PowerBus Messenger with Control) has been developed and implemented in the system.

Keywords Sensor/actuator network · power line communication · control · data messaging

1 Introduction

Power Line Communications (PLC) is an emerging technology for data transfer and control over the existing power lines in sensor-actuator networks. PLC provides narrowband (and in some cases broadband) solutions by utilizing the ubiquitous power lines infrastructure and thus has exempted the need to install new wiring network. Among the most well-known power line communication technologies are X10 [1], Insteon [2], HomePlug [3], CEBus [4], UPB [5] and LonWorks [6]. A large number of institutions and companies world-wide are involved in research, development and implementation of new PLC solutions as well as refining and furthering the existing ones. Among the newfangled PLC developments is the PowerBus (PB) Technology developed by Domosys® [7]. The company pursued research outside of the scope of the CEBus standard to enhance the performance of control

Moi-Tin Chew
School of Engineering and Technology, Massey University, New Zealand,
e-mail: M.Chew@massey.ac.nz

Serge Demidenko
School of Engineering and Technology, Massey University, New Zealand; School of Engineering, Monash University, Australia-Malaysia

Yoke-Fei Ng
School of Engineering, Monash University, Australia-Malaysia

networks in five significant areas: device interoperability, product development time, production costs, communication performance, and compliance with worldwide regulatory standards. The PB-based systems are quite effective in electrical control related tasks. However, they lack of data communication functions.

The presented solution enhances the PB technology by adding data communications to it without introducing any significant changes to the existing hardware operating on the AC power supply lines. The entire system architecture offers 'plug-and-play' capability and requires only simple set-up. The system is based on the use of one convenient package - the PowerBus Starter Kit [8]. The hardware of the kit consists of two fully configurable PowerCard™ Modules and one PowerGate™ Modem.

The central microcontroller (U-Chip™) is the kernel of PB hardware. It is employed in both the PowerCard™ Module and PowerGate™ Modem. The microcontroller is responsible for the data transmission and reception, control of the integrated devices remotely or locally, activation of the programmed scenarios, and communication with the host personal computer (PC) through the serial interfaces (UART or SPI Extension Bus) thus enabling the power line based remote control and monitoring tasks [7, 9].

The PB network system is assembled with the PowerGate™ modem as the network controller (Master), and the PowerCard™ modules as the configurable sockets for the electrical appliances and actuators (Slaves).

In the reported system, custom-made PB nodes are built by utilizing these PowerBus Starter Kit's components (Fig. 1). The PB nodes are also used as communication

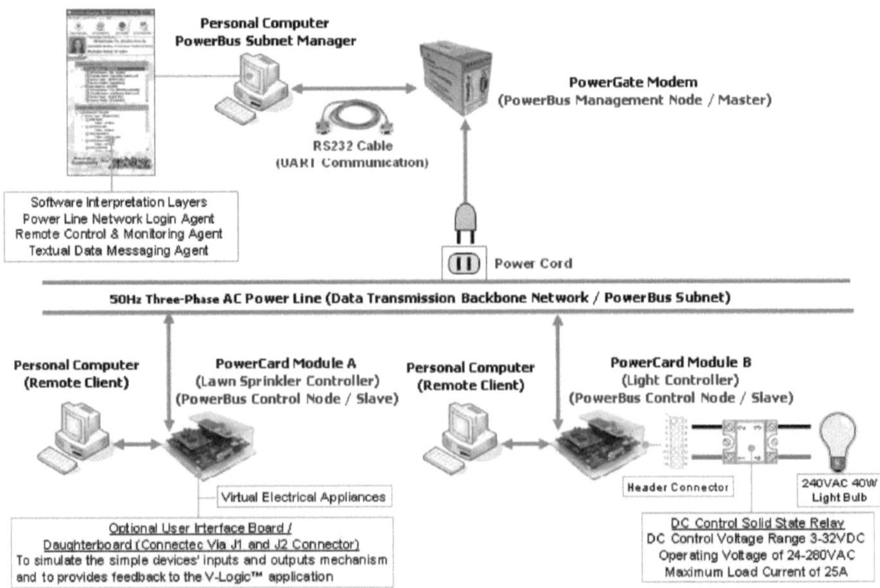

Fig. 1 Example of the PowerBus network architecture

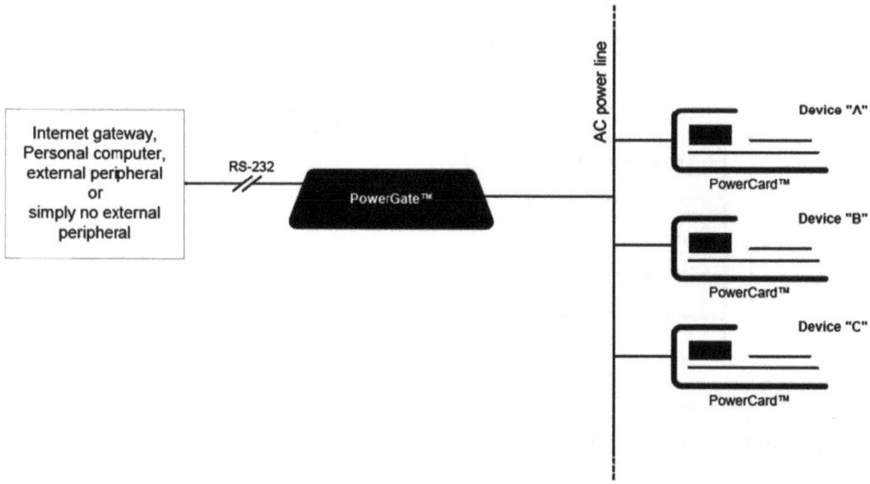

Fig. 2 Example of PowerBus Network connection [Courtesy: Domosys Corp.]

devices. The connected to the PC nodes can send required information to the main supervising unit, present device data summaries, network activities and control functionalities (thus displaying a good level of intelligence). Furthermore, each particular node is programmed to behave independently in the absence of control instructions through the communication medium. Therefore, if no host PC or Master is present in the network, the node still operates as an autonomous system and manual control is possible through the input switches.

Since each node in the network has local intelligence, the system has no central point of failure. PowerBus nodes and the network (Fig. 2) can be easily installed and configured by using the vendor software to facilitate the rapid system implementation [7].

2 System Hardware

The PowerBus protocol stack (Fig. 3) is a derived subset that is originated from the OSI 7-layer model, with exclusively the physical, data link, network, transport and application layers being conformed. These five layers are implemented and hosted on a single PowerBus U-Chip^{TM} IC. The fact that the entire protocol stack resides in the flash memory of the U-Chip^{TM} and can be upgraded in the field over the power line, ensures that a PB device can be upgraded to the latest released firmware easily at any time.

PowerBus also embeds security features (such as 80-bit encryption keys) and can access the Internet through the OPC or OSGi drivers since every chip has a unique, worldwide MAC address, which is useful for addressing PB nodes over the Internet [7].

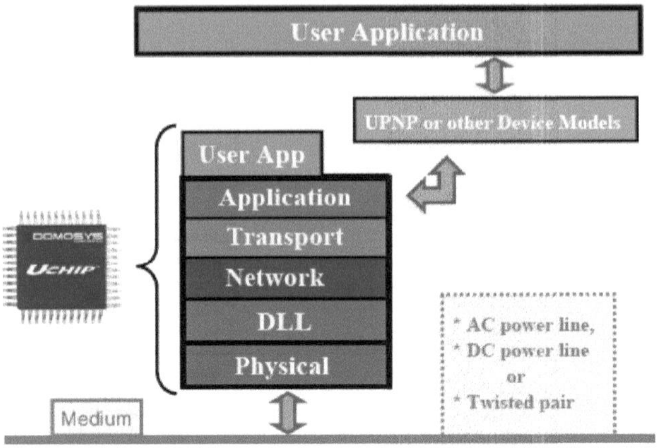

Fig. 3 Protocol Stacks Layer of PowerBus™ [Courtesy: Domosys Corp.]

U-Chip™ is a RISC8052 u-processor combined with peripheral circuits thus incorporating both control and communication features in a single IC. From the control point of view, it incorporates 43 I/O pins with different I/O firmware programmable functions (digital 1, 8, 16-bit I/O), a triac driver, 2 PWM outputs, 8-bit ADC input, and a pulse counter. From the communication point of view, the U-Chip™ (Fig. 4) incorporates serial communication interface via UART, SPI bus

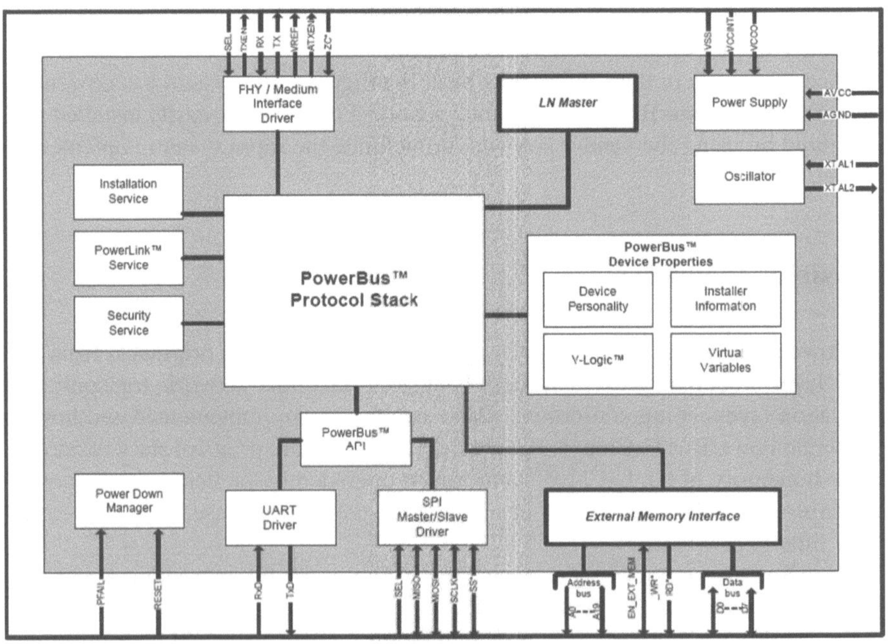

Fig. 4 U-Chip™ ICs block diagram [Courtesy: Domosys Corp.]

and the elementary firmware to implement the aforementioned PB protocol, in such a way that communications procedures inside the network are transparent to the user [9].

Each local point of control in the PB network (called PB node or PB device) contains a U-Chip™ for control and communication, I/O interface connectors for physical device connection, a medium interface card (MIC) transceiver providing physical interface to the AC power line medium, coupling transformer and a DC power supply (Fig. 5). Nodes are objects that interact with physically attached I/O devices and communicate with other nodes over a network by using the PB protocol [7]. Nodes sense, monitor, count, measure time, manage switches and relays, and respond to conditions reported by other nodes.

Transceiver of a PB node is the Medium Interface Card (MIC). It provides a simple and a cheap connection of the node to the PB network. It is a DIP-packaged analog module delivering signal amplification and filtering functions [11].

To superimpose the I/O communication signals of U-Chip™ through MIC to the mains two different signals have to coexist there: the powers supply signal (230Vrms/50Hz) and the communication signal (3.5Vp-p/110kHz). The coupling circuit mounted in the DC power supply adaptor is a high frequency transformer to provide the necessary galvanic isolation with a series capacitor used to block the low frequencies while allowing the high frequency signals to pass through it. Some other miscellaneous elements are added to minimize the coupling impedance and to protect the communication circuit against possible electrical perturbations, which frequently appear across power lines [12].

The reliable physical layer elements coupled with its PowerLink Service (an automatic repeating and routing mechanism embedded in the core of every PB

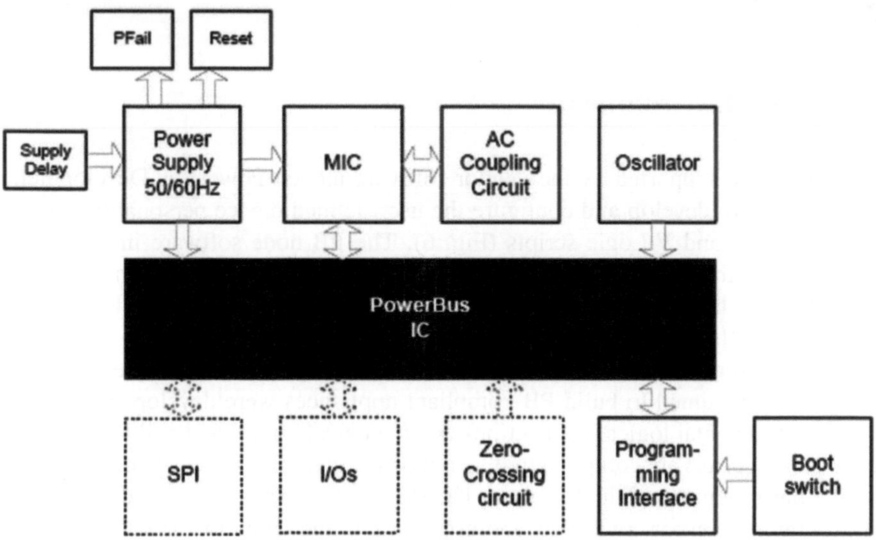

Fig. 5 Generic PowerBus device block diagram [Courtesy: Domosys Corp.]

product) ensure the required robust communication in the harsh power line environments [13]. In order to increase resistance to disturbances and interruptions inherent to electrical installations, PB employs the proprietary narrowband single-carrier modulation scheme of RYSK (Reduced Yaw Shift Keying) enabling reliable data transmission [14].

A PB network constituted by a set of intelligent nodes is responsible for the co-ordination, control and management of PB-enabled appliances' functions. It also provides facilities for the appliances and system components to exchange data messages and functional requests. The PB protocol can be exploited for either master-slave or peer-to-peer network topologies.

This presented system employs the master-slave topology as it is aimed for communications from top to bottom or vice versa (whereas the peer-to-peer topology is used for multidirectional communications between all devices of the network) [7]. In the implemented network architecture, the slaves (PowerCard^{TM} Modules) are the only agents that access the electrical appliances directly and execute the control operations locally. For remote control the slaves will broadcast their available services and will wait for the service request message to be generated from the master in the PB network. When they receive a request of service, they will execute a list of actions to achieve the required task [10]. In general, for the master-slave configuration, a particular slave is only concerned with the events in its immediate locale. Therefore a slave is rarely required to communicate with the neighbouring PB nodes. This greatly reduces the required bandwidth of the communication system.

The master (PowerGate^{TM} Modem) is responsible for the network coordination, maintain registered information of all the slaves, remote control function, initiate and route messages between the networked nodes [15]. This reduces the level of encapsulation of the networked nodes since the master will organize and schedule the communications with the other nodes.

3 Software Environment

PB systems are supported by the vendor software named PowerBus Development Tool enabling to develop and configure the user-defined device personality, virtual variables (V^2s) and V-Logic scripts (Fig. 6). The PB node software implementation must use the V-Logic language – an interpreted programming language designed for U-Chip^{TM} embedded applications and directly supporting the U-Chip PB firmware [9].

In the presented system the sets of autonomously executed sequences with V-Logic scripts aimed to build PB-compliant appliances were developed to establish the fundamental logical connections and to control virtual and real PB devices. The V-Logic scripts are compiled first, and then the compiled codes are downloaded into the flash memory of the PB node. The PB node's kernel will periodically scan the programmed event conditions and ascertains if and when to carry out the pre-programmed and scheduled action logics. V-Logic language is able to provide a

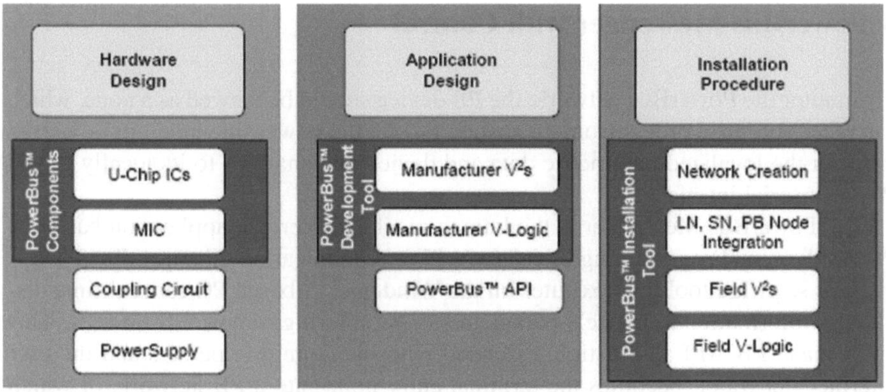

Fig. 6 PowerBus Device Development Cycle [Courtesy: Domosys Corp.]

framework, through which the local PB devices may communicate with, and control objects linked to other remote PB devices. The V-Logic functionality can only be implemented through PowerCard^{TM} module or any slave devices with embedded U-541 IC.

The V-Logic uses a particular syntax whereby each I/O objects or control parameters of a PB device has a set of properties associated with it, known as Virtual Variable or V^2. These V^2s can represent data in a variety of formats, such as Boolean, 32-bit floating point, unsigned char, signed integer and the like. The named condition or the enumerated representation of the V^2 completely defines the I/O object in question and they can be Read-Only, Write-Only or Read/Write. A PB device is implemented with such a collection of V^2 and it is operated upon the V-Logic scenarios. In essence V^2s are the logical representation of the underlying physical processes, of which the control network nodes drive or measure. The virtual variable interprets the low level device data to a high level meaningful descriptive context describing the network environment in the V-Logic script. If an input and an output V^2 are defined for a specific application, they could be bounded and associated together in the V-Logic script and linked to represent a variety of PB devices. This is achieved through a bi-directional mapping of V^2 entities, possible by virtue of object-oriented methodology inherent in the V-Logic specification [7, 9].

Each PB node or PB device can be associated with one or more V^2s. Each V^2 represents a mere single datum of the appliance. It can correspond to an aspect of the physical world, such as the state of a switch, or to be of more abstract nature, such as a preset temperature set point. In addition the V^2s are required to be specified and to be mapped to the I/O pins of U-Chip^{TM}. The discovery and the interrogation of all networked PB nodes to determine the V^2 parameters and statuses of each PB devices of the network are achievable through the use of PowerBus Installation Tool. The tool offers a simplest platform for monitoring and controlling V^2s. In terms of program interaction with the users, however, it is rather complicated and multifarious.

4 PowerBus Messenger with Control

To monitor the PowerBus network, the PB device should be viewed as a node, which can receive control orders from the distant PC via the power line medium as well as transmit the localized parametric data and device personalities to its locally linked PC via a serial interface.

The PowerBus Messenger is the human-machine interface application based on PB Application Programming Interface (API) communication schemes (Fig. 7). It is the host software tool that executes on the Windows^{TM}–based PC for real-time displaying informational device's parameters or monitoring remote PB devices. This tailor-made PB API application, extended with the definition parsers and the user interface generator, becomes the terminal entry point into a PB network, of which the enumerated service requests messages can be inputted and outputted through the connected PB device. It can automatically generate the user interfaces according to the type of the discovered PB devices as well as the information provided by the target devices. The software aims at supporting connection and instant access to various networked PB devices. It offers also a responsive user-to-user messaging method. It employs the GUI blueprint of the well-known instant messaging software – the Microsoft MSN Messenger as an implementation platform [16, 17]

The main objective of the PowerBus Messenger software development is to provide an easy and a simplified access to the high-level functionality of the PB technology, while shielding the PB users from the complexities of the low-layer protocols. PB API is a part of the application layer functional units and it provides access the PB Device's Properties, reads or writes V^2s and subscribes to networked PB Events. By using PB API all network activities can easily be monitored and communication statistics accessed. It enables also to obtain an entire view of the network including the detailed properties for each PB device, as well as send and/or receive raw data packets. The PB API message, a string of bytes with a maximum data length of 250 bytes, consists of a Primitive Type Byte followed by other informational bytes. It carries the instructions of which the PowerBus Messenger application attempts to match from its categorized lists of service types and procedural responses interpretation databases. If matched, the selected procedures are executed and the API Primitive Arguments Bytes that followed the Primitive Type Byte will provide the necessary information to proceed with the execution. The destination of the API Message can be a PB Node or group of PB Nodes that is identified by the Destination/Source Address Byte that is assembled in it [9].

General primitive format

First Byte........									........Last Byte
1	2	3	4	5	6	7	8	9...N	N + 1
Control Byte	Data Length (MSB first)		Destination/Source Address			Transport Layer Header	Primitive Type	Primitive Arguments	Checksum

Fig. 7 API Communication scheme

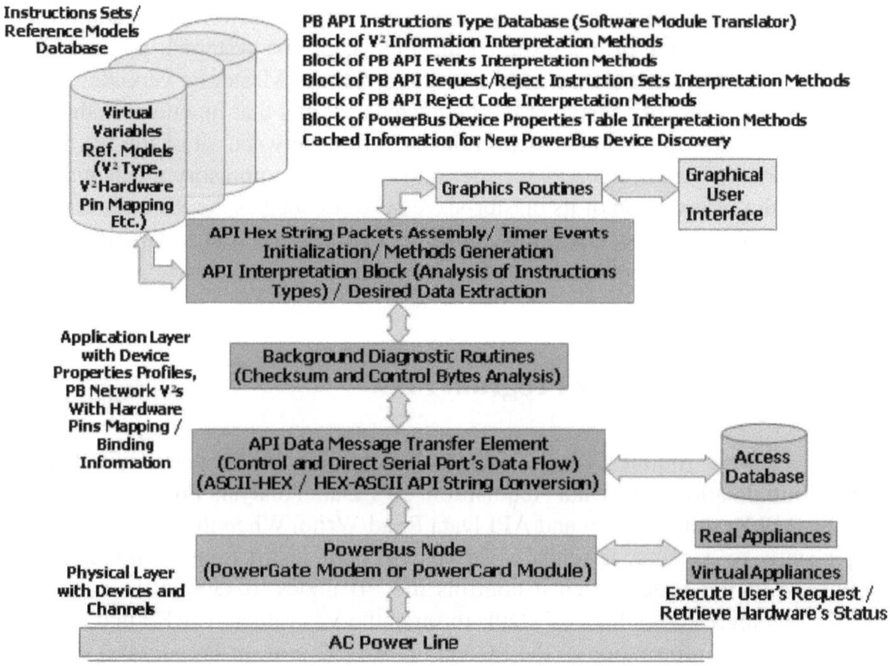

Fig. 8 PowerBus Messenger architecture

The PowerBus Messenger (Fig. 8) uses a significant number of service requests primitives to transfer data messages and to convey control information. The name of a service requests primitive denotes the direction of the information transfer and the purpose of transfer. The specific tasks performed by the service requests primitive is determined by a value contained within the PB API message string.

There are four types of API service request messages: GET (01h), SET (02h), REPORT (05h) as well as USER-DEFINED (40h). For instance, GET is used to read the value of a property or a V^2, SET is used to write the value of a property or a V^2, while REPORT is used to report the value of a V^2 or to report an Event. Conversely, the USER-DEFINED service request, which is used by the host (PC or Microcontroller) to transfer data to a remote PB device, is being exclusively entrenched in PowerBus Messenger to implement the textual data transfer function. It enables PB users to promptly and easily use the instant messaging on virtually any PowerBus-enabled PC. PB Messenger has sets of these API service request message generation entities. Sequences of service requests are pre-programmed in this application to repeat periodically and random service requests generation is also supported, such as the instant messaging function.

In order to make the PowerBus Messenger more practical, usable and a functional a total number of 56 types of device property table reference models (programmable data fields of U-Chip^{TM}), 20 types of API events interpretational reference models, 18 types of API reject codes, etc., have been implemented in it.

To make the messenger to perform and maintain the networked operation appropriately, a minimum of one terminal must exists on the system at all time. This terminal (Logic Network Manager) is associated with the Master PowerGate^{TM} Modem. The Logic Network Manager is a dedicated terminal that functions as the Network Administrator providing synchronous API flow between all of the PB nodes, categorize PB devices or PB API services and executes compulsory mechanisms to retrieve all the information of its master-slave network such as the PB node address, subnet address, device list, URL, look-up table, firmware version and the like, as all these information is crucial for PowerBus Messenger users.

5 PowerBus Messenger Program Flow

The PB software routines can be subdivided into six major parts, specifically: System Initialization, API Data Acquisition, API Data Analysis Process, API Data Storage, API Results Display and API Data Read/Write. When the PowerBus Messenger starts to execute, it sends a various PB API instructions to initialize the parameters of PB nodes. Then it controls the PB nodes to sample their existing U-Chip^{TM} integrated-appliances' status through the V^2s values. The PB nodes transmit the resultant information to the PC via the RS232 port, or broadcast it through the power line medium to other remote PC-based PowerBus Messenger users. At the PC side, the information is stored in the Microsoft Access database or the messenger program's static array variables for the need of inquisition in the upcoming execution. Graphical user interface is then dynamically generated according to the acquired services description embedded in the target PB devices. The users could then control PB devices plugged on the power line and get the status feedback [17, 18].

Initially, PowerBus Messenger checks up the Device Properties (Fig. 9) of the locally attached PB Device (master or slave) via the RS232 port. The information of the PowerBus Protocol Layered Structures programmed in the local embedded U-Chip^{TM} IC is retrieved by executing different GET and SET PB API Request Instructions to read the complete Device Property Tables sequentially. Next, the gathered information is saved in the Static Arrays Variables, which serve as an intermediate program database, to be used for configuration purposes.

It should be noted that for the master-slave topology, PB API messages are only transmitted on the PB network when requested by the master [9]. Slaves are never self-triggered to transmit an API message without first receiving an acknowledged request from their master.

To cater the dissimilarity in between the master and slave operational properties, upon the completed analyses on the aforesaid Device Property Tables, the PowerBus Messenger is able to identify the role of the local PC-connected PB device. Consequently, it triggers then different sets of program subroutines for the master and slave, respectively, to carry out the appropriate functions. Without the presence of the master in the PB network, the slave can merely control its locally connected external appliances through the PowerBus Messenger and communication with other

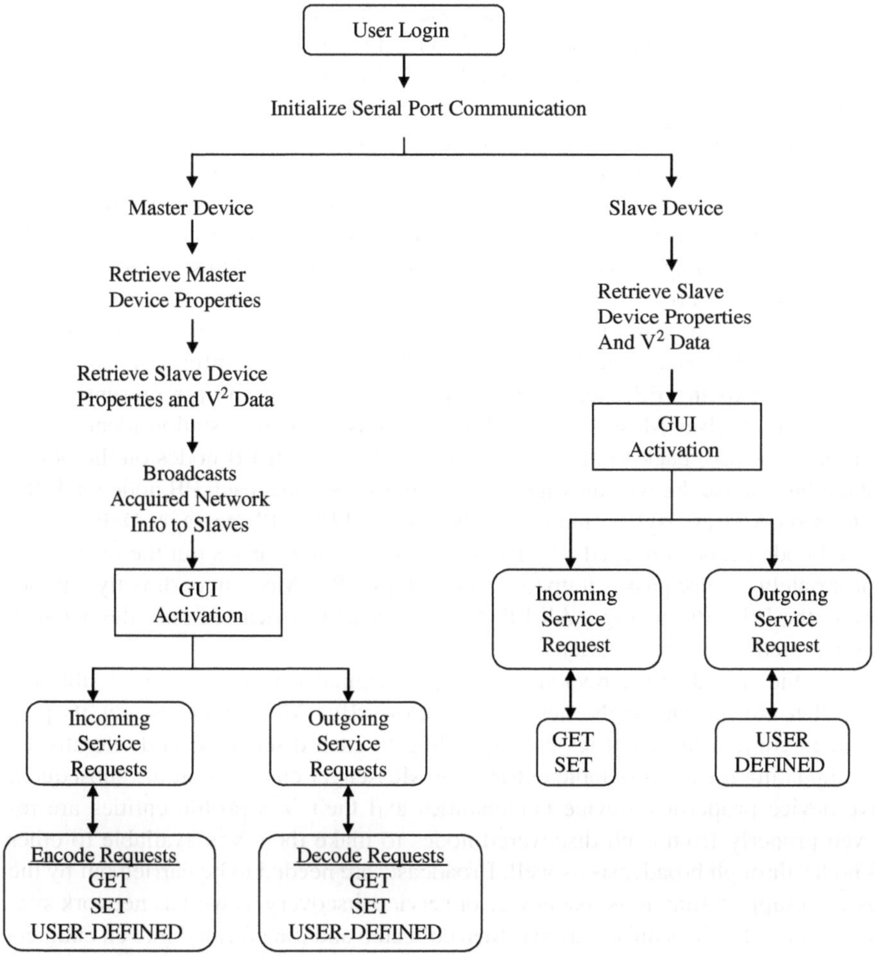

Fig. 9 Simplified PowerBus Messenger program flow chart

PB nodes or slaves is not achievable by any means. Thus, the instant messaging feature is disabled when the master is absent in the network. For the master the management and diagnosis services, which are organized by PB Messenger software routines entirely, include querying the content type of the V^2s, the PB Nodes status, querying and updating addressing information, PB device identification, configuring routers and the like.

All the devices in the PB network share a common power line communication channel. Therefore identical Logic Network (04h) and Subnet (03h) addresses are allocated to them. For the PowerBus protocol to propagate the PB API messages to the exact destined node, it needs the PB Node Address (02h) of the destination node. This address is stored in one of the 56 Device Property Table entries onboard the master U531-Chip^{TM} – the Look-Up Table (7Ah), and is extracted in the previous initial process catered for the master.

The dynamic network configuration process, executed by the master with accordance to the PB API instructions issued out from the PowerBus Messenger in favour of address acquisition and distribution service, is to discover and validate other specific PB nodes in the network. PowerBus Messenger constructs iterative API instructions according to the number of previous acquired addresses (addresses of slave nodes) and it broadcasts them on the medium via the master. This caused other slave PB nodes to transmit the feedback response messages containing their unique 6-byte visible ID (B2h) and their corresponding URL (12h) information (used to preserve the existing PB users' names (who are signed-in and are presently activating the PowerBus Messenger), and it is used as a pointer/location identifier on a remote database) back to the master. The entire group of PB nodes on the power line medium sense the API message at about the same time, each PB node reads the Destination Address Byte contained in the acquired PB API message. Only the addressed node decodes it, reads the message contents and carries out the instruction appropriately, or else passes it up to its host's PowerBus Messenger directly without parsing the PB API message. All PB API Messages broadcasted onto the network are encrypted to ensure security.

Upon the arrival of the response messages originated from other accessible PB nodes that are present in the network, the PowerBus Messenger updates its programmed address table entries corresponding to each discovered node iteratively. After updating the address table entries, PB Messenger ensures that the subsequent slave device properties, device personalities and their V^2s profile entities are retrieved properly from each discovered nodes to make their V^2s available to other PB nodes through broadcasts as well. Broadcasts are needed to be carried out by the master to support functions like device or service discovery as well as network synchronization. The V^2s information is then collected and transformed into an enumerated representation suitable for system processing. The V^2s information is retrieved continuously to reduce the dependences on external, possibly inaccurate out-of-date database, of which this cited inaccuracy is induced by prolonged program execution period as the V^2s values may not be constant all the time. Any discovered Slaves/PB nodes are then integrated as part of the PowerBus Messenger networking system without any further delay [19, 20].

Once the configuration processes are completed, the GUI elements are generated and activated to display the acquired network information. PB Messenger GUI presents the PB network resembling the Windows^{TM} Resource Explorer. The messenger displays and manages the data items/properties in tree-view (Fig. 10). Data items are classified as PB nodes and their controllable V^2s. The GUI merely shows the PB Messenger user a list of discovered PB nodes that are being registered in the configuration query process. The GUI representation is as simple as possible, but powerful enough to describe the essential features of a discovered PB device and interact with it.

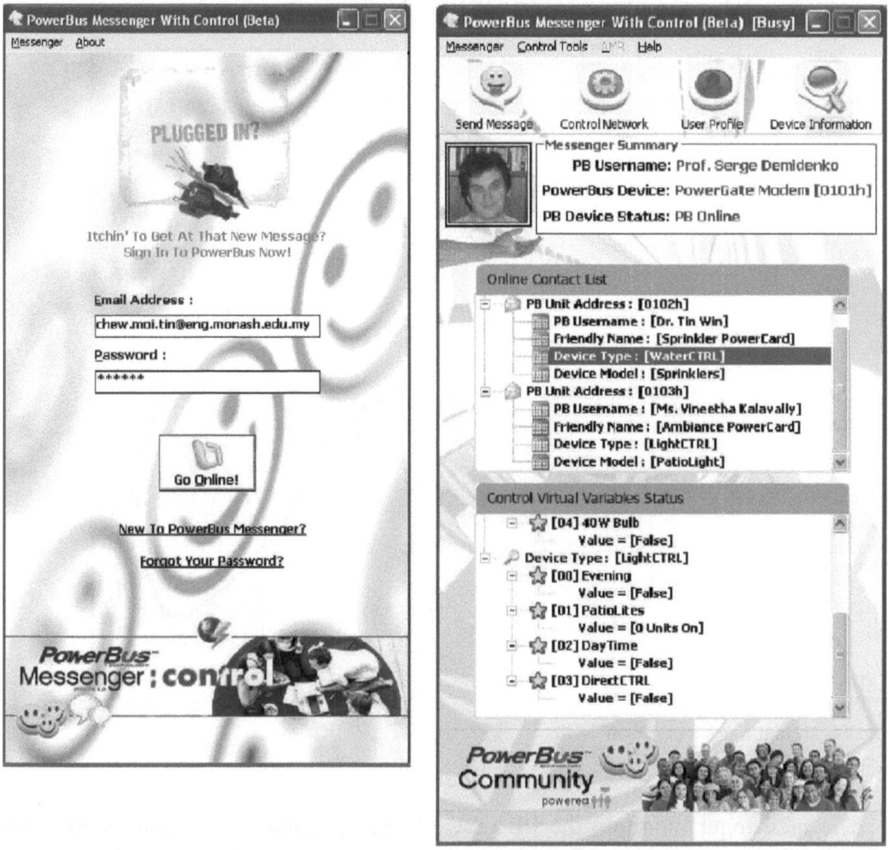

Fig. 10 User's sign-in form and the main form of the PowerBus Messenger

PB Messenger users can make use of the accessible textual messaging feature to deliver short-text notes to other available remote PB users (Fig. 11). The maximum length of ASCII texts that could be delivered at once is 250 characters/bytes (without overhead). Messages up to 32 bytes are contained in a single PowerBus packet. Messages longer than 32 bytes are divided into multiple PB packets using the segmentation service of the application layer that supports up to 250 bytes. A PowerBus message can theoretically be as long as 1.8Mbytes.

Users have full access to the individual V^2s of each networked PB devices. In addition when Direct-Network Control function is enabled (manual override), it allows the PB users to change the state of certain V^2s (Read/Write) or monitor the values of other V^2s (Read-Only). Controlling the V^2s via the PB Messenger invokes the hardware-level responses, and the physical PB devices in respond to the generated PB API instructions carry out the tasks accordingly. If the PB Messenger user disables the Direct-Network Control function, the slave can carry on with its pre-programmed V-Logic scripts and act as a stand-alone PB device to execute developed scenarios [14].

Fig. 11 Instant messaging form

The messenger monitors the PB network input/output traffics continually and periodically requests other PB nodes to send their latest device properties records and any changes in the state of their V^2s through cyclically polling. The most recent broadcasted/received PB API messages and events, along with their enclosed information are analyzed to determine the API message categories. Next, the desired information is extracted with a suitable categorized decode method, and the previously acquired configuration records are overwritten for updates.

Because the patterns of the PB network environment are dynamic, the program must be able to maintain and update the network statistics for dynamic user interface generation. For example, on every change of a specially designated V^2, the PB Messenger software automatically initiates the necessary API message exchange to propagate the updated V^2 value to other appropriate slave receivers.

6 Structure of PB Messenger

The PB Messenger includes an inner multi-task scheduler and it adopts an event-driven structure. This built-in multi-task scheduler allows the Messenger program to express logically and to execute numerous event-driven tasks in an organized way, as well as to control the prioritized execution of these tasks. Examples of such tasks include timer events activating per-defined duties on expiration, run-time library modules of function calls to carry out events checking, manage program's input/output activities, send and receive API messages across the network, to control miscellaneous functions of the U-Chip^{TM} and so forth. Table 1 summarizes major program routines that are employed in PowerBus Messenger.

Table 1 Routines and functions of PB Messenger

Subroutine/Function	Functional roles
MSComm_OnComm()	Initialize the serial communication port settings. Trapping serial communication events and errors. Transmit/Receive PB API Messages to/from other PB Devices.
TIMERPROC()	Scheduled event that is triggered to refresh the entire network configuration procedures. Reset all the static parametric variables and other programmed events. Retrieve and update the entire program parameters.
MASTERPSEUDOPING() SLAVEPINGTIMEPROC()	Scheduled event that validates the presence of Slaves on the PB Network, as well as the connectivity to each Slave. Circularly invokes additional processes to remove the former GUI items, of which representing the PB Devices that are no longer exists in the PB Network, from the PowerBus Messenger. Plug and Unplug Detection over Power Line.
OnOffValidationControl()	Scheduled event that will validate the On or Off condition of connected PB Device. Triggers supplementary processes if PB Device is still being turned On, otherwise disabled the entire sets of GUI controls.
PBAPIParsing()	API Message Parser that contains lists of PB API contexts and predefined methods (Hex-ASCII / ASCII-Hex Translation) to handle any incoming PB API Messages. Decode and identify types of PB Service Requests and trigger the corresponding activities to perform the Request. Several GUI components reside at this function as an observer.
GETVirtualVariableResult() ReportPrimitive()	Reference Models and Data Structures that are characterized for the attributes of V^2s. Methods to manipulate and interpret V^2s data reside in this function.
GETPropertyResult()	Reference Models and Data Structures that are characterized for the attributes of PB Device Properties Tables. It includes a range of methods to manipulate and to interpret PB Device Properties, maintains records of contents for protocol layered structures, and preserves the application-dependent information of which the PB Device will be needed to further process the data transmissions routine.
tvwContactList_Click() tvwControlStatus_Click()	GUI elements that allow the PB Network information to be browsed and displayed graphically. Able to trigger and complete the requested device control activities.
cmdUpdateValue_Click()	GUI elements used to control the remote appliances or to modify the writable V^2s enumerated values.
PBMessageReception()	GUI elements used to activate the Instant Messaging feature, of which is assembled with reference to the USER-DEFINED Service Request. Conversation History is logged.

Polling method is used to retrieve the required data from the PB network and to interrogate each PB node about its status remotely or locally by PowerBus Messenger. Although polling method has the benefit of no address tables entries of the nodes on the PB network need to be modified frequently, nonetheless, every PB node has to be polled individually (requests and responses) causing a considerable amount of unnecessary network traffic and driving the PB Messenger to enter an exhaustively execution stage. Therefore, an extended delay interval in between each polling request instruction is adopted to relieve the traffic and to allow the program to have sufficient time to deal with the other events.

Despite the robustness of the PB technology, potential errors could still impact on the network performance. Such errors could generate increasing network traffic, incomplete packets transmission or even excessive usage of the IACK (interrupt acknowledge) call-backs. Insufficient delay time between the API packets transmissions could further prop up the aforesaid errors. Therefore careful diagnostics of network performance is required to avoid overload in network channels, prolonged transaction times and repetitive unstable processes [21, 22].

To achieve the objective, each critical communication error detected by PB device or distinguished by the PowerBus Messenger is prompted to notify the PB users through different types of alert dialog boxes. The time-stamped detailed error descriptions are logged for further diagnosis. By doing this, the potential bottlenecks and instabilities in the PB network can be identified. It allows also carrying out detailed investigation of the abnormal system behaviour. As a result faults can then be detected, localized and corrected at an early stage with minimum efforts. This is particularly important when the access to the PB nodes is offered through a far-off remote connection.

7 Conclusion

This paper presents implementation of a simple power line control and communication system that satisfies the fundamental dependability requirements. The system is built on the PowerBus technology enabling implementation of the large-scale network. It provides a good practical balance of flexibility, transparency, and expandability. The system can employ different physical media and network topologies that are supported by the PowerBus standard. In addition to the standard control functions earlier developed for the PowerBus technology, the system is capable of providing such functions as text messaging, file transfer, automated meter reading, audio control network applications and a number of others.

References

1. Boone K (2006) Using X10 for home automation, UK Automation, http://www.uk-automation.co.uk/article_info.php?articles_id=7
2. Darbee P (2006) INSTEON Compared, SmartLabs Technology, Irvine, CA, USA, http://www.smartlabsinc.com/files/INSTEONCompared20060102a.pdf

3. Gardner S, Markwalter B, Yonge L (2000) HomePlug Standard Brings Networking to the Home, Communication Systems Design, 6 (12), http://www.intellon.com/pdfs/article_CommDesign1105.pdf
4. Wacks K P (1997) Home Automation and Utility Customer Services, Aspen Publishers, USA
5. UPB™ Universal Powerline Bus. The UPB System Description, Ver. 1.4 (2007), Powerline Control Systems, Northridge, CA, USA, http://www.pulseworx.com/downloads/upb/UPBDescriptionv1.4.pdf
6. Building Automation Technology Review (2003), Paper 005-0162-01, Echelon Corp. San Jose, CA, USA
7. PowerBus™ Technology Overview (2002), Domosys Corp., Canada, http://www.domosys.com/support/download/dnld_manager.asp?fileid=5
8. Up and Running in 30 Minutes (2003), Document AN-PWD-0400-01, Domosys Corp., Canada
9. PowerBus™ ICs Datasheet: Master/Slave (2007), Document D-PWB-0100-11, Domosys Corp., Canada
10. PowerCard™ Module User's Manual (2002), Document UM-PWC-0100-05, Domosys Corp., Canada
11. Medium Interface Card Data Sheet (2007), Document D-MIC-0100-08, Domosys Corp., Canada
12. PowerBus DC Adapter Data Sheet (2002), Document D-PWA-0100-02, Domosys Corp., Canada
13. PowerBus™ Technology Glossary (2002), Document G-PWB-0100-04, Domosys Corp., Canada
14. PowerBus™ Quick Reference Guide (2004), Document OR-PWB-0100-05, Domosys Corp., Canada
15. DIN PowerGate™ Modem User Manual (2007), Document UM-DPWG-0100-11, Domosys Corp., Canada
16. Windows™ Live Messenger (2006), Microsoft Corp., USA http://get.live.com/messenger/overview
17. Corcoran P M, Desbonnet J (1997) Browser-Style Interfaces to a Home Automation Network. IEEE Trans. on Consumer Electronics, 43(4): 1063–1068
18. Neugebauer M, Plonnigs J, Kabitzsch K, Buchholz P (2004) Automated Modeling of LonWorks Building Automation Networks. In: Proc. IEEE Workshop on Factory Communication Systems, pp 113–118
19. Liang N-S, Fu L-C, Wu C-L (2002) An integrated, flexible, and Internet-based control architecture for home automation system in the Internet era. In: Proc. IEEE International. Conference ICRA, 2, pp 1101–1106
20. Nunes R J C (2004) A Web-based approach to the specification and programming of home automation systems. In: Proc. 12th IEEE Mediterranean Electrotechnical Conference, 2, pp 693–696
21. Shahnasser H, Wang Q (1998) Controlling industrial devices over TCP/IP by using LonWorks. In: Proc. IEEE Global Telecommunications Conference, 2, pp 1309–1314
22. Lusted K, Humborg K, Nolan P J, Corcoran P M (1995) An Active-Node Fault Diagnosis System for CEBus Networks. IEEE Trans. on Consumer Electronics, 41(3): 884–888

Part II
Gas/Bio Sensors

Remarks on Emotion Recognition Using Breath Gas Sensing System

Kazuhiko Takahashi and Iwao Sugimoto

Abstract This paper proposes a smart gas sensing system to achieve emotion recognition using breath gas information. A breath gas sensing system is designed by using a quartz crystal resonator with a plasma-polymer film as a sensor. For computational experiment of emotion recognition, the machine learning-based approaches, such as neural network (NN) and support vector machine(SVM), are investigated. In emotion recognition experiments by using gathered breath gas data under psychological experiments, the obtained maximum average emotion recognition rates are 75% using ANN and 83% using SVM for three emotions: pleasure, displeasure, and no emotion. Experimental results show that using breath gas information is feasible and the machine learning-based approach is well suited for this task.

Keywords Emotion · breath gas · quarts crystal resonator · plasma-polymer film · neural network · support vector machine

1 Introduction

Emotion recognition is an interesting but difficult task. People can recognize emotional speeches with about 60% accuracy and emotional facial expressions with about 70–98% [1]. Studies on emotion recognition with computers differ on the number of categories and the kinds of categories to use. Some emotion recognition systems in speech or facial expressions which have been used include several emotional states such as joy, teasing, fear, sadness, disgust, anger, surprise, and neutral. In those studies, emotions that are consciously and purposefully expressed by the subjects are treated since consciously expressed emotions are easier to recognize,

Kazuhiko Takahashi
Department of Information Systems Design, Doshisha University, Kyoto, Japan,
e-mail: katakaha@mail.doshisha.ac.jp

Iwao Sugimoto
School of Bionics, Tokyo University of Technology, Tokyo, Japan,
e-mail: sugimoto@cc.teu.ac.jp

control and significantly to gather data on, however recognition rates are 50–60% in emotional speech recognition [2, 3] and 80–90% in facial expressions [4, 5]. Physiological indexes are also useful to evaluate emotions since they can be measured physically and objectively and can be easily applied to engineering approaches. Physiological changes according to exciting emotions can be observed on changes of the body surface and/or autonomic nervous system [1]: e.g., skin conductivity, electrocardiogram, electromyogram, and blood volume pressure. Using brain waves seems to be effective [6, 7] since emotions are excited in the limbic system and are deeply related to cognition process. In these cases, however, sensors have to be attached to the human body to collect physiological data and expert techniques of handling the sensors are also required. On the other hand, it might be considered that there is a relationship between mental states such as emotion and stress, and physical states of the internal organs or the oral cavity. Since feeling emotions affect the autonomic nervous system, studies of a relevance between breathing patterns and emotions have been carried out [8]. From a viewpoint of relevance between illness and emotions such as a laughter and stress [9], it is shown that changes of the autonomic nervous system based on feeling emotions affect physical states of the internal organs recently. In the field of remoteness medical care, tele-medicine, and tele-care, studies on a disease diagnosis by breath gas test [10, 11, 12, 13] have been carried out actively since it is clear that internal organs disease on a liver, lungs, and blood has bad breath (chemical features) from experience and an analytic chemistry point. As the breath gas test is a non-invasive method of biomarkers without pain, it protects the human body off from the risks of inserting electrodes or sensors. Thus, in a medical field, the breath gas test is expected as diagnosis technology to inspect an internal change of human body. On the other hand, oriental medicine explains the relationship between emotions and internal organs by using five elements (water, fire, wood, metal, and earth) theory of traditional Chinese philosophy as shown in Table 1: liver vs anger, heart vs joy, spleen vs worry, lungs vs sadness, and kidney vs fear. Although it is not commonly guaranteed that there exists such a relationship clearly yet, it might be possible to estimate emotions by using information of internal organs' states indicated by breath gas indirectly. Furthermore, if an evaluation method of mental states by using change of breath gas states can be achieved, information from the breath gas might provide a new man-machine interface. However,

Table 1 Five elements of Chinese philosophy and biogas components of disease

Five elements	Five organs	Five emotions	Disease	Biogas components
Wood	Liver	Anger	Liver disease, Cirrhosis, Hyper-cholesterolemia	Ammonia, Isoprene, Volatile amines, Volatile sulfur compounds
Fire	Heart	Joy	Myocardial infarction	Pentane
Earth	Spleen	Worry	Diabetes mellitus	Acetone
Metal	Lungs	Sadness	Asthma	Carbon Monoxide
Water	Kidney	Fear	Renal failure	Volatile amines

to date, few study of estimating mental states such as emotions from breath gas information has been undertaken. Therefore, we assume that emotions affect changes in the components of breath gas and investigate a possibility of using biogas emitted from a human body as a new channel of estimating human emotions.

In this paper, we propose a smart gas sensing system for breath gas test and investigate emotion recognition by using machine learning-based approach as a first step to achieve emotion recognition from breath gas information. In Sect. 2, the basic design of breath gas sensing system utilizing a quartz crystal resonator sensor that is a type of mass transducer is described. In Sect. 3, collecting breath gas data using the sensing system under psychological experiments is carried out and some results of computational emotion recognition are presented.

2 Breath Gas Sensing System

Figure 1 shows a block diagram of the breath gas sensing system that consists of a sensor module, a power supply, and a personal computer. The sensor module utilized a sensitive sensor device based on a quartz crystal resonator (QCR) with a plasma-polymer film (PPF) where the sensor device can detect volatile organic compounds (VOCs) with parts-per-billion (ppb) levels under the dry air conditions [14, 15, 16]. The PPF was prepared by radio-frequency (RF) sputtering with an organic solid target, which was suitable for mass production and needed no reactive or toxic reagents for processing. A 9 MHz AT-cut QCR, which is 8.5 mm diameter and 0.1 mm thick, was used. As shown in Table 2, the PPFs with different chemical structures were prepared by using various target materials including biomaterials (s1 ~ s5) and synthetic polymers (s6 ~ s7). The electrode's area of the sensor device was 0.13 cm^2. The relationship between the frequency shift of the QCR Δf (Hz) and mass variation of the organic thin film Δm (ng) is defined with Sauerbrey equation:

$$\Delta m = -1.05 \Delta f \quad (1)$$

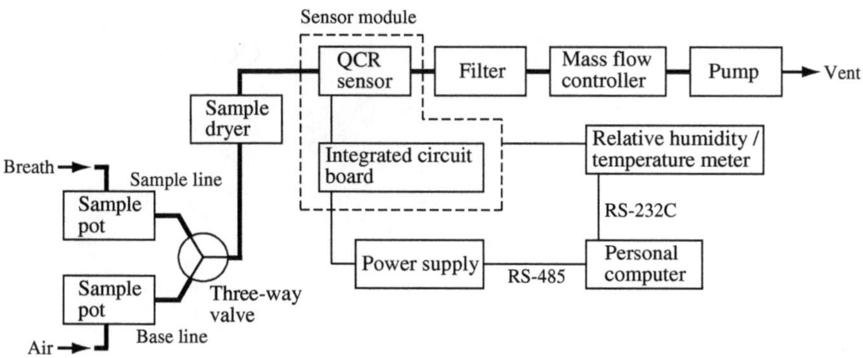

Fig. 1 Diagram of breath gas sensing system

Table 2 Materials for organic thin film of sensor cell

Sensor cell	Material
s1	D-phenylalanine
s2	D-tyrosine
s3	D-glucose
s4	DL-histidine
s5	Adenine
s6	Polyehylene (PE)
s7	Polychlorotrifluoroethylene (PCTFE)
s8	D-phenylalanine (Sealing)

The PPF-QCRs were placed in a flow sensor cell and simultaneously attached to an integrated circuit board equipped with a custom LSI for oscillation and resonant frequency measurements and a multiport serial interface as shown in Fig. 2. This sensor module can mount eight PPF-QCRs.

The sensor module was placed in a freeze sealing case. The relative humidity and temperature in the case were measured by a humidity sensor (HMT337, Vaisala, Finland, accuracy: ±1.0% for 0 to 90% relative humidity). As shown in Figure 1, the breath measurement system consisted of two gas flow lines: a sample line carries breath gas and a reference line carries base gas. In our sensing system, air in indoor atmospheric condition was used as a reference or cleaning gas for establishing the initial state of the QCR sensors. These flow lines were switched by a three-way valve to connect with the gas line of the sensor cell. A sample dryer (Desiccant/Membrane gas dryer DM-110-24, Perma Pure, U.S.A.) was inserted between the valve and the sensor module to dehumidify gas. The flow rates of all flow lines were controlled around 300 mL/min generally by a mass flow controller (SEC-E40MK3, HORIBASTEC, Japan, accuracy: ±1.0% F.S., control range: 2 ~ 100% F.S., flow range: 10/20/50/100/200/500 SCCM 1/2/3/5/10 SLM) and a dry vacuum pump (Linicon LV-125, NITTO-KOHKI, attainment vacuum level: −33.3 kPa,

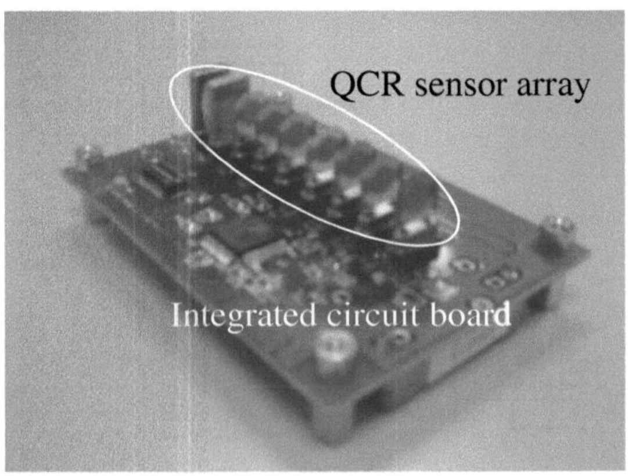

Fig. 2 Configuration of sensor module

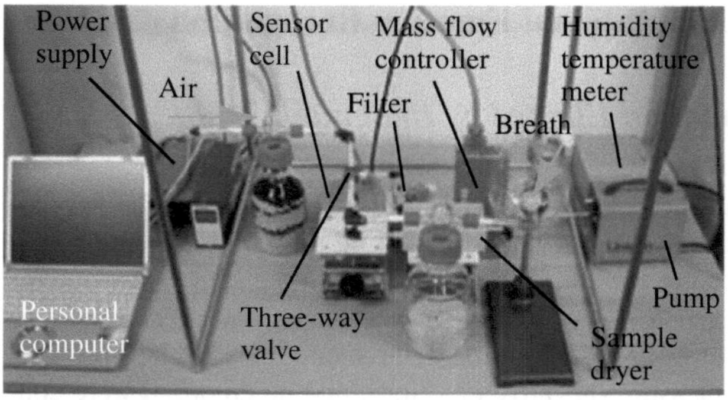

Fig. 3 Experimenal setup

exhalation: 5 L/min). To remove minuteness mist in the gas line, an inline filter (FT4-5, GL Science, Japan, diameter: 5 μm) was placed between the sensor module and the mass flow controller.

Figure 3 shows the experimental setup for measuring and analyzing breath gas using arrays of PPF-coated QCRs in the laboratory. As shown in Fig. 3, dehumidification by a molecular sieve and deodorization by a carbon were performed before atmosphere of the room was introduced into the reference line, and a sample pot was used as a buffer before introducing breath gas into the sample line. The molecular sieve was also in the sample pot in order to dehumidify the sample gas. A subject covered her/his mouth with a nose pad and blew breath as shown in Fig. 4. The breath gas was introduced into the sensor cell by switching the three-way valve after the baseline fluctuation had been suppressed to below ± 0.1 Hz for 1minute. Data were collected at 1-second interval for the QCR sensor array and the relative humidity/temperature sensor.

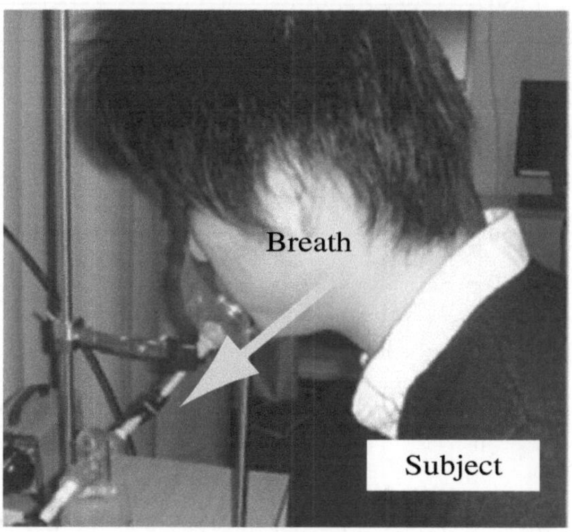

Fig. 4 Collection of breath gas

3 Experiments of Computational Emotion Recognition

3.1 Recognition of Emotion Elicited by Imaging [17, 18]

The responses of the breath gas sensing system were collected under the psychological experiments. By considering a guide to gathering physiological data for affective recognition described in the affective recognition work [1], the experiments were carried out following an *event-excited, lab setting, feeling*, *open-recording*, and *emotion-purpose* methodology. In this experiment, three emotions, such as pleasure (positive emotion) and displeasure (negative emotion), and no emotion, were considered. A subject imaged situations of feeling the emotion to excite the emotion, then the subject blew breath into the sample line of the sensing system. The experiment was carried out in our laboratory where the illumination, sounds, and room temperature were controlled to maintain uniformity. The subject was a male, native Japanese and the breath gas was collected four times, one time for each of the three emotions. Figure 5 shows examples of responses from each sensor cell. The horizontal axis is time and the vertical axis is the frequency shift. The subject blew breath to the sample line around 30 seconds, and the flow line was changed into the

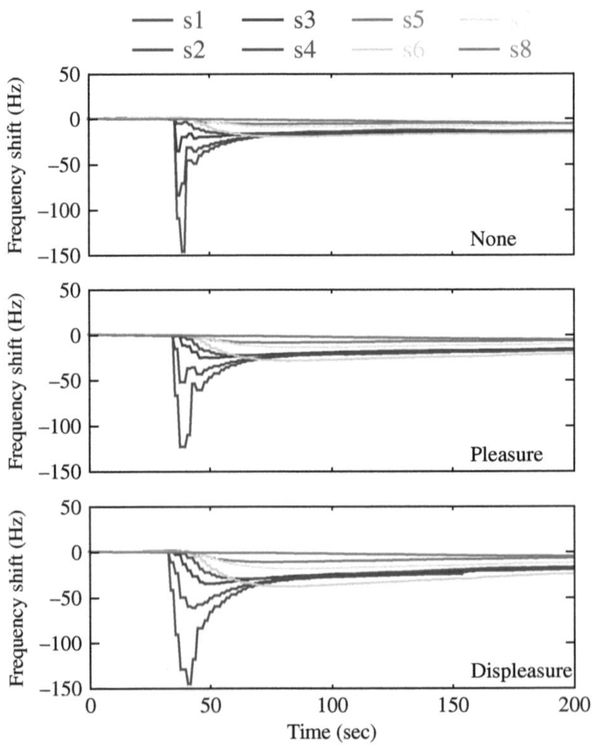

Fig. 5 Examples of responses from breath gas sensing system

Fig. 6 Feature extraction

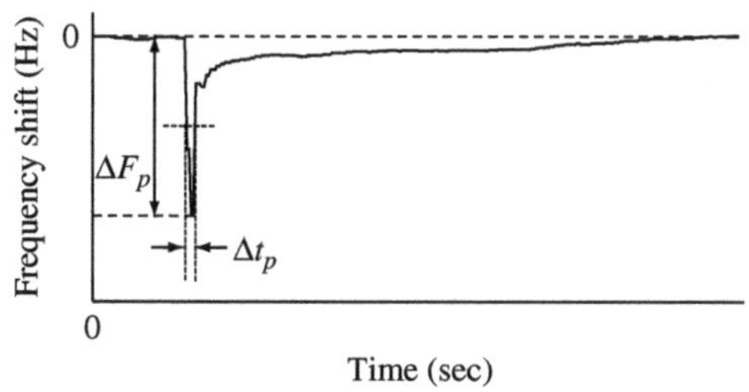

reference line around 90 seconds by the three-way valve. Except for the sensor 8, the response of the frequency shift shows a local minimum after breath introduction and then converges to around zero as shown in Fig. 5. The amplitude of each response curve depends on how strongly the subject blows breath to the sample line, however, the shape of the response curve is not affected.

Two features, such as peak Δf_p and pulse width Δt_p, can be extracted from each response curve of sensor. As shown in Fig. 6, the pulse width is defined by the interval that the output of the sensor crosses over the half of the peak. Average value of each feature is calculated by using collected data for every emotional state in every sensor as shown in Fig. 7, but remarkable differences between sensors are not observed by t-test with 5% level of statistical significance. Therefore, the machine learning-based approach is introduced for the emotion recognition task.

The emotion recognition system was composed of artificial neural network (ANN), which is a multi layer feedforward network, and decision logic as shown in Fig. 8. The relationship between inputs and outputs of the ANN is given as follows.

$$y_i = \sum_j w_{3_{ij}} f_j \left\{ \sum_k w_{2_{jk}} f_k \left(\sum_l w_{1_{kl}} x_l \right) \right\} \tag{2}$$

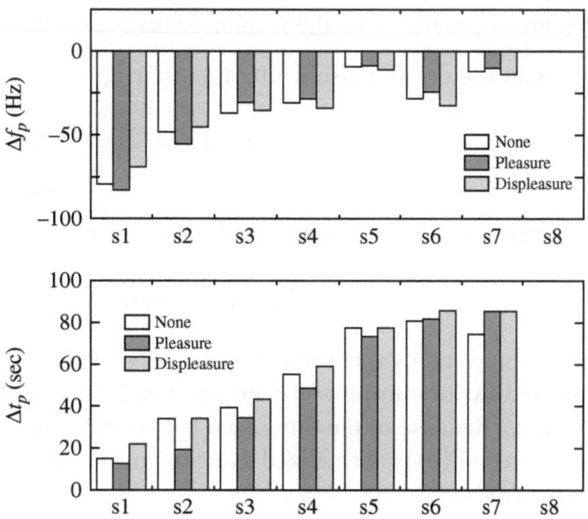

Fig. 7 Average features under emotional states

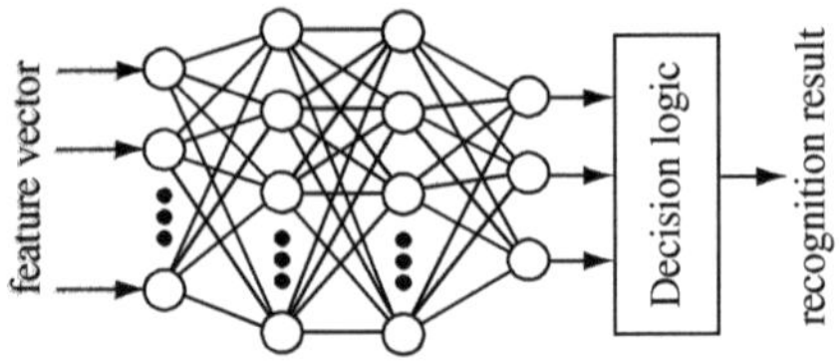

Fig. 8 Artificial nueral network-based emotion recognition system

where x_l is the input to the lth neuron in the input layer, y_i is the output of the ith neuron in the output layer, $w_{m_{qr}}$ ($m = 1,2,3$) is the weight that includes threshold, and $f(\cdot)$ is a sigmoid function. The learning of the ANN is carried out by the back propagation algorithm to minimize the cost function J as follows.

$$J = \frac{1}{2}\sum_p \sum_i \left(y_{d_i} - y_i\right)^2 \tag{3}$$

$$w_{m_{qr}}(t+1) = w_{m_{qr}}(t) - \eta \frac{\partial J}{\partial w_{m_{qr}}(t)} + \mu \Delta w_{m_{qr}}(t) \tag{4}$$

where η is the learning factor, α is the momentum factor, $\Delta w_{m_{qr}}(t)$ is the weight increments at the t-th iteration, y_{d_i} is the teaching signal, and p is the total number of the training data. The learning factors η and μ are tuned by the following rules.

$$\eta(t+1) = \begin{cases} (1+\beta)\eta(t) & J(t) < J(t-1) \\ \eta(t) & J(t) = J(t-1) \\ (1-\gamma)\eta(t) & J(t) > J(t-1) \end{cases}$$

$$\mu(t+1) = \mu(t) + \Delta\mu$$

where β, γ, and $\Delta\mu$ are positive constant values.

In the experiment, two types of input vector to the ANN were composed of the features extracted from the response curves as follows.

- input vector (a): pulse width $\Delta t_{p_i} (i = 1,2,\cdots,7)$

$$\boldsymbol{x} = \left[\Delta t_{p_1}\ \Delta t_{p_2}\ \cdots\ \Delta t_{p_7}\right]$$

- input vector (b): peak Δf_{p_i} and pulse width $\Delta t_{p_i} (i = 1,2,\cdots,7)$

$$\boldsymbol{x} = \left[\Delta f_{p_1}\ \Delta f_{p_2}\ \cdots\ \Delta f_{p_7}\ \Delta t_{p_1}\ \Delta t_{p_2}\ \cdots\ \Delta t_{p_7}\right]$$

The input vector (a) is used only the pulse width information since it is hard to be affected by how to blow the breath. On the other hand, the vector (b) utilizes all features. Because of the difference between the units of two features, each element of the vector (b) is normalized by using mean and standard deviation when it applies to the ANN.

Table 3 Emotion recognition results using ANN with input vector (a) [%]

In \ Out	Pleasure	Displeasure	None
Pleasure	75.0	25.0	0.0
Displeasure	50.0	25.0	25.0
None	25.0	25.0	50.0

There were 3 neurons in the output layer of the ANN. In the learning process of the ANN, the teaching signals $y_{d_1}y_{d_2}y_{d_3}$ were defined as follows: the set of [100] indicated displeasure, the set of [001] indicated pleasure, and the set of [010] indicated no emotion. The learning was quitted if the cost function achieved 10^{-4}. In the emotion recognition process, the output from the ANN was investigated in the decision logic that selects the best emotion; the output neuron whose output value closes '1' is chosen, and it indicates the recognition result. The leave-one-out cross-validation method was used to evaluate the recognition ability of ANN.

Table 3 shows the emotion recognition result by using the input vector (a) where a (7-11-11-3) network was used. The number of neuron in the hidden layers and the learning factors were determined by trial and error in order to converge the learning of ANN: $\eta(0) = 10^{-4}$, $\beta = \gamma = 10^{-5}$, $\mu(0) = 0.1$, and $\Delta\mu = 10^{-4}$. In the tuning of the learning factors, the upper and lower values of η were 0.1 and 10^{-5}, respectively, and upper value of α was 0.9. Though the recognition rate of 100% is achieved for the teaching data after the learning of ANN is completed, the averaged recognition rate defined with the average of the diagonal element is lower than 50% for the test data. Displeasure and no emotion are hard to recognise correctly. Table 4 shows the emotion recognition result by using the input vector (b) where a (14-14-6-3) network was used. The averaged recognition rate of 75% is achieved, however, misrecognition of pleasure increases.

As an alternative of using the ANN, support vector machine (SVM) was applied to the emotion recognition experiment. The SVM is originally designed for two-class classification and is finding the optimal hyperplane that minimizes the expected classification error of test samples by using statistical learning theory. Given a labelled set of training data ($\boldsymbol{x}_i$, u_i) where $\boldsymbol{x}_i$ is input vector ($\boldsymbol{x}_i \in R^N$) and u_i is the associated label ($u_i \in (-1,1)$), the optimal hyperplane $\boldsymbol{wx} + b = 0$ can be found by minimizing $||\boldsymbol{w}||^2 + C\sum_i \xi_i$ constrained by: $\xi_i \geq 0$ and $u_i(\boldsymbol{w}_i \cdot \boldsymbol{x} + b) \geq 1 - \xi_i$. Here ξ_i is the slack variable that is introduced to account for non-separable data, C is the margin parameter that quantifies the trade-off between training error and system capacity. Solving the quadratic programming problem, the optimal hyperplane can be defined as follows.

Table 4 Emotion recognition results using ANN with input vector (b) [%]

In \ Out	Pleasure	Displeasure	None
Pleasure	50.0	50.0	0.0
Displeasure	25.0	75.0	0.0
None	0.0	0.0	100.0

$$g(\boldsymbol{x}) = \sum_i u_i \alpha_i^* K(\boldsymbol{x}, \boldsymbol{x}_i^*) + b^* \tag{5}$$

where $K(\cdot,\cdot)$ is a kernel function, $\boldsymbol{x}_i^*$ is a support vector that corresponds to a nonzero Lagrange multiplier α_i^*, and b^* is a bias parameter. While the kernel function is a simple dot product for a linear SVM, a nonlinear function of the kernel function projects the data to high dimensional feature space in a nonlinear SVM and the optimal hyperplane is found in that space. Several kernel functions such as Gaussian and polynomial functions have been used in the nonlinear SVM.

$$K(\boldsymbol{x}, \boldsymbol{y}) = exp(-\gamma |\, \boldsymbol{x} - \boldsymbol{y}|^2) \tag{6}$$

$$K(\boldsymbol{x}, \boldsymbol{y}) = (\boldsymbol{x} \cdot \boldsymbol{y} + 1)^d \tag{7}$$

where γ is spread of a Gaussian cluster and d is the degree of freedom in a polynomial kernel. To apply the SVM for multiclass classification, the one-vs-all method [19] as shown in Fig. 9 was implemented. Three SVMs that correspond to each of the three emotions were used. In the emotion recognition process, the feature vector was simultaneously fed into all SVMs and the output from each SVM was investigated in the decision logic that selects the best emotion; the class of the SVM indicates the recognition result.

The input vector of the SVMs was the input vector (a). Training and testing of the SVMs were carried out by the leave-one-out cross-validation method. In the SVMs, the 3rd order polynomial function was used as the kernel function and the margin parameter C was chosen as 1 in the case of the input vector (a) and was set to 10 in the case of the input vector (b). The other parameters were defined by trial and error in order to achieve complete classification rate for training data. Tables 5 and 6 show the emotion recognition result. The averaged recognition rate of 83% is achieved with the input vector (a), while the recognition rate decreases with the input vector (b).

These results indicate that the machine learning technique such as ANN and SVM is suited for the emotion recognition task using breath gas information, but the recognition results are obtained by using only one subject's data. Therefore, the result obtained in this experiment depends greatly on person. More number of sample data should be required in order to evaluate the emotion recognition system in details.

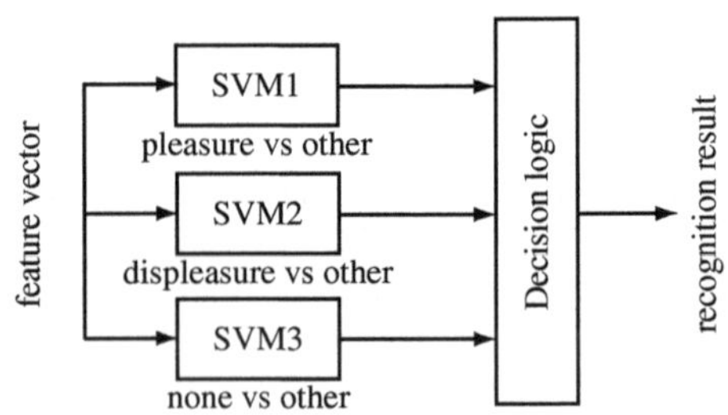

Fig. 9 SVM-based emotion recognition system

Table 5 Emotion recognition results using SVM with input vector (a)[%]

In\Out	Pleasure	Displeasure	None
Pleasure	100.0	0.0	0.0
Displeasure	25.0	75.0	0.0
None	25.0	0.0	75.0

Table 6 Emotion recognition results using SVM with input vector (b)[%]

In\Out	Pleasure	Displeasure	None
Pleasure	50.0	25.0	25.0
Displeasure	50.0	50.0	0.0
None	0.0	50.0	50.0

3.2 Recognition of Emotion Excited with Dental Rinse

In this experiment, two emotions, such as comfortableness (refreshing emotion) and no emotion, were considered. By using a dental rinse as a stimulus for exciting emotion, the responses of the breath gas sensing system were gathered. The experiments were carried out in our laboratory where the illumination, sounds, and room temperature were controlled to maintain uniformity. The experiments were done by 10 subjects of our laboratory (males, native Japanese). In the experiment, the subject first blew his breath into the gas sensing system under no emotion. After 10 minutes interval, the subject washed his mouth with the dental rinse and then blew his breath into the gas sensing system. The dental rinse is a commercial product (http://www.teamgum.net/eng/index.html). The emotions of subjects were evaluated by using questionnaires based on the multiple mood scale [20].

In order to compose input vector of the emotion recognition system, principle component analysis (PCA) is introduced to extract features from each response curve of sensor. The data matrix is defined by using all sensor data that are obtained during the interval of 30 seconds after the flow line was changed into the sample line. The PCA is applied to the data matrix, and two principle components are calculated in the 8 sensors. Figure 10 shows the PCA score plot. The horizontal axis is the first principle component (PC1) while the vertical axis is the second principle component (PC2). The top of Fig. 10 shows the PCA score plot obtained before using dental rinse. The bottom of Fig. 10 shows the PCA score plot obtained after using dental rinse. Here the results of all subjects are illustrated. The distribution of the PC2 is changed after using the dental rinse, however it is very small and depends on the subject.

In the questionnaires, 8 emotions, such as depression/anxiety, hostility, boredom, liveliness, well being, friendliness, concentration, and startle, were investigated before and after using the dental rinse. A semantic profile was utilized to evaluate the questionnaires. The semantic profile shows that the degrees of negative emotions (depression/anxiety, hostility, boredom) decrease while the degrees of positive

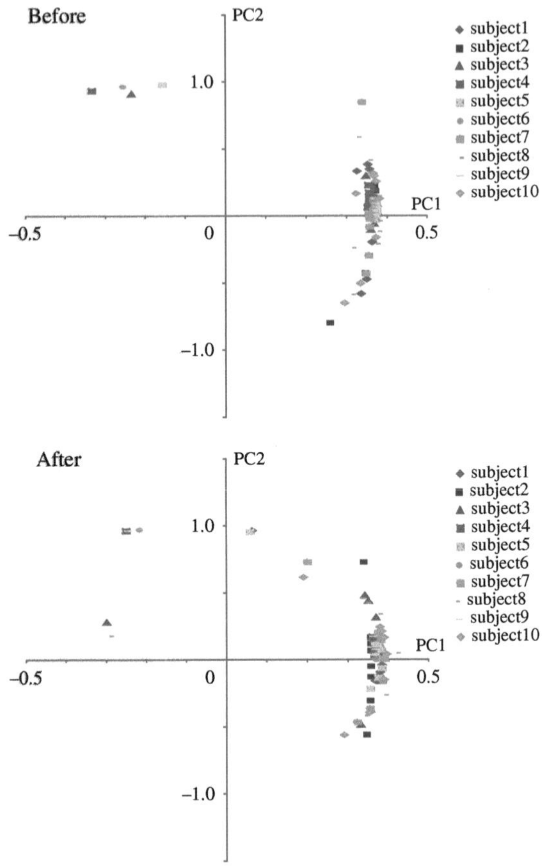

Fig. 10 Score plot of PCA (*top*: before using dental rinse, *bottom*: after using dental rinse)

emotions (liveliness, friendliness, concentration) increase after using the dental rinse. Therefore, we assume that the subject feel comfortableness/refreshing emotion by using the dental rinse.

To achieve computational emotion recognition, the ANN was used in the emotion recognition system. The input vector of the ANN was composed with the first and second principle components of PCA results for all sensors as follows.

$$\boldsymbol{x} = \left[PC_1^{S_1} \; PC_2^{S_1} \; PC_1^{S_2} \; PC_2^{S_2} \; \cdots \; PC_1^{S_8} \; PC_2^{S_8} \right]$$

where $PC_i^{S_j}$ is the ith principle component of the sensor j. Training and testing of the ANN were carried out by the leave-one-out cross-validation method. Table 7 show the emotion recognition result where a (16-12-8-6-2) network was used. The averaged recognition rate of 70% is achieved. This result shows the feasibility of our emotion recognition system based on the ANN using breath gas information.

Table 7 Emotion recognition results using ANN [%]

In\Out	None (before)	Refreshing (after)
None (before)	50.0	50.0
Refreshing (after)	10.0	90.0

4 Conclusions

This paper proposed a smart gas sensing system to achieve emotion recognition using breath gas information. A quartz crystal resonator with a plasma-polymer film was used as a sensor, and breath gas sensing system was designed under air in indoor atmospheric condition. The machine learning-based approach was conducted for computational emotion recognition and its characteristics were investigated. Experimental results demonstrated that using breath gas information is feasible and that the machine learning-based approach such as artificial neural network and support vector machine is well suited for this task.

There are many works to be done in the emotion recognition from breath gas information. In the sensing system, stabilization of the reference is required since the reference is easily affected from the air condition of the room. Improving how to introduce breath gas into the sensing-system is also important. Possible features of breath gas have to be selected in the feature extraction process, and further trials with different recognition methods may help improve recognition performance.

Acknowledgements This work was supported by JSPS Grant-in-Aid for Scientific Research (C) 18500142.

References

1. Picard R W, Vyzas E, Healey J (2001) Toward Machine Emotional Intelligence: Analysis of Affective Physiological State. IEEE Transactions on Pattern Analysis and Machine Intelligence, 23(2):1175–1191
2. Dellaert F, Polzin T, Waibel A (1996) Recognizing Emotion in Speech. In: Proceedings of Fourth IEEE International Conference on Spoken Language Processing, 3, Philadelphia, U.S.A., pp 1970–1973
3. Nicolson J, Takahashi K, Nakatsu R (2000) Emotion Recognizing in Speech Using Neural Networks. Neural Computing and Applications, 9(4):290–296
4. Essa I A, Pentland A P (1995) Facial Expression Recognition using a Dynamic Model and Motion Energy. In: Proceedings of International Conference on Computer Vision, Cambridge, U.S.A., pp 360–367
5. Yacoob Y, Davis L (1996) Recognizing Human Facial Expressions from Log Image Sequences Using Optical Flow. IEEE Transactions on Pattern Analysis and Machine Intelligence, 18(6):636–642

6. Ishino K, Hagiwara M (2003) A Feeling Estimation System Using a Simple Electroencephalograph. In: Proceedings of 2003 IEEE International Conference on Systems, Man, and Cybernetics, Washington, D.C. , U.S.A., pp 4204–4209
7. Takahashi K (2005) Remarks on Emotion Recognition from Multi-Modal Bio-Potential Signals. The Japanese Journal of Ergonomics, 41(4):248–253
8. Masaoka Y, Honma I (2004) The Effect of Pleasant and Unpleasant Odors on Levels of Anticipatory Anxiety: Analysis Observing Respiratory Patterns. AROMA RESEARCH, 5(1):44–49
9. Yoshino S, Kurai T (2001) A Laughter and Immunity. Journal of Clinical and Experimental Medicine, 197(12):916–917
10. Dubowski K M (1974) Breath Analysis as a Technique in Clinical Chemistry. Clinical Chemistry, 20:966–972
11. Manolis A (1983) The Diagnostic Potential of Breath Gas. Clinical Chemistry, 29:5–15
12. Phillips M (1992) Breath Tests in Medicine. Scientific American, 267:74–79
13. Yatagai M, Kawasaki M (2003) AROMA SCIENCE Series 21(4). FRAGRANCE JOURNAL Ltd., Tokyo
14. Sugimoto I, Nakamura M, Ogawa S, Seyama M, Katoh T (2000) Petroleum Pollution Sensing at Ppb Level Using Quartz Crystal Resonators Sputtered with Porous Polyethylene Under Photo-Excitation. Sensors and Actuators, B64:216–223
15. Seyama M, Sugimoto I, Miyagi T (2002) Application of an Array Sensor Based on Plasma-Deposited Organic Film Coated Quartz Crystal Resonators to Monitoring Indoor Volatile Compounds. IEEE Sensors Journal, 2(5):422–427
16. Sugimoto I, Nagaoka T, Seyama M, Nakamura M, Takahashi K (2007) Classification and Characterization of Atmospheric VOCs Based on Sorption/Desorption Behaviors of Plasma Polymer Films. Sensors and Actuators, B124:53–61
17. Takahashi K, Sugimoto I (2007) Neural Network Based Emotion Recognition Using Breath Gas Sensing System. In: Proceedings of the 2nd International Conference on Sensing Technology, Palmerston North, New Zealand, pp 467–472
18. Takahashi K, Sugimoto I (2006) Remarks on Breath Gas Sensing System and Its Application to Man-Machine Interface. In: Proceedings of the 3rd International Conference on Autonomous Robots and Agents, Palmerston North, New Zealand, pp 361–366
19. Chapelle O, Haffner P, Vapnik V (1999) Support Vector Machines for Histogram-based Image Classification. IEEE Transactions on Neural Networks, 10(5):1055–1064
20. Terasaki M, Kishimoto Y, Koga A (1992) Construction of a Multiple Mood Scale. Shinrigaku kenkyu : The Japanese Journal of Psychology, 62(6):350–356

Electrochemically Synthesised Pd- and Au-Nanoparticles as Sensing Layers in NO_x-Sensitive Field Effect Devices

K. Buchholt, E. Ieva, L. Torsi, N. Cioffi, L. Colaianni, F. Söderlind, P.O. Käll and A. Lloyd Spetz

Abstract An environmental pollutant of great concern is NO_x (nitrogen monoxide and nitrogen dioxide). Here we report the utilisation of electrochemically synthesised gold and palladium nanoparticles as catalytically active gate material on gas sensitive field effect sensor devices. The synthesised nanoparticles have been characterised by TEM and XPS, and the morphology of the thermally treated nanostructured sensing layers has been investigated using SEM and XPS. Measurements on the gas response of the palladium as well as the gold nanoparticle sensors towards a number of analytes found in automotive gas exhausts were performed and their response patterns were compared. The initial gas response measurements show interesting sensing properties for both the gold and the palladium nanoparticle sensors towards NO_x detection.

Keywords NO_x- sensor · gold · palladium · nanoparticle · field effect sensor

---

K. Buchholt
Department of Physics, Chemistry and Biology, Linköping University, Linköping, Sweden,
e-mail: kribu@ifm.liu.se

E. Ieva
Dipartimento di Chimica Università degli Studi di Bari, Bari, Italy

L. Torsi
Dipartimento di Chimica Università degli Studi di Bari, Bari, Italy

N. Cioffi
Dipartimento di Chimica Università degli Studi di Bari, Bari, Italy

L. Colaianni
Dipartimento di Chimica Università degli Studi di Bari, Bari, Italy

F. Söderlind
Department of Physics, Chemistry and Biology, Linköping University, Linköping, Sweden

P.O. Käll
Department of Physics, Chemistry and Biology, Linköping University, Linköping, Sweden

A. Lloyd Spetz
Department of Physics, Chemistry and Biology, Linköping University, Linköping, Sweden

1 Introduction

Fuel consists of hydrocarbons which break down into smaller molecules during combustion. For diesel engines some of the fuel ends up as particulate matter, one of the main pollutants in exhaust gases. Another main pollutant is nitrogen oxides (NO_x), which forms in the combustion chamber at high temperatures. NO_x react with hydrocarbons in sunlight and produce ozone and photochemical smog. Ozone can cause breathing difficulties and damage to plant life and NO_x in itself can increase respiratory illnesses and contributes to acid rain [9]. Due to the increased knowledge of the negative environmental and health effects of NO_x, emission legislations in both Europe and the United States have become much stricter. A great deal of research interest is therefore focused on developing NO_x-sensors.

Metal Insulator Semiconductor (MIS) field effect devices used as gas sensors can be of different types: transistors, Schottky diodes or capacitors, with transistors being preferred for commercial devices. The gas sensing principle for field effect sensors is based on molecules adsorbing and dissociating on a catalytically active gate material on the sensor. These interactions create a change in the electric charges on the semiconductor surface, which in turn results in a shift in the sensor output voltage. The interactions of the gas molecules with the gate material depend on the operating temperature and the morphology and chemical characteristics of the gate material [19]. The importance of the gate morphology has been shown for the detection of ammonia with a field effect sensor, where it influences both the sensitivity and the selectivity of the sensor [13]. Modifying the gate morphology by decreasing film thickness and optimising grain size has also been shown to give increased sensor responses for SnO_2 films [20].

The hydrogen response of MIS field effect devices is the most studied and is caused by dissociation of hydrogen on the catalytic metal and trapping of atomic hydrogen at the metal insulator interface where a polarised layer is formed. This dipole layer is observed as a voltage shift of the electrical characteristics of the device as described in [4]. The results of this study support a response mechanism where atomic hydrogen species, formed by the dissociation of hydrogen molecules on the catalytic metal surface, spill over to the silica surface and interact with the oxygen atoms of the insulator surface creating O-H bonds, as has been shown by DRIFT experiments for exposure of Pt/SiO_2 to H_2 and NH_3 [4, 21].

The use of nanostructured films as gate material has the potential to give sensors with increased sensitivity, and faster response and recovery due to the larger surface area available for interaction with the gas molecules, as compared to conventional thin film sensing layers. The nanostructure of the films also has the potential to introduce new selectivity properties of the sensors and the use of electrochemically synthesised nanoparticles gives the possibility to vary the size of the particle core, and/or the composition of the organic shell [8, 19].

MIS sensors with palladium as gate material that are sensitive towards hydrogen were described already during the 70s, reviewed in [11], and palladium has been the subject of intense research interest for hydrogen sensors ever since. Field effect sensors with palladium films as sensing layer have been reported to show response also towards other gases like ammonia and ethanol [12, 15], as well as being able to classify mixtures of H_2 with NO_2 and O_2, when used together with Pd/Ni sensors and a pattern recognition algorithm [7]. The use of electrochemically synthesised palladium nanoparticles in a chemiresistor-type sensing device for the detection of ethanol vapours is reported in [3].

Field effect gas sensors with dense polycrystalline gold films as gate material have previously been reported to show response towards NO_2, with faster response times and larger voltage shifts for thinner gates with smaller grain sizes [5]. A review made by Haruta on the catalytic properties of gold shows that its catalytic nature can be tuned by controlling the particle size and the architecture of the metal-support interaction [6], and the adsorption interaction of NO_2 with gold has been investigated in [10].

We have reported elsewhere the use of gold nanoparticles as gate material for NO_x field effect gas sensors [8]. Here a comparison is made between the use of electrochemically synthesised palladium and gold nanoparticles as catalytically active gate materials in capacitive field effect sensors capable of NO_x detection, as first reported and discussed in [1], with extended XPS characterisation of the sensing layers.

2 Experimental

2.1 Nanoparticle Synthesis and Characterisation

2.1.1 Electrochemical Synthesis

Palladium (or gold) nanoparticles were electrochemically synthesised using the Reetz sacrificial anode electrolysis [17, 18]. For the synthesis, a three-electrode cell was used with an $Ag/AgNO_3$ reference electrode, a palladium (or gold) sacrificial anode, and a platinum cathode. Electrolyte solution was composed of 0.1 M quaternary ammonium halide dissolved in a 3:1 mixture of tetrahydrofuran and acetonitrile. Quaternary ammonium salt acted both as supporting electrolyte and as stabiliser of the nanoparticles. During the synthesis the cell was kept under nitrogen atmosphere. Palladium nanoparticles were synthesised in the presence of tetra-butyl-ammonium bromide (TBAB), while gold nanoparticles were stabilised by tetra-octyl-ammonium chloride (TOAC). For further details regarding the synthesis conditions see [8].

2.1.2 TEM Characterisation

The nanomaterial morphology was characterised by means of transmission electron microscopy (TEM). The samples were observed at 100 kV using a PHILIPS 400 T TEM for the Au nanoparticle material and a Jeol JEM2010 for the Pd nanoparticle material.

2.2 Sensor Device Configuration and Nanoparticle Deposition

2.2.1 Device Configuration

The sensor devices employed were MIS capacitors with a catalytically active layer of Pd- or Au-nanoparticles as the gate material. The ohmic backside contact consists of evaporated, annealed Al. The semiconductor material used was p-doped Si. On this semiconductor surface a thermal oxide of SiO_2 was grown. Bonding pads of evaporated Cr/Au were then deposited on the insulator. The sensor chip, a ceramic heater, and a Pt-100 element for temperature control, were mounted on a 16-pin holder and electric contacts made from the sensor to the pins with gold bonding. A schematic view of the sensor device is shown in Fig. 1.

2.2.2 Nanoparticle Sensing Layer Deposition

A fixed volume of palladium or gold nanoparticles was drop deposited as a catalytically active gate material on the insulator of the sensor, partly overlapping the bonding pad. The solvent was evaporated through annealing in air at 300°C for the palladium nanoparticles, and at 200°C for the gold nanoparticles. The thermal treatment was performed at a higher temperature than the gas sensing measurements to avoid reconstruction of the nanomaterial sensing layer during the measurements.

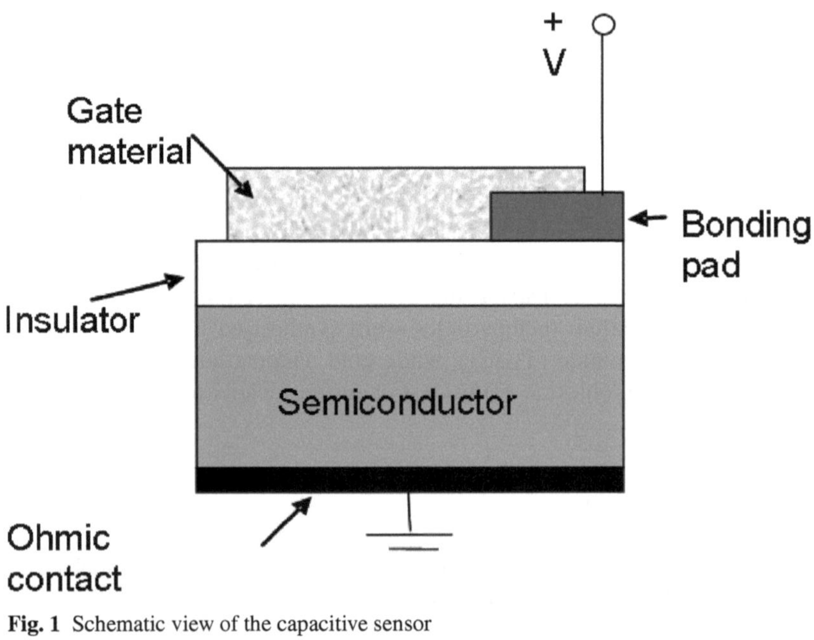

Fig. 1 Schematic view of the capacitive sensor

2.3 Chemical Composition and Sensing Layer Morphology

2.3.1 XPS Analysis

Surface chemical characterisation of the nanomaterials was performed by means of X-ray Photoelectron Spectroscopy (XPS) using a Thermo VG Theta Probe spectrometer equipped with a micro-spot monochromatic Al Kα source. Both survey and high-resolution spectra were acquired in fixed analyser transmission mode with a pass energy of 200 eV and 150 eV, respectively.

2.3.2 SEM Characterisation

Surface morphological study was carried out on annealed samples by means of Scanning Electron Microscopy (SEM) using a Zeiss Cross Beam 1540 EsB and a Leo 1550 VP Field emission SEM.

2.4 Gas Measurement Setup

During the gas sensing measurements the holders were mounted in aluminium blocks connected to a gas flow line. A computer-controlled gas mixing system was used for a flow of the test gases over the sensor surfaces at 100 ml/min. The measurements were performed using an MCM, (Multi Capacitor Meter) equipment, which makes it possible to measure the voltage, at a constant capacitance, of several devices in a time sharing mode. The output signal of the device is the voltage across the structure at a constant capacitance while exposing the sensor to gases. During the measurements the sensors were exposed to the following series of test gases: NO, H_2, CO, NH_3, and C_3H_6. The same test series was performed with NO_2 replacing the NO. The NO_x response was specifically investigated through exposing the sensors towards different concentrations of NO_x. Nitrogen with 10% O_2 was used as the carrier gas. Gas sensing measurements were performed while keeping the sensors at 150 and 175°C using a temperature control setup.

3 Results and Discussion

3.1 TEM Characterisation of the Synthesised Nanoparticles

Typical Transmission Electron Spectroscopy micrographs of palladium and gold nanoparticles are reported in Fig. 2a and b respectively. In the case of Pd-NPs, the average core diameter is equal to 2.4±0.5 nm, while Au-NPs have an average core

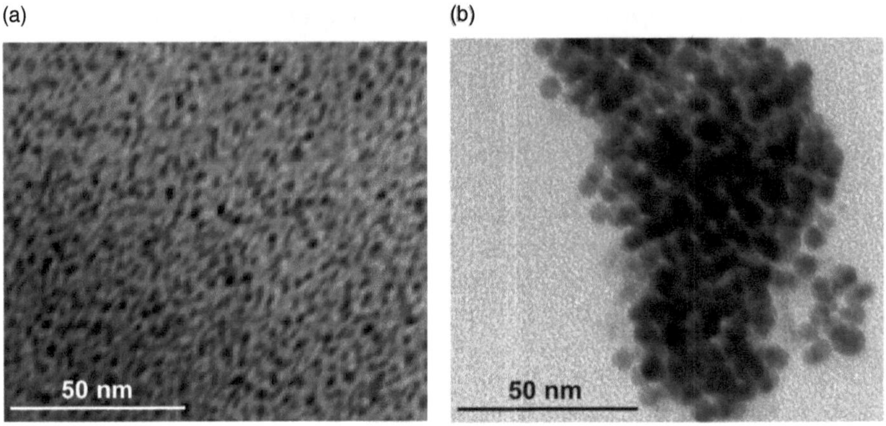

Fig. 2 TEM micrograph of Pd-nanoparticles (**a**), Au-nanoparticles (**b**)

diameter of 5.0±1.1 nm. In both cases, a very narrow size dispersion of the particle population was observed, the core diameter standard deviation being around 20% of the average core size.

3.2 Chemical Composition and Morphology

3.2.1 Chemical Composition

The surface chemical composition of both the as-synthesised and annealed nanomaterials was determined by X-ray Photoelectron Spectroscopy (XPS). For both the pristine metal nanocolloids, the surface carbon/metal atomic ratio was found to be close to 10^3; carbon, nitrogen and halides (bromide in case of palladium, chloride in case of gold particles) are usually the most abundant elements, this is due to the large excess of surfactant used during the synthesis. Palladium and gold were detected at lower concentrations. The surface chemical speciation of the nanostructured metals is reported in Fig. 3 (Pd3d high resolution XP spectra) and Fig. 4 (Au4f high resolution XP spectra).

The spectrum of pristine palladium nanoparticles (upper panel of Fig. 3) was composed of two doublets, i.e. two chemical states were detected.

The first one is attributed to nanostructured elemental palladium ($BE_{Pd3d5/2}$ = 335.3±0.1 eV), and is slightly shifted, as compared to the peak position of bulk metallic Pd ($BE_{Pd3d5/2}$ = 335.1±0.1 eV). Such evidence has already been demonstrated to be strictly correlated to the small size of the particles [2]. The second doublet ($BE_{Pd3d5/2}$ = 337.3±0.1 eV) is attributed to $Pd^{(II)}$ species, mainly in the form of bromides ($BE_{Pd3d5/2}$ = 337.5±0.1 eV), although the presence of low amounts of PdO ($BE_{Pd3d5/2}$ = 336.5±0.1 eV) cannot be excluded.

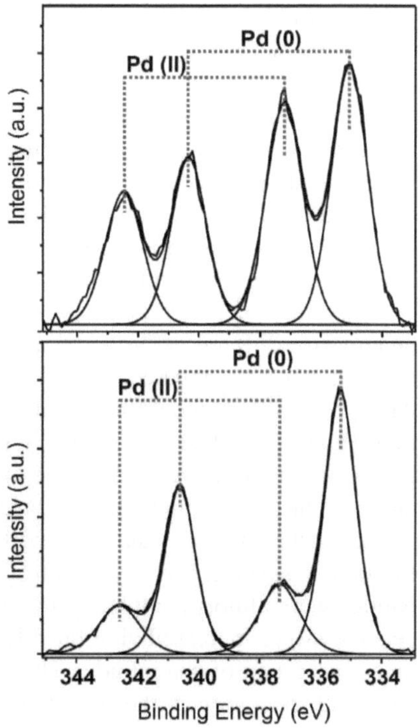

Fig. 3 XPS characterisation of Pd-nanoparticles stabilised by TBAB: high resolution XP spectra of pristine (*upper panel*) and annealed (*lower panel*) nanoparticles

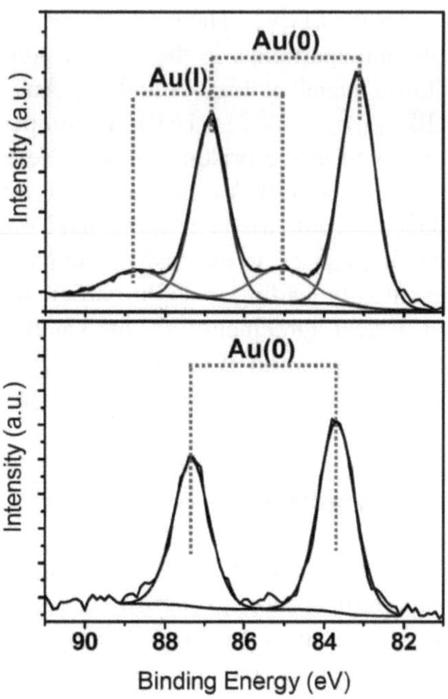

Fig. 4 XPS characterisation of Au-nanoparticles stabilized by TOAC: high resolution XP spectra of pristine (*upper panel*) and annealed (*lower panel*) nanoparticles

The spectrum of pristine gold nanoparticles (upper panel of Fig. 4) is composed of two doublets, that were attributed to nanostructured $Au^{(0)}$ ($BE_{Au4f7/2} = 83.0 \pm 0.1\,eV$) and $Au^{(I)}$ ($BE_{Au4f7/2} = 85.0 \pm 0.1\,eV$). The peak position of nanostructured elemental gold showed a marked BE shift, as compared to bulk gold ($BE_{Au4f7/2} = 84.0 \pm 0.1\,eV$). Noteworthy, the influence of particle size effects on the nano-Au XP spectra lead to a BE decrease, while, on the contrary, a size-related BE increase was observed in case of Pd nanoparticles. A deeper insight into the fundamental aspects leading to this well known chemical shift can be found in reference [16]. Finally, $Au^{(I)}$ doublet was attributed to NR_4AuCl_2 species, according to reference [14].

A comparison of XPS spectra of pristine Pd and Au nanoparticles shows apparent quantitative differences in the relative abundance of the species with higher oxidation number, as compared to the elemental chemical environment (nanostructured $Au^{(0)}$, or $Pd^{(0)}$). It should be noted that in pristine colloids, the oxidized/elemental ratio shows a marked sample-to-sample dependence, being affected by several experimental parameters (such as the electrosynthesis time, the halide concentration, and even the nanoparticle deposition conditions).

Thermal annealing caused a marked removal of the organic substances from the sample surface: carbon and nitrogen percentages were greatly decreased, while halides were almost completely removed. Gold and palladium surface atomic percentages were increased significantly and the C/metal ratio was lowered by more than one order of magnitude. The thermal treatment significantly modified the chemical speciation in the palladium and gold high resolution regions. The spectrum of annealed nano-palladium (lower panel of Fig. 3) was composed of two doublets, namely nanostructured $Pd^{(0)}$ ($BE_{Pd3d5/2} = 335.4 \pm 0.1\,eV$) and $Pd^{(II)}$ ($BE_{Pd3d5/2} = 337.4 \pm 0.1\,eV$). Their attribution is the same as described in the case of pristine nanoparticles. In the case of gold nanoparticles, the $Au^{(I)}$ doublet was absent (lower panel of Fig. 4) and the $Au^{(0)}$ doublet was shifted towards higher values ($BE_{Au4f7/2} = 83.7 \pm 0.1\,eV$), as compared to the *nano*-$Au^{(0)}$ chemical environment observed in the pristine sample. Nevertheless, the $Au^{(0)}$ BE value was still lower than that of *bulk* Au. Independently of the *initial* oxidized/elemental relative abundance, a comparison of upper and lower panels in Fig. 3 or 4 shows that thermal annealing causes an increase of the nanostructured $Au^{(0)}$ or $Pd^{(0)}$ relative abundance, eventually leading to slight particle aggregation and/or growth phenomena. This is in perfect agreement with the smaller size-dependent chemical-shift observed for the peak position of nano $Au^{(0)}$ and also matches with the results of Scanning Electron Microscopy (SEM) investigations reported in the next paragraph.

3.2.2 Sensing Layer Morphology

To study the effect of the annealing treatment on the morphology of the palladium nanoparticles, and to confirm that the particles were still nanostructured the samples were investigated using Scanning Electron Microscopy (SEM). Figure 5 shows a micrograph of palladium nanoparticles subjected to thermal annealing at 300°C. The SEM characterisation indicates that the annealing procedure has induced a size

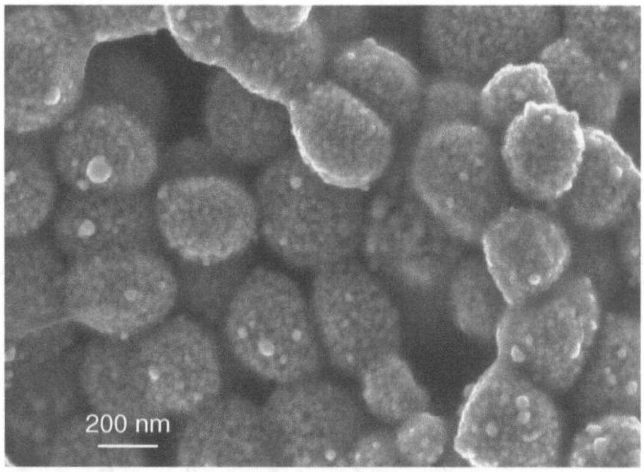

Fig. 5 SEM image of Pd nanoparticles annealed at 300°C

increase of the Pd nanoparticles. The material however appears to consist of larger clusters of smaller Pd nanoparticles, and remains nanostructured in its character.

The same investigation was made of the morphology of the gold nanoparticle sensing layers after the annealing treatment, as shown in Fig. 6. The annealing treatment causes an increase in the particle size as compared to the as-synthesised nanoparticles, also in this case, but the gold nanoparticle film remains nanostructured after the annealing.

3.3 Gas Sensing Measurements

In Fig. 7 a typical response curve at 175°C for a capacitive sensor with palladium nanoparticles as catalytically active gate material is shown.

These initial measurements show that the sensor has response towards hydrogen as would be expected from the numerous reports in literature of hydrogen sensors with palladium as catalytically active material [11, 12, e.g.]. The sensor also displays

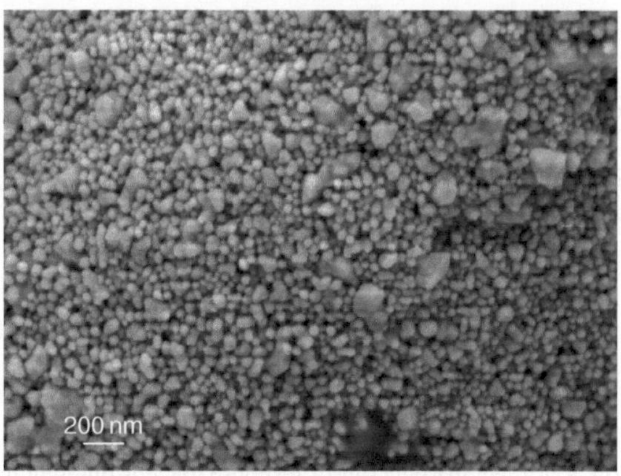

Fig. 6 SEM image of Au nanoparticles annealed at 200°C

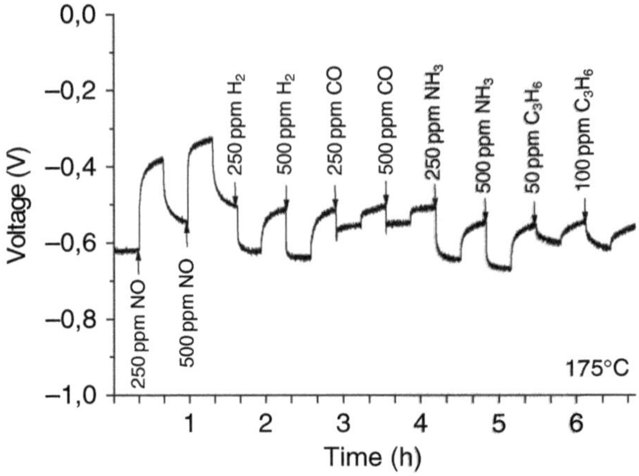

Fig. 7 Pd-sensor response towards different gases

some response towards CO, NH_3, and C_3H_6. The most interesting feature of the sensor is the relatively high response towards NO_x.

In Fig. 8 a typical Au-nanoparticle sensor response at 175°C is shown. The sensor has the largest response towards NO_x with some sensitivity towards H_2 and NH_3. Unlike the palladium sensor there is no measured response towards CO or C_3H_6 at the tested concentrations. The voltage shift when exposing both types of sensors to NO_x gas pulses, is towards positive values, which is consistent with the findings of Filippini et al. for MOS capacitors with dense polycrystalline gold as gate material, and attributed to the oxidising character of NO_2 [5].

Calibration curves for NO_2 at 175° C are shown in Figs. 9 and 10 respectively. Both the Au- and the Pd-nanoparticle sensors were able to detect NO_x in the tested concentration range between 50 and 200 ppm. For both the sensors with gold

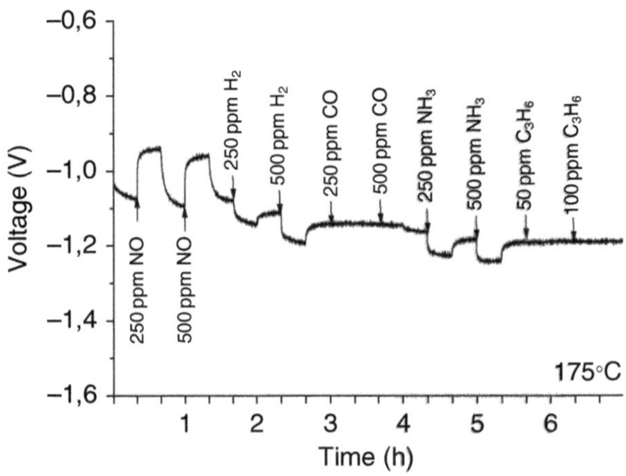

Fig. 8 Au-nanoparticle sensor response towards different gases

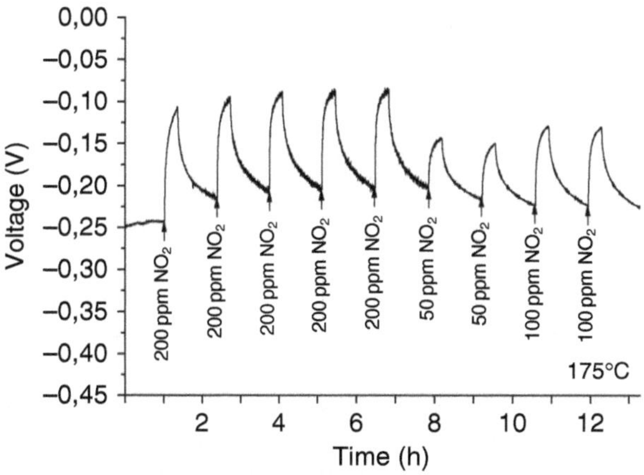

Fig. 9 NO_2-response of Pd-nanoparticle sensor

nanoparticles as sensing layer, and the sensors with palladium nanoparticles as sensing layer, similar responses were obtained for both NO and NO_2.

For low temperature applications the Au-nanoparticle sensing layers remain an interesting possibility for NO_x-detection sensors. However, the sensors with palladium nanoparticles as sensing layer are the most interesting candidates for further investigation related to applications which require higher operating temperatures during the measurements, due to the higher melting point of palladium and thereby higher resistance towards reconstruction of the sensing layer. The gas sensing measurements shown in Figs. 7 to 10 were also performed at 150°C, displaying a somewhat slower speed of response. Thus, higher operating temperatures above 200°C may speed up the gas response as shown in [22], and for the realisation of measurements at a higher operating temperature SiC could be used as the semiconductor material. This is however not a possibility for the Au-nanoparticle sensors since prolonged thermal treatment at 200°C causes particle growth phenomenon [8].

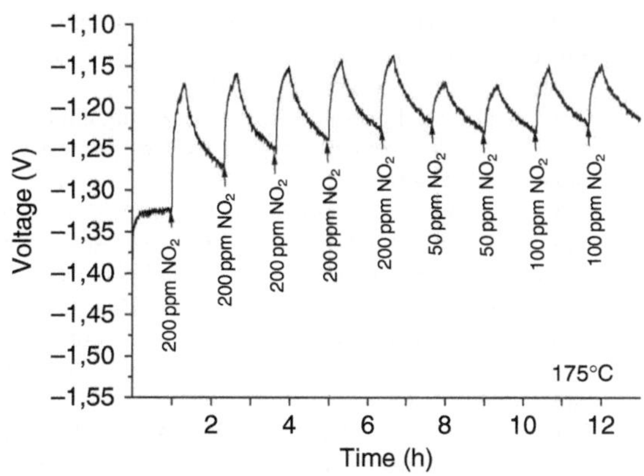

Fig. 10 NO_2-response of Au-nanoparticle sensor

Further investigations are needed of the sensitivity, selectivity, and cross-sensitivity of these sensors, as well as elucidation of the interaction mechanism between the nanomaterial sensing layers and the gas molecules.

3.4 Summary and Conclusions

Palladium and gold nanoparticles were electrochemically synthesised and characterised with XPS and TEM. The synthesised particles were then deposited as catalytically active sensing layers on capacitive field effect sensors and subjected to a thermal annealing procedure. The annealed sensing layers were investigated with XPS and SEM, which confirmed that the layers were still nanostructured. Measurements on the gas sensitivity of the palladium as well as the gold nanoparticle sensors towards a number of gases found in automotive gas exhausts were performed as well as NO_x calibration measurements. The sensor with Au as sensing layer showed the largest response towards NO_x, and smaller responses towards H_2 and NH_3. The sensor with Pd as sensing layer also displayed the largest response towards NO_x with smaller response for H_2, CO, NH_3 and C_3H_6.

The NO_x-sensor with gold nanoparticles as gas sensing gate material has slightly better selectivity towards NO_x than the sensor based on palladium nanoparticles. This difference in sensor response patterns gives the possibility to use the two types of sensors in a sensor array where they would give complementary information.

Acknowledgements The authors thankfully acknowledge the skilful assistance of Evald Mildh in mounting of the sensors and Ingemar Grahn for electronics support and development. Grants from the Swedish Research Council and CeNano, Linköping University are also acknowledged.

References

1. Buchholt K, Ieva E, Torsi L, Cioffi N, Colaianni L, Söderlind F, Käll PO, Lloyd Spetz A (2007) A comparison between the use of Pd- and Au-nanoparticles as sensing layers in a field effect NOx-sensitive sensor. In: Proc. 2007 ICST 2nd International Conference on Sensing Technology, New Zealand, pp 87–92
2. Cioffi N, Torsi L, Losito I, Sabbatini L, Zambonin PG, Bleve-Zacheo T (2001) Nanostructured palladium–polypyrrole composites electro-synthesised from organic solvents. Electrochimica Acta 46: 4205–4211
3. Cioffi N, Traversa L, Ditaranto N, Taurino AM, Epifani M, Siciliano P, Bleve-Zacheo T, Sabbatini L, Torsi L, Zambonin PG (2006) Core-shell Pd nanoparticles embedded in SnOx films. Synthesis, analytical characterisation and perspective application in chemiresistor-type sensing devices. Microelectronics Journal 37: 1620–1628
4. Eriksson M, Salomonsson A, Briand D, Åbom A (2005) The influence of the insulator properties on the hydrogen response of field-effect gas sensors. Journal of Applied Physics 98:034093
5. Filippini D, Weiss T, Aragón R, Weimar U (2001) New NO_2 sensor based on Au gate field effect devices. Sensors and Actuators B: 78: 195–201

6. Haruta M (1997) Size- and support-dependency in the catalysis of gold. Catalysis Today 36(1): 153–166
7. Hughes RC, Osbourne GC, Bartholomew JW, Rodriguez JL (1995) The detection of mixtures of NOx's with hydrogen using catalytic metal films on the Sandia robust sensor with pattern recognition. In: The 8th International Conference on Solid-State Sensors and Actuators, and Eurosensors IX, Sweden, pp 730–733
8. Ieva E, Buchholt K, Colaiannia L, Cioffi N, Sabbatini L, Capitani GC, Lloyd Spetz A, Käll PO, Torsi L (2007) Au nanoparticles as Gate Material for NO_x Field Effect Gas Sensors. Accepted to the special issue of sensor letters, EMRS 2007 spring meeting
9. Lloyd AC, Cackette TA (2001) Diesel engines: environmental impact and control. Journal of Air & Waste Management Association 51: 809–847
10. Lu X, Xu X, Wang N, Zhang Q (1999) Bonding of NO_2 to the Au atom and Au (111) surface: A quantum chemical study. Journal of Physical Chemistry A 103: 10969–10974
11. Lundström I, Sundgren H, Winquist F, Eriksson M, Krantz-Rülcker C, Lloyd Spetz A (2007) Twenty-five years of field effect gas sensor research in Linköping. Sensors and Actuators B 121: 247–262
12. Lundström I, Armgarth M, Lloyd Spetz A, Winquist F (1986) Gas Sensors based on catalytic metal-gate field effect devices. Sensors and Actuators 10: 399–421
13. Löfdahl M, Utaiwasin C, Carlsson A, Lundström I, Eriksson M (2001) Gas response dependence on gate metal morphology of field-effect devices. Sensors and Actuators B 80: 183–192
14. McNeillie A, Brown DH, Smith WE, Ginson M, Watson L (1980) X-Ray photoelectron spectra of some gold compounds. Journal of Chemical Society Dalton Transactions, 767–770
15. Peterson LG, Danetun H, Fogelberg J, Lundström I (1986) Oxygen as promoter or poison in the catalytic dissociation of H_2, C_2H_4, C_2H_2, and NH_3 on palladium. Applied Surface Science 27: 275–284
16. Radnik J, Mohr C, Claus P (2003) On the origin of binding energy shifts of core levels of supported gold nanoparticles and dependence of pretreatment and material synthesis. Physical Chemistry Chemical Physics, 5: 172–177
17. Reetz MT, Helbig W (1994) Size-selective synthesis of nanostructured transition metal clusters. Journal of American Chemical Society 116: 7401–7402
18. Reetz MT, Helbig W, Quaiser SA, Stimming U, Breuer N, Vogel R (1995) Visualization of Surfactants on nanostructured Palladium Clusters by a Combination of STM and High-resolution TEM. Science 267: 367–369
19. Salomonsson A, Roy S, Aulin C, Cerdà J, Käll PO, Ojamae L, Strand M, Sanati M, Lloyd Spetz A (2005) Nanoparticles for long-term stable, more selective MISiCFET gas sensors. Sensors and Actuators B 107: 831–838
20. Vuong D, Sakai G, Shimanoe K, Yamazoe N (2005) Hydrogen sulfide gas sensing properties of thin films derived from SnO_2 sols different in grain size. Sensors and Actuators B 105: 437–442
21. Wallin M, Grönbeck H, Lloyd Spetz A, Eriksson M, Skoglundh M (2004) Vibrational Analysis of H_2 and D_2 adsorption on Pt/SiO_2. Applied Surface Science 235: 487–500
22. Wingbrandt H, Lundström I, Lloyd Spetz A (2003) The speed of response of MISiCFET devices. Sensors and Actuators B 93: 286–294

Phthalocyanine Functionalized Hybrids as Receptors for Enhanced Gas/Bio Sensing

Al. Palaniappan, Shabbir Moochhala, Francis E.H. Tay, Moawia O. Ahmed and Nicky C.L. Phua

Abstract Cobalt Phthalocyanine (CoPc) functionalized silica matrix has been tested as receptors for gaseous bio-markers detection. Bio-marker such as Nitric Oxide (NO) is identified and quantified by tetra amino phthalocyanine (CoTAPc) functionalized silica matrix. CoPc is anchored chemically on to a mesoporous silica network which is deposited on a Quartz Crystal Microbalance (QCM). Time of Flight Secondary Ion Mass Spectroscopy (Tof-SIMS) is used to characterize the CoPc/silica hybrids. The sensor response towards the bio marker is tested in a sealed gas chamber. The sensor response of QCM functionalized with CoTAPc and non-modified QCM is also recorded. The experiment results show that sensitivities up to 25 ppb/Hz could be achieved by using the developed CoTAPc/silica hybrid on QCM.

Keywords Mesoporous silica · QCM · cobalt phthalocyanine · gas/bio sensor

1 Introduction

Over the past few decades, gas/bio sensing has been one of the interesting areas of research. There have been several reports on gas/bio sensors using several methodologies each of them having their own pros and cons [1]. Numerous applications

Al. Palaniappan
Department of Mechanical Engineering, National University of Singapore, Singapore,
e-mail: alps@nus.edu.sg

Shabbir Moochhala
DSO National Laboratories, 27 Medical Drive, Singapore 117510

Francis E.H. Tay
Department of Mechanical Engineering, National University of Singapore, Singapore

Moawia O. Ahmed
School of Materials Science and Engineering, Nanyang Technological University 50 Nanyang Avenue, Singapore 639798

Nicky C.L. Phua
Department of Mechanical Engineering, National University of Singapore, Singapore

S.C. Mukhopadhyay, G.S. Gupta (eds.), *Smart Sensors and Sensing Technology*,

such as e-noses, exhaled breath analysis, environmental monitoring, etc., demands gas/bio sensors that are highly sensitive, selective and field deployable. For applications such as breath analysis, the fluctuations in the concentrations of biomarkers are low and can only be significantly noticed in the sub-part per million level (sub-ppm) level. Shortcomings of current detector systems include high costs, complicated procedures and low sensitivities. Therefore, there is a need to develop inexpensive and simple-to-use sensor systems that can monitor the concentration of bio-markers at sub-ppm levels. In this paper, we report a methodology to detect bio markers such as Nitric Oxide (NO) at sub-ppm levels using Cobalt Phthalocyanine (CoPc) functionalized mesoporous silica matrix deposited on a Quartz Crystal Microbalance (QCM).

Metal phthalocyanines are chemically stable centro-symmetric planar organometallic molecules with an extensively delocalized two-dimensional conjugated π-electron system [2, 3]. In order to enhance the affinity of CoPc to the target analytes, the peripheral groups of the phthalocyanine molecules could be substituted with functional groups such as amino, carboxylic and nitro side groups. These groups can be broadly classified as electron-donating or electron-withdrawing groups. By substituting electron-donating side groups, the central metal atom becomes electron rich and hence improves the sensitivity towards oxidizing gases. Besides improving the affinity of CoPc molecules the side groups also serve as a link to covalently bond CoPc to the mesoporous silica matrix. Therefore, for NO detection, a stable CoPc-Silica matrix could be achieved by functionalizing CoPc with amino side groups.

QCM consists of a piezoelectric quartz disk sandwiched between two electrodes and vibrates at its resonant frequency when biased with an alternating current. The resonant frequency shift that is proportional to the change in the mass of the adsorbed layer deposited on its electrodes is continuously monitored to identify and quantify target analytes. The detection sensitivity of the QCM could be enhanced

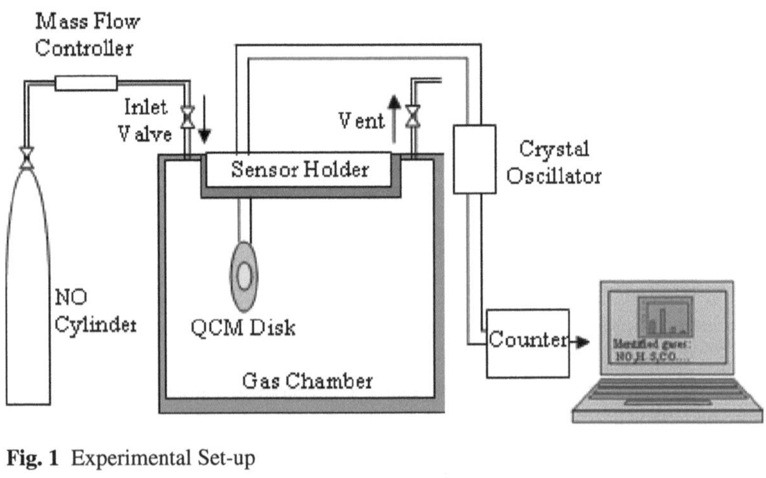

Fig. 1 Experimental Set-up

by immobilizing the sensitive receptor molecules on the mesoporous silica matrix as a three-dimensional sensing platform is created. CoPc functionalized with amino groups, tetra amino phthalocyanine (CoTAPc), is then chemically bonded to the silica matrix deposited on QCM electrodes. This covalent method of immobilization could be advantageous when compared to physis-sorption due to the increased stability of the CoPc-Si matrix and less susceptibility to selective leaching or deactivation [4]. The QCM carrying the CoTAPc-Si hybrid is then introduced in a sealed gas chamber (Fig. 1) containing the target analytes and the sensor response is then monitored continuously to identify and quantify the target analyte concentrations.

2 Materials and Apparatus

10 MHz, AT-cut QCM with polished gold electrodes (diameter 5.1 mm) were purchased from International Crystal Manufacturer (ICM, Oklahoma US) and are used as the sensing platform. Tetraethylorthosilicate (TEOS), ethanol, hydrochloride acid, Isopropyl Alcohol (IPA) and Dimethyl Sulfoxide (DMSO) were purchased from Sigma-Aldrich and used without further purification. These chemicals were used for preparing the silica films. Silica sol was prepared by an acidic catalysis process as described before [5]. Cobalt phthalocyanine, nitrophthalonitrile, ammonium chloride, sodium methoxide, cobalt (II) sulfate heptahydrate, sodium sulfide, 3-chloropropyl trimethoxysilane and xylene is purchased from Sigma-Aldrich and used without further purification.

CoPc could be covalently bonded to the silica matrix deposited on QCM based on the procedures described before [6, 7]. Briefly, one of amino side groups of the CoTAPc synthesized according to the previous report [6] is covalently bonded to 3-chloropropyl trimethoxysilane [7]. The resulting compound is then covalently bonded to the silica matrix deposited on the QCM by elimination of a methanol molecule, i.e., by eliminating methyl group from 3-chloropropyl trimethoxysilane attached to CoTAPc and an OH group on the surface of the silica matrix. In order to evaluate the sensitivity enhancement, CoTAPc is also directly immobilized on the QCM electrodes. CoTAPc is dissolved in DMSO and spin/drop coated on QCM electrodes and subsequently calcined to obtain the planar receptor film (planar QCM).

The QCM deposited with CoTAPc functionalized silica matrix is introduced to a sealed gas chamber (~13 L) containing the target analytes. Flexible hoses are used to connect the sensing chamber to the NO and N_2 cylinders. Mass flow controllers control the concentration of the analytes that are fed to the chamber. Upon attaining the stable resonant frequency, N_2 was flushed in to remove any absorbed moisture. The target analytes were then introduced into the chamber and the resonant frequency is simultaneously recorded by a computer with the aid of LabVIEW interface software. The resonant frequency shift is then correlated to the target analyte gas concentrations.

3 Results and Discussion

Immobilizing CoPc in a mesoporous silica matrix enhances the detection sensitivity of the sensor by creating a three-dimensional sensing platform. The detection sensitivity could be further increased by substituting amino electron-donating side groups. After substitution, the central metal atom of CoTAPc becomes electron rich and hence improves the sensitivity towards oxidizing gases. Since the method of bonding between NO molecule and CoTAPc molecule is an electron transfer from the central metal atom to the NO molecule, it is reasonable to postulate that the substituent side groups need to be electron-donating for improved bonding between CoTAPc and NO. The amino group is strongly electron donating and therefore enhances the NO detection sensitivity.

Matrix assisted laser desorption/ionization time-of-flight (MALDI-TOF) mass spectra of CoTAPc deposited on QCM electrodes were obtained on a Bruker Autoflex TOF/TOF instrument. Figure 2 shows that a sharp peak at ~631 is noticed for the QCM deposited with CoTAPc. A molecular weight of 631 corresponds to the molecular weights (MW) of CoPc and tetra amino groups [MW ($CoPc+4^{*}NH_2$)] fragments. This peak confirms that amine groups are attached to all the four side of CoPc. Despite maintaining the sensing sites, this CoPc derivative enables covalent bonding to the silica matrix.

The QCM deposited with CoTAPc is then placed in a sealed chamber and exposed to known concentrations of NO. Figure 3 shows the NO response of QCM

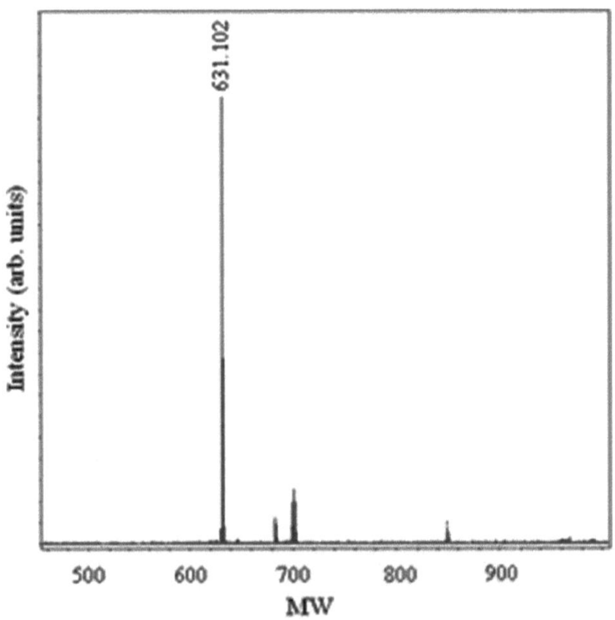

Fig. 2 Mass spectra of the CoTAPc deposited on QCM

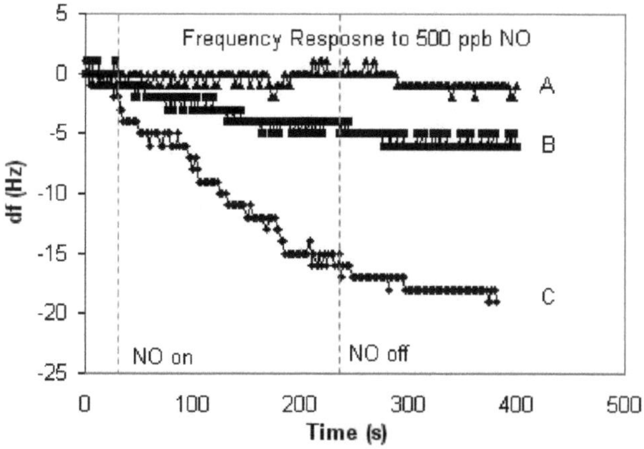

Fig. 3 Sensor response to 500 ppb NO of non-modified QCM (curve A), planar QCM (curve B), and CoTAPc/silica hybrid (curve C).

functionalized by CoTAPc (Planar QCM; QCM without mesoporous silica film - Curve B) and QCM functionalized with CoTAPc/silica hybrid (curve C). The QCM deposited with CoTAPc/silica hybrid effectively adsorbs NO and there is negligible adsorption of NO on a non-modified QCM (Curve A). The CoTAPc in the silica network adsorbs NO due to the affinity between the central metal atom (Co) and NO, thereby increasing the mass of the crystal. As a result, the resonance frequency of the QCM decreases continuously with time till a saturated adsorption is reached. Saturation occurs when all the available bonding sites in the CoTAPc-Si matrix are occupied by NO molecules. QCM deposited with the CoTAPc/silica hybrid shows better frequency response around 16 Hz to 500 ppb NO and when compared to planar QCM and thereby greater detection sensitivity. For a test concentration of 500 ppb, the detection sensitivity of QCM functionalized with CoTAPc/silica hybrid is approximately three times higher than that of planar QCM.

Figure 4 shows the preliminary results for sensor regeneration studies. It can be seen from Fig. 4 that the adsorbed NO desorbs on heating the QCM for 1 hour at 250°C. The tracing back of resonant frequency values to initial values suggest that the sensor regeneration could be achieved by heat treatment process. Figure 4 shows that the resonant frequency does not reach the initial resonant frequency value after heat treatment for 1 h, however, full regeneration is expected after heating 12 h [8]. Figure 4 also shows that almost similar frequency shift values for same concentration are obtained for subsequent measurements, however the sensor is saturated due to the occupation of bonding sites by NO. Therefore, the adsorbed NO must be desorbed to obtain repeatability, which could be achieved by the suggested heat treatment process [8].

A random test concentration of 0.3 ppm (parts per million) is chosen to evaluate the detection sensitivity of the CoTAPc/Si functionalized QCM. Upon obtaining stable resonantfrequency, 0.3 ppm NO is introduced in the test chamber containing

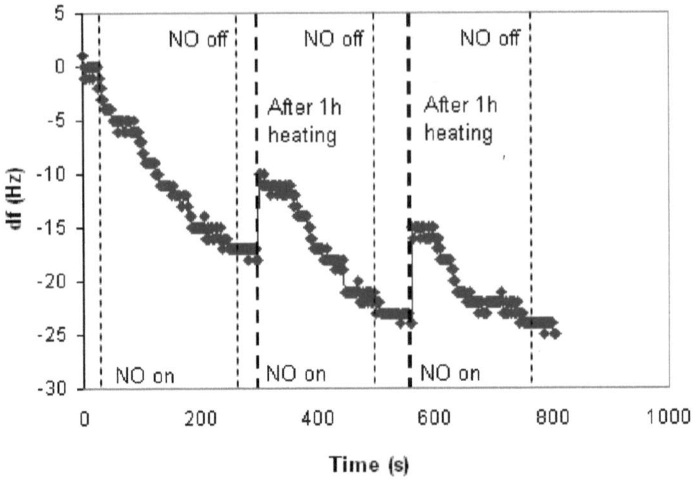

Fig. 4 Sensor regeneration

the QCM functionalized with Si-CoTAPc. A maximum frequency shift of ~12 Hz is observed (in approximately 200 s), which shows that the NO detection sensitivity of the QCM functionalized with Si-CoTAPc is ~25 ppb/Hz (parts per billion/Hz). Figure 5 confirms that the frequency shift of the CoTAPc deposited QCM is only due to NO adsorption as there is no significant change in the sensor response of the uncoated QCM. Furthermore, it could be concluded the there is no significant interference of ambient temperature or pressure on the sensor response.

Figure 6 shows the frequency response summary of QCM deposited with the Si/CoTAPc hybrid to different concentrations of NO. The frequency response of the QCM to the same concentrations of CO is also recorded in order to evaluate the

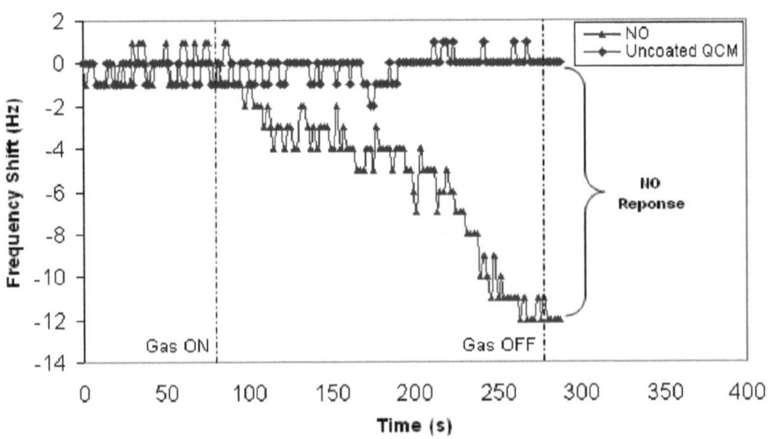

Fig. 5 Sensor response to 300 ppb NO

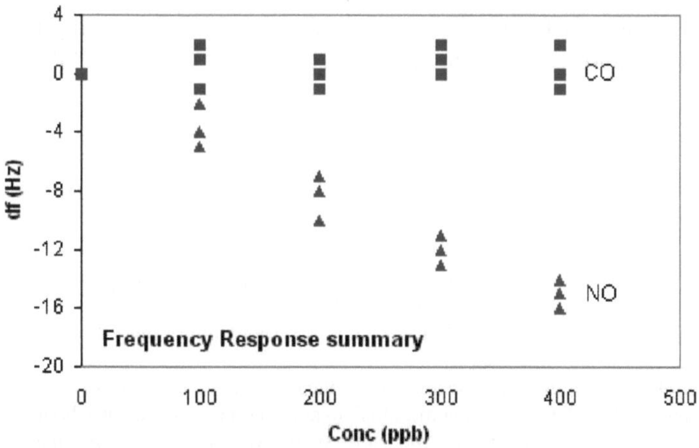

Fig. 6 Frequency response summary

selectivity of the developed sensor. It can be seen from Fig. 6 that QCM deposited with the Si/CoTAPc hybrid exhibits much higher response to NO than CO, showing that the developed Si/CoTAPc hybrid selectively adsorbs NO. The higher adsorption of NO could be attributed to the fact that NO molecules have unpaired electrons in the valence shell in contrast to CO. Figure 6 also shows that NO frequency shift – concentration relationship is fairly linear for a wide range of concentrations.

For all the test concentrations, the NO response is higher when compared to CO, which shows that selective NO detection systems could be developed by incorporating Si/CoTAPc hybrid with QCM. The selectivity of the developed sensors could be further enhanced by incorporating appropriate filters to eliminate contaminants and moisture. Soda lime absorbents could also be used to remove interfering species such as CO_2 [9]. The experiments were repeated several times to confirm the reproducibility of the sensors. As shown in Fig. 6, similar frequency shift values are obtained for three measurements, which confirms the reproducibility of the developed sensor.

4 Conclusion

A QCM based, simple to use, NO detection system has been proposed. A strategy to functionalize the silica matrix deposited on QCM with CoPc has been proposed. The CoPc could be bonded to the mesoporous silica matrix via covalent linkages there by creating a stable silica-receptor matrix for NO detection. The Si/CoTAPc hybrid has been tested as receptors for NO detection. Experiment results shows that the CoTAPc/Si hybrid exhibits poor sensitivity to CO. Sensitivities up to 25 ppb/Hz have been achieved by the developed system. Further studies will focus on selective bio-marker detection strategies. Further studies will also focus on optimizing the CoPc-Si functionalization and on the regeneration and reusability of the CoTAPc/Si hybrids.

References

1. K. Arshak E. Moore, G.M. Lyons, J. Harris, and S. Clifford (2004) A review of gas sensors employed in electronic nose applications. Sensor Review, 24: 181–198
2. C.C.Leznoff, A.B.P. Lever (Eds.) (1989, 1992, 1993, and 1996) Phthalocyanines, Properties and Applications. Vol. 1–4, VCH, New York
3. A.L. Thomas (Ed.) (1990) Phthalocyanine research and applications. CRC Press, Boston
4. X.-B. Lu, H. Wang, and R. He (2002) Aluminum phthalocyanine complex covalently bonded to MCM-41 silica as heterogeneous catalyst for the synthesis of cyclic carbonates. Journal of Molecular Catalysis A: Chemical, 186: 33–42
5. K. Alexander, G. Martijn, S. Vera, S.I. Nico, A.J.M. Sommerdijk, J.A. Jansen, and R.J.M. Nolte (2001) Silica-based hybrid materials as biocompatible materials for glucose sensors. Sensors & Actuators B, 81: 68–75
6. B.N. Achar and K.S. Lokesh (2004) Studies on tetra-amine phthalocyanines. Journal of Organometallic Chemistry, 689: 3357–3361
7. T. Buck, D. Wiihrle, G. Schulz-Ekloff, and A. Andreev (1191) Structure and mercaptan oxidation activity of cobalt (II) phthalocyanines covalently bonded to silica of low surface area. Journal of Molecular Catalysis, 70: 259–268
8. P. Ballirano, R. Caminiti, C. Ercolani, A. Maras, and M.A. Orru (1998) X-ray powder diffraction structure reinvestigation of the α and β forms of cobalt phthalocyanine and kinetics of the $\alpha \rightarrow \beta$ phase transition. Journal of American Chemical Society, 120: 12798–12807
9. M. Teresa, S.R. Gomes,* P. Sergio, T. Nogueira, A.C. Duarte, and Joao A.B.P. Oliveira (1999) Development of a methodology for the determination of carbon monoxide using a quartz crystal microbalance. Analyst, 124: 1449–1453

A Comparison of the Gas Sensing Properties of Purified and Platinum Decorated Chemical Vapour Deposition Grown Multi Walled Carbon Nanotubes

R.E. Pearce, M. Andersson, J. Belmonte, I. Gracia, M.M. Stevens, K. Buchholt, J. Shaw, M.S.P. Shaffer and A. Lloyd Spetz

Abstract Multi walled carbon nanotubes (MWCNTs) are known to respond well to a range of gases and vapours. Metallic additives are often introduced to improve the sensitivity and selectivity to some gases. Here the difference in response between MWCNT sensors with and without a metal additive is discussed for a range of gases. Resistive sensors fabricated from chemical vapour deposition (CVD) grown multi walled carbon nanotubes (MWCNTs) collected between gold microelectrodes by dielectrophoresis are presented as a cheap, scaleable and facile method of producing carbon nanotube gas sensors. The MWCNTs sensors were exposed to a series of test gases including NO_2, NH_3, CO and H_2 and exhibited low ppm. detection at room temperature. Increasing the temperature not only reduced the recovery time of the sensors, but also increased the sensitivity to some gases whilst a decrease in the sensitivity was seen for other gases. A method for decorating MWCNTs with

R.E. Pearce
Department of Chemistry, Imperial College London, UK; Department of Physics, Chemistry and Biology, Linköping University, Sweden, e-mail: ruth@ifm.liu.se

M. Andersson
Department of Physics, Chemistry and Biology, Linköping University, Sweden

J. Belmonte
Department of Materials, Imperial College London, UK

I. Gracia
National Centre for Microelectronics (CNM), Bellaterra, Spain

M.M. Stevens
Department of Materials, Institute of Biomedical Engineering, Imperial College London, UK

K. Buchholt
Department of Physics, Chemistry and Biology, Linköping University, Linköping, Sweden

J. Shaw
Tyco Safety Products, UK

M.S.P. Shaffer
Department of Chemistry, Imperial College London, UK

A. Lloyd Spetz
Department of Physics, Chemistry and Biology, Linköping University, Linköping, Sweden

S.C. Mukhopadhyay, G.S. Gupta (eds.), *Smart Sensors and Sensing Technology*,

nanometer sized Pt particles is presented and compared as a sensing material to purified MWCNTs which have undergone acid reflux and base washing. The role of defects and the sensing mechanisms are discussed.

Keywords Multiwalled carbon nanotubes · CVD · NO_2 · NH_3 · CO · H_2 · resistive sensors

1 Introduction

CNT sensors show response for ppm levels of a range of gases at room temperature, making them ideal for low temperature applications or applications where low power consumption is required. While at low temperatures recovery times are long for many gases, running the sensors at temperatures above room temperature reduces the recovery time significantly.

Temperature affects not only the recovery time but also the sensitivity and selectivity of the MWCNT sensors. It is demonstrated that higher temperatures can enhance or decrease the response of a MWCNT sensor to a gas. For gases which only weakly physisorb to the MWCNTs lower temperatures are required for maximum sensitivity, as at higher temperatures this mechanism becomes unfavourable.

Processing and purification of CNTs also has an effect on the gas sensitivity of MWCNT sensors as the tube walls can be modified with acid groups during the process, and defects in the tube walls may be increased. Modifications such as the decoration of the MWCNTs with metal nanoparticles, also affect the dominant sensing mechanism. The mechanisms of sensing are elucidated by comparison of the magnitude of response of modified MWCNTs at different temperatures.

The work presented here is based on a conference paper presented at the 2nd International Conference on Sensing Technology (ICST), Palmerston North, New Zealand [12].

2 Experimental

2.1 Synthesis of Carbon Nanotubes

The CVD equipment used is similar to that described by Andrews et al. [1]. The growth parameters were varied until long aligned mats of MWCNTs were grown. The growth parameters used are based on those described by Singh et al. [15]. A brief description of both CVD apparatus and growth parameters follows.

A solution of 3wt% ferrocene in toluene was injected into a two-stage tube furnace (Lenton PTF15) using a motorised syringe pump. The first stage of the furnace was maintained at 200°C where the solution vapourises. The mixture then flows into the main tube furnace which is maintained at 760°C by a temperature controller

Fig. 1 SEM image showing a mat of long, entangled and aligned MWCNTs

(Eurotherm 2216). The growth substrate was a quartz tube with internal diameter of 46 mm. A mixture of argon and 10% hydrogen was used as the carrier gas. The total gas flow rate was maintained at 2000 ml/min, and the injection rate of ferrocene in toluene was 5 ml/hour. The ferrocene decomposes to provide the iron catalyst required to nucleate the CNT growth with an Fe particle found at the root of each CNT, whereas toluene acts as the carbon feedstock.

Figure 1 shows an SEM image of these MWCNTs which in long (~ 500 mm), entangled, aligned mats perpendicular to the walls of the quartz furnace tube.

2.2 Purification and Modification of Carbon Nanotubes

Acid refluxes are commonly used to improve the purity and dispersibility of CNTs in aqueous solution. It has recently been shown [17] that including a basewashing step after the acid reflux removes residual impurities and contaminants associated with this process. Figures 2a and 2b show the CNTs before and after purification respectively.

The purification process which has been used in this work has been discussed elsewhere [17], however a brief description will be given. The acid reflux continued as has frequently been described in the literature [6]; MWCNTs grown by CVD were refluxed in a 3:1 concentrated H_2SO_4/HNO_3 mixture at 120°C for 30 min, and then thoroughly washed with distilled water, until the filtrate was colourless and pH neutral. The refluxed CNTs formed a stable dispersion in water. A base washing step using 0.01M NaOH initially produced a yellow/brown filtrate, the washing was continued until the filtrate was colourless. The resulting CNT dispersion was then washed with distilled water until pH neutral. Finally, the product was washed with 0.01 M HCl, then washed until pH neutral with distilled water.

This process not only cleans the surface of the CNTs and removes much of the amorphous debris, but it also enables some modifications to be carried out by

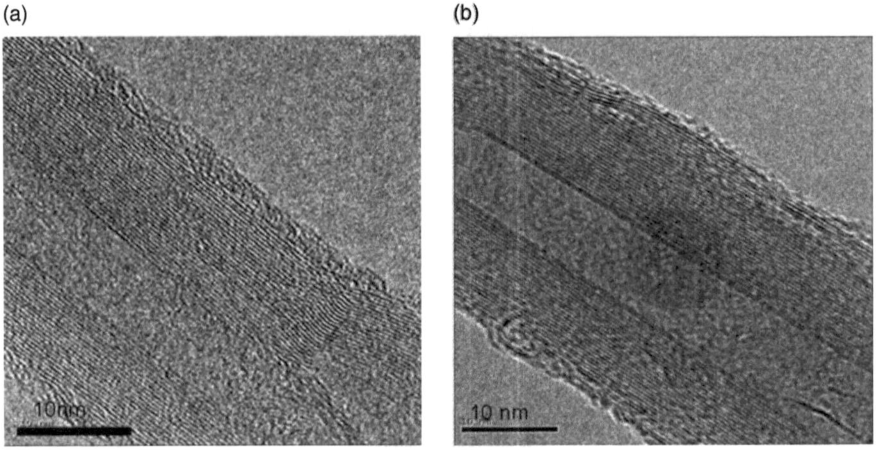

Fig. 2 TEM images **a** before and **b** after purification and vacuum annealing

facilitating the formation of a more stable aqueous solution. To remove any residual acid groups on the surface of the CNTs a vacuum annealing process was undertaken; The CNTs were placed into a tube furnace and heated to 1100°C for 4 hours under vacuum. After the annealing process the MWCNTs were no longer as readily soluble in water due to the desorption of polar groups from the surface of the outer walls which were introduced during the growth and purification process. An ultrasonic probe tip was used in combination with an ultra sonic bath to disperse the CNTs. These processes shorten the length of the CNTs and cause an increase in defects seen in the tube walls. The resulting suspension was only stable for a few hours.

After purification a platinum additive was introduced to a sample of the CNTs, by mixing an aqueous solution of $KPtCl_4$ with an aqueous suspension of purified CNTs. The suspension was then rapidly dried in an oven. The ratio of carbon to platinum was 98 to 2wt%. The platinum salt was then reduced to the metal by reduction under hydrogen: The dried CNT/Pt salt mixture was placed into a tube furnace which was heated to 100°C for 20 minutes under a flow of argon with 10% hydrogen. Small particles between 1 and 5 nm in diameter were formed on the outer tube wall as observed by TEM (Fig. 3). The Pt particles were well separated with very little visible agglomeration of particles seen. Figure 4 shows the size distribution of the Pt particles.

2.3 Sensor Device

The resistive sensor device consisted of gold interdigitated microelectrodes with spacing of 5 μm on SiO_2/Si (fabricated at CNM, Spain) which were glued to a ceramic (Al_2O_3) substrate, with a thin resistive-type Pt heater wire, to enable heating of the sensor device (Fig. 5). The ceramic heater was mounted on to a 16 pin header with gold plated contacts. The heater is suspended above the header to enable rapid

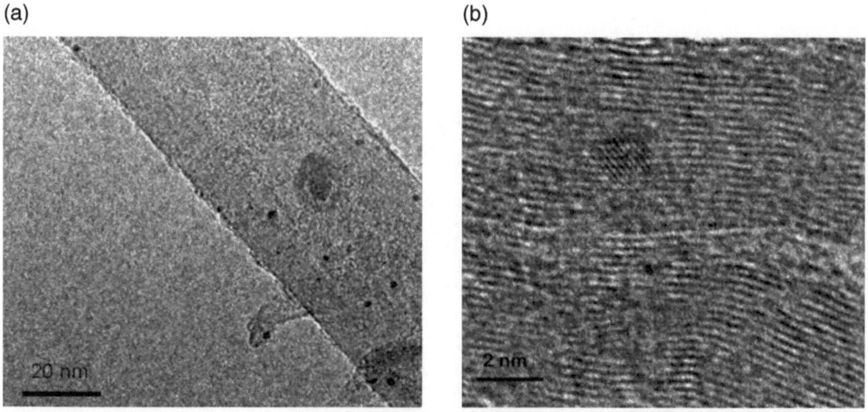

Fig. 3 **a** TEM image of Purified MWCNT decorated with a catalytic platinum additive at 2wt% **b** High magnification image of a Pt particle

changes in temperature to be made. A Pt100 temperature sensor was also mounted onto the ceramic heater for control of the operating temperature. The interdigitated microelectrodes are bonded through gold wires to gold contacts on the pins of the header.

2.4 CNT Deposition on to Gold Microelectrodes

Aqueous solutions (containing 0.01 μl of 1 mg in 100ml) of modified CNTs were deposited on to the gold microelectrodes and collected between the electrodes by dielectrophoresis with an alternating current of 1MHz frequency and a peak to

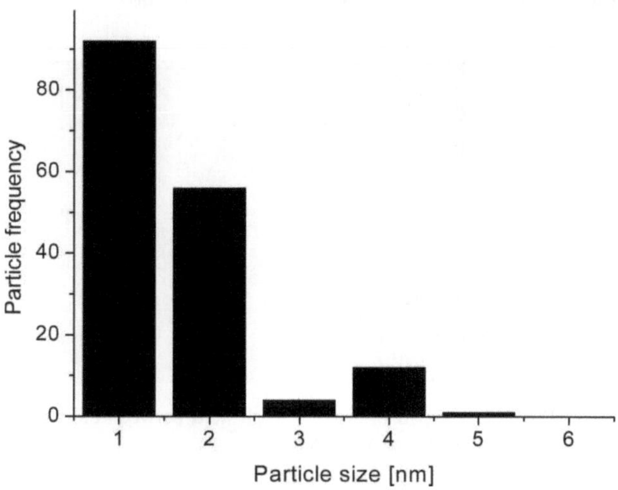

Fig. 4 Size distribution of platinum particles

Fig. 5 The gold microelectrodes are bonded to the contacts on a header with gold wires. A Pt100 temperature sensor is also mounted on to the heater

peak voltage of 0.6V. SEM images showed that the MWCNTs were not aligned as SWCNTs have been shown to be in such a field [10], but became trapped in areas of high field before drying occurred (Fig. 6a). This was also observed by Suehiro et al. [16]. Earlier experiments where electric fields were not applied did not give conductivity even if the concentration of the solution was increased dramatically. SEM images showed that under these conditions the CNTs dried in a ring pattern around the edge of the droplet and did not bridge the electrodes (Fig. 6b).

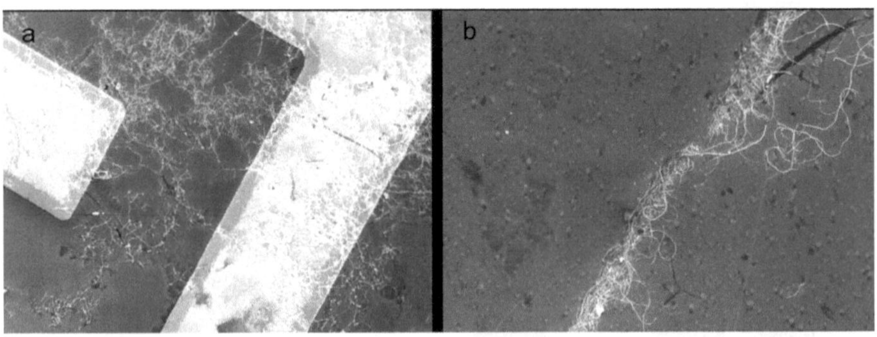

Fig. 6 (**a**) SEM image of Purified MWCNTs trapped in areas of high field between electrodes (**b**) purified MWCNTs dried in a ring around the edge of a droplet

2.5 Gas Measurement Equipment

The headers were mounted into an aluminium gas test chamber with a small $\sim 1\,cm^3$ gas cavity connected to a gas flow line. A computer controlled gas mixing system using Labview was used to control the gas flow over the sensor. The gas flow was maintained at $100\,mlmin^{-1}$ throughout the tests and synthetic air was used as the carrier gas. Power was supplied to the sensor with a source meter (Keithley 2601) and the data was processed using Labview. Temperature was controlled using the Pt100 temperature sensor bonded directly on to the ceramic heater near to the Si substrate, power was supplied to the Pt100 by a regulated DC power supply (Kenwood) and the resultant resistance of the Pt100 was measured with a multimeter (Fluke 8840A).

3 Results and Discussions

The sensors were sequentially exposed to NO_2, CO, NH_3, and H_2, at 25, 100 and 500 ppm. The pulses of test gas were 1200 seconds (20 minutes) in duration. The carrier gas (dry air) was pulsed before and after each test gas pulse for 3600 seconds (1 hour). The gas flow was maintained at 100 ml/min thorough out the tests. The same gas test sequence was repeated at operating temperatures of 40, 100 and 200°C.

3.1 Response of Purified Sensor

Figure 7 shows the typical sensitivities $((RTEST_{GAS} - R_{AIR})/R_{AIR})^{*}100$ of the purified but otherwise unmodified MWCNT sensor to the test gases at 40, 100 and 200°C.

The sensor shows sensitivity to both electron withdrawing and electron donating gases CNTs being amphoteric in nature, with NO_2 demonstrating the greatest sensitivity. It can be generally observed that for NO_2 and CO increasing the temperature increases the magnitude of response, whereas an increase in temperature appears to lead to a decrease in the magnitude of response for NH_3.

Sensing occurs in CNTs by an alteration of the density of states close to the Fermi level. MWCNTs are semimetallic and so do not have a band gap, however the valence and conduction band of graphene narrow as they near the Fermi level, forming a Fermi point at which the density of states is zero. This means that even a small change in the Fermi level will have a large effect on the conductance of a sample.

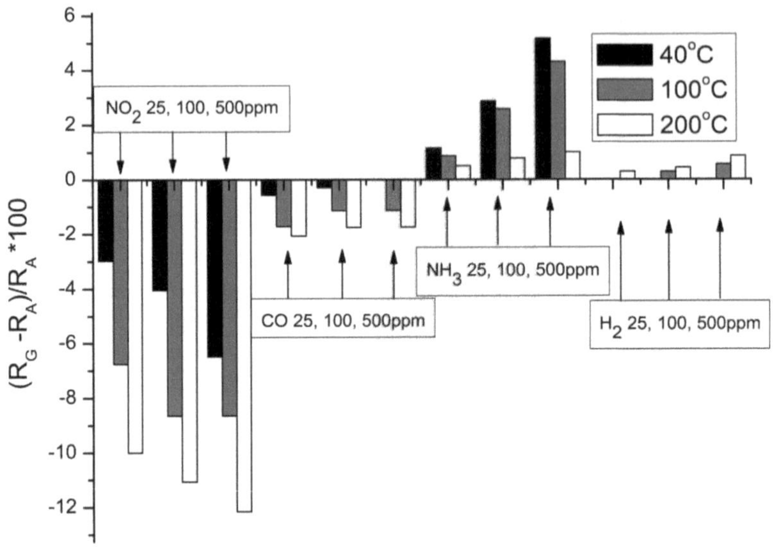

Fig. 7 Response of purified MWCNT sensor to NO_2, NH_3, CO and H_2 at 25, 100 and 500 pm in air at 40, 100 and 200°C

3.2 Response of MWCNT Sensor with Pt Additives

The same test gas series was then applied to the MWCNT sensor with Pt additives at 2wt% see Fig. 8. The response of the sensor with Pt additives appears similar to the purified sensor with responses seen to all gases however, some differences due to the Pt additive can be seen, and these will be discussed for each gas in turn.

Catalytic metal additives are added to increase the sensitivity and decrease the temperature at which the sensitivity maximum is observed. Catalytic metal additives are thought to have two possible mechanisms of altering the sensor response of a metal oxide sensor [3], and these methods may be applicable to a CNT based sensor. The first method is an electronic method whereby the catalytic additives change their oxidation state on encounter with a gas, and all reactions take place on the metallic particles, in the case of CNT sensors, the CNT would then have a role only as a transducer. The second method of altering a sensor response is by a chemical mechanism, platinum is thought to work via this chemical mechanism [2]. In this method, the additives increase reaction rate by first absorbing the gases then in the case of a CNT sensor transferring them on to the CNT walls.

Kong et al. [7] were the first to add metal particles to their CNT sensors. They decorated an individual SWCNT FET device with Pd particles to improve the response towards H_2, claiming that as-prepared SWCNTs were insensitive to this gas, the mechanism which was used to explain this sensing was molecular dissociation of H_2 to H atoms on the Pd surface which then dissolve into the Pd particles, lowering the work function of the metal and transferring electrons to the p-type CNT.

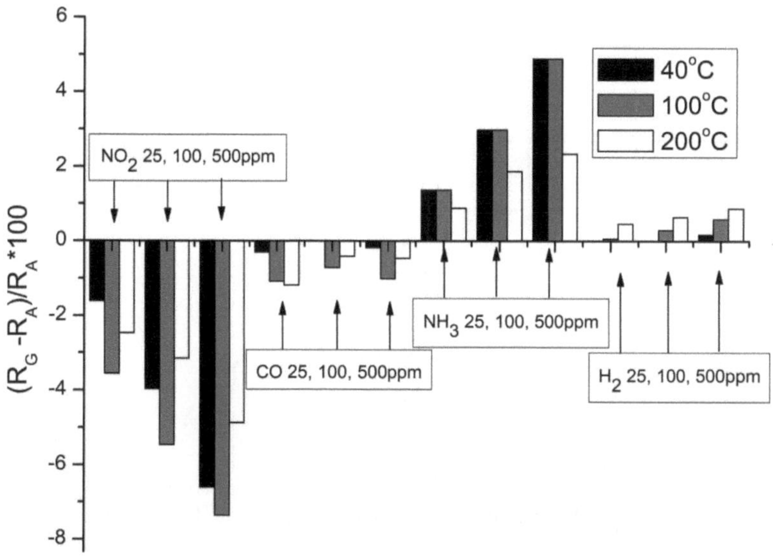

Fig. 8 Response of Pt decorated MWCNT sensor to NO_2, NH_3, CO and H_2 at 25, 100 and 500 pm in air at 40, 100 and 200°C

3.3 NO_2 Sensitivity

The purified sensor's response to NO_2 is seen to increase with temperature. At 40°C the response of the two sensors is very similar in magnitude (See Fig. 9). With increasing temperature the purified sensor continues to demonstrate an increase in sensitivity whereas the sensor with Pt additives does not show the same increase at 100°C as is seen for the purified sensor. At 200°C a decrease is seen in the sensitivity of the sensor with Pt additives as the sensitivity decreases below the value at 40°C.

With increasing temperature the differences in sensitivity between the two sensors increases as the platinum particles become more catalytically active at higher temperatures. Contrary to the expected increase in sensitivity with addition of catalytic metal additives a decrease in sensitivity is seen with the addition of Pt decoration.

The recovery time after NO_2 exposure is improved for the Pt decorated sensor with increasing temperature. For the purified sensor, an improvement is seen from 40 to 100°C however on increase in temperature to 200°C a smaller percentage recovery is seen. The percentage recovery is very similar for the purified and Pt decorated sensor at 40°C, slightly faster recovery is seen for the Pt decorated sensor at 100°C, and at 200°C the percentage recovery is much greater for the Pt decorated sensor than for the purified sensor. This again indicates that the Pt particles do not play an active role in sensing at room temperature. On running the Pt decorated sensor at higher temperatures there is a trade off between faster response times but lower sensitivity.

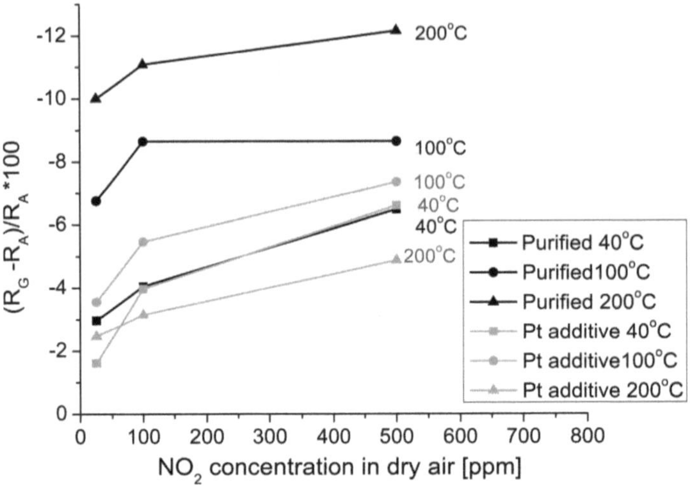

Fig. 9 NO_2 response of purified MWCNT sensor and MWCNT sensor with Pt additive

3.4 NH_3 Sensitivity

The most striking difference in the response towards NH_3 between the two sensors (Fig. 10) is at 200°C, where the response of the purified sensor appears to almost disappear. A small decrease in response is also seen at 100°C. This reduction in sensor response at 200°C is probably due to weak physisorption of the NH_3 molecule becoming unfavourable at this temperature. Theoretical simulations have suggested

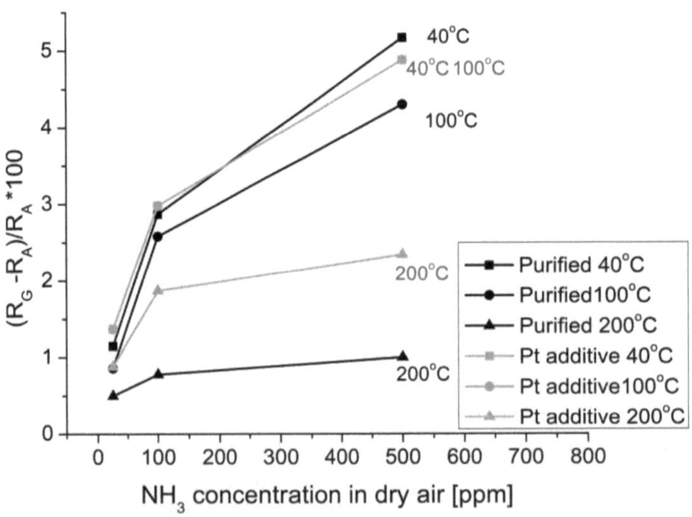

Fig. 10 NH_3 response of purified MWCNT sensor and MWCNT sensor with Pt additive

that there is no binding affinity between a CNT and an NH_3 molecule [8], however NH_3 is known to interact strongly with preabsorbed oxygen on graphite [4], and IR studies found that heating under vacuum reduced the NH_3 absorption on SWCNTs, but heating in air regenerated the NH_3 absorption [5].

Contrary to the response of the purified sensor, it is seen that for the Pt decorated sensor NH_3 response does not decrease drastically with temperature. At 40°C the response is similar in magnitude to that at 40°C for the purified sensor, which indicates that the Pt particles are not playing an active roll in the sensing at this temperature. At 100°C the response in very similar in magnitude as it was at 40°C, and at 200°C a slight decrease in magnitude of response is seen, however the response at 200°C is much greater than that of the purified sensor, so a different sensing mechanism appears to be operating at higher temperatures.

At elevated temperatures the physisorption sensing mechanism of the NH_3 may be replaced by dissociative adsorption and hydrogen abstraction on the Pt nanoparticles decorating the tube walls [11]. This increase in catalytic type behaviour of the Pt particles at 200°C is supported by the increase in the sensing of hydrogen seen for the Pt decorated sensor at higher temperatures see Sect. 3.5.

The recovery times for NH_3 are long at lower temperature with very little recovery seen for both the Pt decorated and the purified sensor at 40°C during the hour recovery under air. These long recovery times have also been reported by Kong et al. [8]. At 100°C there is a small improvement in recovery times for both the Pt decorated and purified sensor. At 200°C the response for the purified sensor is very small and does not recover during the recovery period, and appears to have permanently changed the resistance of the sensor, this may be caused by NH_3 chemisorbing at defect sites on the MWCNT walls. For the Pt decorated sensor at 200°C the recovery is improved with a recovery of approximately 90% for a pulse of 500 ppm concentration.

3.5 H_2 Response

Figure 11 shows the response of the sensors towards hydrogen. The H_2 response is small but comparable in magnitude for both sensors at 40 and 100°C, indicating that the Pt particles are not playing a large role in the H_2 detection at lower temperature. The Pt decorated sensor response to hydrogen is much greater than that of the purified sensor at 200°C, indicating that at this temperature the mechanisms of sensing are different and the Pt particles are playing a much greater active role. Dissociative absorption of hydrogen on the Pt particles is likely to be a favourable mechanism at this temperature. After dissociation on Pt nanoparticles H_2 is thought to then spill over on to defect sites on the MWCNT wall. The sensing of hydrogen with MWCNTs without metal additives does not have an immediately obvious mechanism, however H_2 is thought to adsorb on defect sites on acid treated SWCNT walls [9]. The H_2 may dissociate at defect points on the CNT walls. The recovery time of both sensors to H_2 is greatly improved with temperature. At 200°C the recovery time is around 5 minutes for the Pt decorated sensor and 10 minutes for the purified sensor.

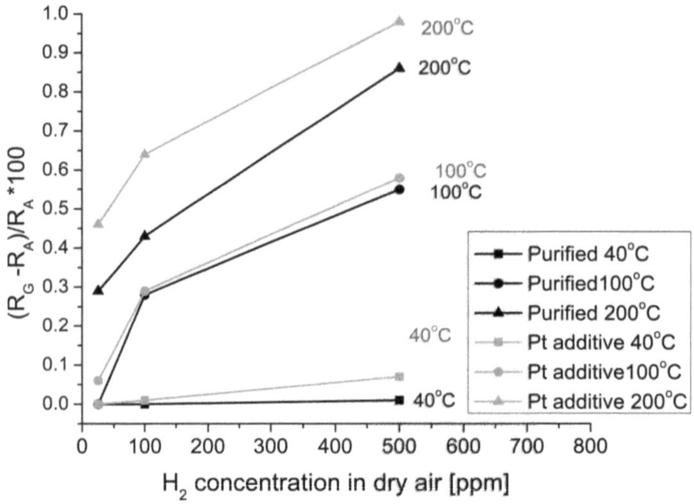

Fig. 11 H_2 response of purified MWCNT sensor and MWCNT sensor with Pt additive

3.6 CO Response

The CO response (Fig. 12) does not scale with concentration, the response decreases with concentration from 25 to 100 ppm. It is also seen that the purified sensor shows a greater response to CO that the Pt decorated sensor. MWCNTs have been reported to show no sensitivity towards CO [13], however graphene sheets have recently shown a small increase in resistance on exposure to CO [14].

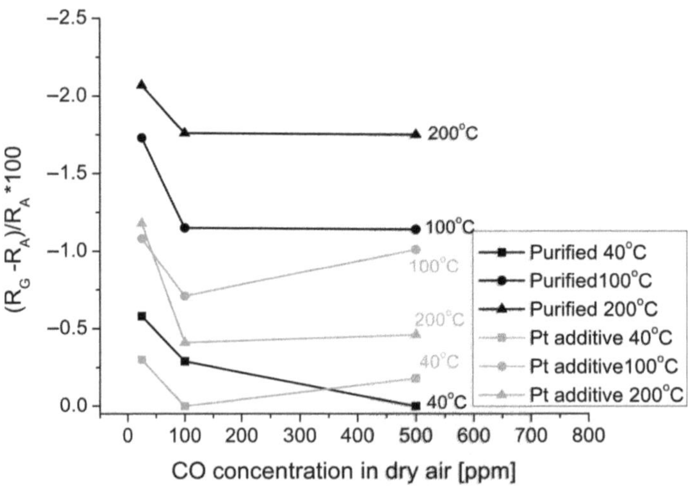

Fig. 12 CO response of purified MWCNT sensor and MWCNT sensor with Pt additive

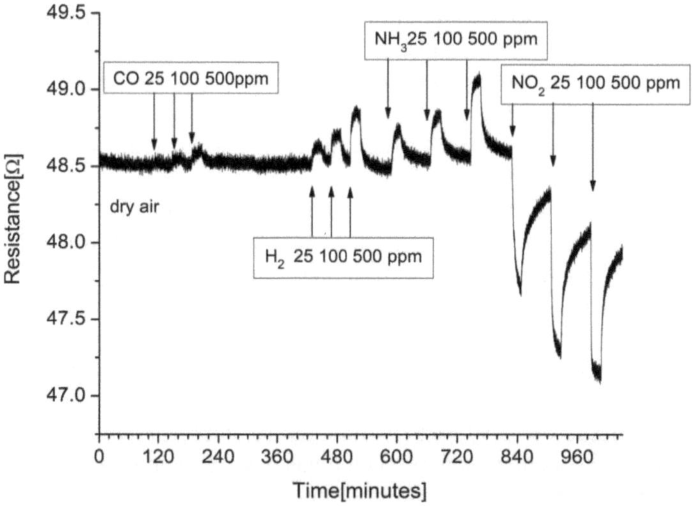

Fig. 13 Response of MWCNT sensor with Pt additive at 200°C to reversed test gas series

A further investigation into whether the CO response was in any way affected by the previous exposure to NO_2 was undertaken; The sensors were heated in air at 200°C for 48 hours, then exposed to the test gases at 200°C in the opposite order to remove the effect that pre-adsorbed NO_2 which may have remained from earlier test pulses see Fig. 13.

After long desorption times ≥48 hours the MWCNT sensor with Pt additives shows a small increase in resistance upon exposure to CO, this response does increase with concentration. After re-exposure to NO_2 the sensor resumes its previous response of a large decrease in resistance upon exposure to CO. For the Pt decorated sensor, this shows that the CO response *is* dependant on the previous gas environment. The experiment was repeated with the purified sensor; a small decrease in response was always seen upon exposure to CO. After very long desorption times of several days, the response to CO became very small, but never reversed in direction.

The mechanism of sensing for the Pt decorated sensor is likely to be oxidation of the CO to CO_2 over the Pt particles causing the reduction at the platinum particles, and so gain in resistance of the Pt decorated MWCNTs. CO has been detected at 0% humidity with Pt impregnated SnO_2 sensor with a maximum sensitivity at around 200°C [3].

4 Summary and Conclusions

MWCNT were synthesised by a CVD method and then decorated with Pt at 2 wt%. The MWCNTs were then drop deposited on to interdigitated gold microelectrodes, and trapped in the high field areas by dielectrophoresis. The sensitivity to a test gas series consisting of NO_2, NH_3, CO and H_2 at concentrations varying between 25 and 500 ppm was investigated at 40, 100 and 200°C.

The temperature at which the largest difference between the two sensors was seen was 200°C where it is seen that the Pt particles are more active. At lower temperatures there is less difference seen between the two sensors, indicating that the Pt particles play less of an active role.

It is likely that there are many competing sensing mechanisms at any temperature for each gas. In the case of NH_3 the physisorption begins to become unfavourable with temperatures above 40°C, and for the Pt decorated sensor, it is thought that a hydrogen abstraction mechanism takes over as the dominant sensing mechanism. For the other test gases the effect of the platinum is seen to improve the speed of recovery, as Pt acts catalytically to increase the reaction rate, which is seen with both H_2 and NO_2.

Cross sensitivity is an issue for CNT sensors and, as has been shown here, the response of the MWCNT sensors does depend on the gas environment to which the CNTs have previously been exposed.

Acknowledgements The authors are greatly indebted to the skilful work of Evald Mildh and Ingemar Grahn. Thanks are also due to Laleh Safina, Hui Qian and Ben Cottam.

The authors would like to acknowledge EPSRC, Tyco and GASG for financial support. This research project has also been supported by a Marie Curie Early Stage Research Training Fellowship of the European Communities Sixth Framework Programme under contract number MEST-CT-2004-504272.

References

1. Andrews R, Jacques D, Rao AM, Derbyshire F, Qian D, Fan X, (1999) Continuous production of aligned carbon nanotubes: a step closer to commercial realization. Chem Phys Lett 303, 5–6:467–74
2. Bond GC, Molloy LR, Fuller MJ, (1975) Oxidation of carbon monoxide over palladium–Tin_IV. oxide catalysts: an example of spillover catalysis. Chem Commun: 796–797
3. Cabot A, Arbiol J, Morante J, Weimar U, Bârsan N, Göpel W, (2000) Analysis of the noble metal catalytic additives introduced by impregnation of as obtained SnO sol–gel nanocrystals for gas sensors. Sensors and Actuators B 70:87–100
4. Cheng A, Steele WA, (1990) Computer simulation of ammonia on graphite. II. Monolayer melting. J Chem Phys 92:3867–3874
5. Feng X, Irle S, Witek H, Morokuma K, Vidic R, Borguet E (2005) Sensitivity of ammonia interaction with single-walled carbon nanotube bundles to the presence of defect sites and functionalities. J Am Chem Soc 127, 30:10533–10538
6. Hu H, Zhao M, Itkis ME, Haddon RC, (2003) Nitric acid purification of single walled carbon nanotubes. J Phys Chem B 107:13838–13842
7. Kong J, Chapline M, Dai H, (2001) Functionalized carbon nanotubes for molecular hydrogen sensors, Adv Mater 13, 18:384–1386
8. Kong J, Franklin NR, Zhou C, Chapline MG, Peng S, Cho K, Dai H, (2000) Nanotube Molecular Wires as Chemical Sensors. Science 287:622–625
9. Kombarakkaran J, Pietraß T, (2008) Electron spin resonance studies of hydrogen adsorption on single-walled carbon nanotubes. Chem Phys Lett 452, 1–3:152–155

10. Krupke R, Hennrick F, vLöhenysen H, Kappes MM, (2003) Separation of metallic from semiconducting single-walled carbon nanotubes. Science 301:344–347
11. Mieher WD, Ho W (1995) Thermally activated oxidation of NH3 on Pt(111): intermediate species and reaction mechanisms. Surface Sci 322, 1–3:151–167
12. Pearce R, Belmonte J, Shaw J, Andersson M, Buckhholt K, Lloyd Spetz A, Shaffer M, (2007) The effect of temperature on the gas sensing properties of CVD grown MWCNTs in: 2nd International Conference on Sensing Technology, Palmerston North, New Zealand pp 455–460
13. Santucci S, Picozzi S, Digregorio F and Lozzi L (2003) NO2 and CO gas adsorption on carbon nanotubes Experiment and theory. J Chem Phys 119 20:10904–10910
14. Schedin F, Geim AK, Morozov SV, Hill EW, Blake P, Katsnelson MI, Novoselov KS, (2007) Detection of individual gas molecules adsorbed on graphene, Nat Mat 6:652–655
15. Singh C, Shaffer MS, Windle AH, (2003) Production of controlled architectures of aligned carbon nanotubes by an injection chemical vapour deposition method. Carbon 41:359–368
16. Suehiro J, Zhou G, Hara M, (2003) Fabrication of a carbon nanotubes-based gas sensor using dielectrophoresis and its application for ammonia detection by impedance spectroscopy. J Phys D; Appl Phys 36:109–114
17. Verdejo R, Lamoriniere S, Cottam B, Bismarck A, Shaffer M, (2007) Removal of oxidation debris from multi-walled carbon nanotubes. Chem Commun:513–515

Electrical Sensing of Biochemicals using Macroporous Silicon

C. RoyChaudhuri, J. Kanungo, R. Dev Das, S.K. Dutta, S. RoyChaudhuri, S. Majhi and H. Saha

Abstract Electrical sensing of biochemical solutions has been reported in this paper using macroporous silicon formed by anodic etching of silicon with DMF and HF solutions. The fabricated porous silicon layers are around 30 µm thick with pore diameter around 1 µm and has been electrically characterised over a wide frequency range. The value of the capacitance can be varied over a wide range from pF to nF by tailoring the nature and dimensions of the electrical contacts over and above the values determined by its morphology. The C-V and I-V characteristics have been studied with two types of contacts—vacuum evaporated aluminium and aluminium paste. An electrical model of the porous silicon along with the contact geometry has been used to explain the basic device characteristics and the large variation of capacitances. The response of the porous silicon layer has been studied after oxidation both by hydrogen peroxide and by thermal method with different concentrations of glucose, potassium and sodium chloride solutions due to the physiological importance of the solutions. The macroporous silicon sensor has also been used to estimate

C. RoyChaudhuri
Department of Electronics and Telecommunication Engineering, Bengal Engineering and Science University Shibpur, Howrah-711103, e-mail: chirosreepram@yahoo.com

J. Kanungo
IC Design and Fabrication Center, Department of Electronics and Telecommunication Engineering, Jadavpur University, Kolkata-700032, e-mail: sahahiranmay@yahoo.com

R. Dev Das
IC Design and Fabrication Center, Department of Electronics and Telecommunication Engineering, Jadavpur University, Kolkata-700032, e-mail: sahahiranmay@yahoo.com

S.K. Dutta
Department of Physics, City College, e-mail: swap9885@dataone.in

S. RoyChaudhuri
Consultant Physician, Anandalok Hospitals, Kolkata, e-mail: soumyocardium@yahoo.com

S. Majhi
Indian Association of Cultivation of Science, Kolkata-700032

H. Saha
IC Design and Fabrication Center, Department of Electronics and Telecommunication Engineering, Jadavpur University, Kolkata-700032, e-mail: sahahiranmay@yahoo.com

S.C. Mukhopadhyay, G.S. Gupta (eds.), *Smart Sensors and Sensing Technology*,

the conditions of dehydration electrically as a function of osmolality of the solution which is essentially a manifestation of the concentration of sodium and potassium ions in solution. The porous silicon layer with thermally grown oxide shows a more significant difference in response with the varying concentrations of the solvents depending on their dielectric constant, dipole moment and molecular dimension.

Keywords Macroporous silicon · stable electrical contact · capacitive detection · biochemical sensing application

1 Introduction

Porous silicon formed by anodic etching of crystalline silicon can be made nano-meso-macro porous as desired by controlling the composition of anodic etching solution and parameters. Its large surface to volume ratio [1] and compatibility with CMOS integration makes it a promising material for use in chemical sensing. Recently porous silicon layers are also being exploited for various biochemical-sensing applications like detection of bacteria [2], DNA hybridization [3] and other biological chemicals. But most of these reports are based on the optical properties of nano and meso-structured porous silicon which suffer from the limitations of rather low sensitivity and non-portability. However to impart portability to such sensors at a relatively low cost, it is necessary to detect the changes electrically. Nanostructured porous silicon has been found to suffer from the problem of establishing stable contact [4] due to the large number of dangling bonds resulting in drift and non-reproducibility. There are some reports on the use of meso and macroporous silicon for electrical sensing of some organic solvents and molecules like oxygen [5]. Different contact geometries on such mesoporous silicon layer have been tried [6, 7, 8, 9] but in almost all of them, the response of the device depends on the characteristics of the electrical contact with the porous silicon. To avoid this problem, contacts have been taken from the underlying bulk silicon [5] in a macroporous silicon sensor but that results in low sensitivity and large response time. Some initial reports of macroporous silicon with only aluminium paste contacts has been found to result in a very high change in capacitance on exposure to solvents [10].

In this paper we have reported in greater details the electrical sensing of biochemical solutions using macroporous silicon. The C-V and I-V characteristics of macroporous silicon has been studied with two types of contacts—vacuum evaporated aluminium and aluminium paste on macroporous silicon to maximize the sensitivity. The value of the capacitance of the sensor is found to be significantly large in case of metal paste. The capacitance can also be changed considerably by tailoring the nature and dimensions of the lateral contacts on porous silicon layer. An electrical model of the macroporous silicon along with the contact geometry has been developed to understand the capacitive and resistive behaviour of the sensor and its variation with frequency and applied bias.

The biochemical solutions which have been used for sensing in this work are different concentrations of glucose, potassium chloride and sodium chloride. The macroporous silicon sensor has also been used to estimate the conditions of dehydration electrically which is essentially a manifestation of the concentration of sodium and potassium ions in solution. The sensor coated with both the different types of oxides has been characterised with the biochemical solutions. The porous silicon layer with thermally grown oxide shows a significant difference in response with solvents depending on their dielectric constant, dipole moment and molecular dimension. These solutions are used owing to their physiological significances. Sodium and potassium are the major ions of the extracellular and intracellular fluids respectively. Both of them play a major role in the maintainance of acid base equilibrium and osmotic pressure of the body. They are vital for cardiac muscle activity and also serve to retain water. Glucose is a monosacharride which serves as a vital source of energy for all physical and physiological processes. Dehydration is one of the commonest paediatric medical problem leading to significant mortality and morbidity. Plasma osmolality is the most reliable method of assessing even small deviation in hydration status of body. But determination of the same is complex. Urine, saliva and sweat osmolality closely follow once dehydration exceeds 5% [11, 12], though there are multiple confounders [13]. A combined approach of determination of change in osmolality in blood plasma (or serum), urine and saliva give a better picture of body fluid content [13] – specially in paired samples for detection of perpetuating dehydration. But determination of osmolality requires an osmometer, which is expensive and not applicable for field situation. Thus development of sensor, which can detect the conditions of dehydration electrically, will be helpful to physicians and applicable for field use. An electrical model of the macroporous silicon along with the contact geometry has been developed to explain the large capacitance of the sensor and its variation with frequency and applied bias.

2 Experimental

2.1 Sensor Fabrication

The porous silicon layers are fabricated by electrochemical etching of p-type silicon (ρ 10–20 Ωcm) under constant current conditions with a current density of 4 mA/cm^2. The electrolyte used is 4 wt% hydrofluoric acid (48 wt%) in N,N dimethylformamide (DMF) [5]. The use of a mild oxidizer such as DMF results in straight and smooth pore walls with pore diameter in the micrometer range. The porous layers are etched for 60 minutes resulting in about 30 μm thick layers. The SEM picture of the porous silicon layer of dimension 1 cm by 1 cm is shown in Fig. 1. The schematic and the picture of the sensor is shown in Fig. 2.

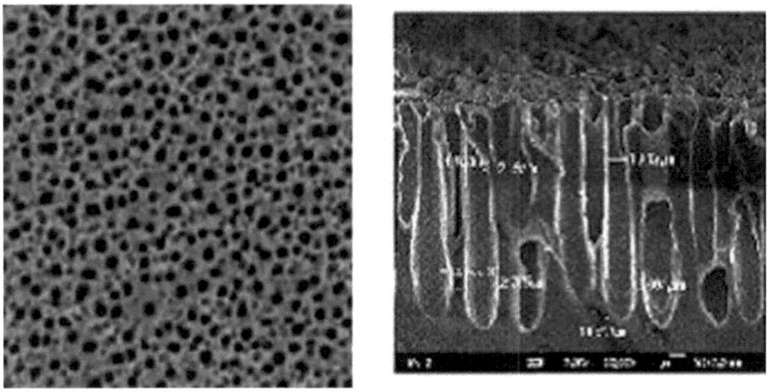

Fig. 1 SEM of macroporous silicon sample

Oxidation of the porous silicon surface is required for stable operation of the devices. The porous silicon layer is oxidized by two methods: chemically oxidized by immersion in 30 wt% hydrogen peroxide (H_2O_2) for a period of one hour at around 70°C and thermally oxidized in a furnace at a temperature of 950°C for 45 minutes. Also the effect of the dipole moment of the solutions are reduced in the presence of an oxide layer. The oxide protects the surface of silicon from potassium and sodium electrolytes. After oxidation, the porous layers are rinsed with deionized water and ethanol and dried under a stream of nitrogen. The metal contacts of dimensions 3 mm by 1 mm with a separation of around 1mm in lateral fashion on porous silicon are then deposited by screen printing with commercially available aluminium paste followed by its firing at 720°C for 45 seconds in nitrogen ambient in a temperature controlled furnace and also by vacuum evaporation of aluminium followed by its firing at 450°C for 45 seconds. Silver paste is then applied by screen printing and baked at a temperature of around 120°C for stable bonded electrical contacts.

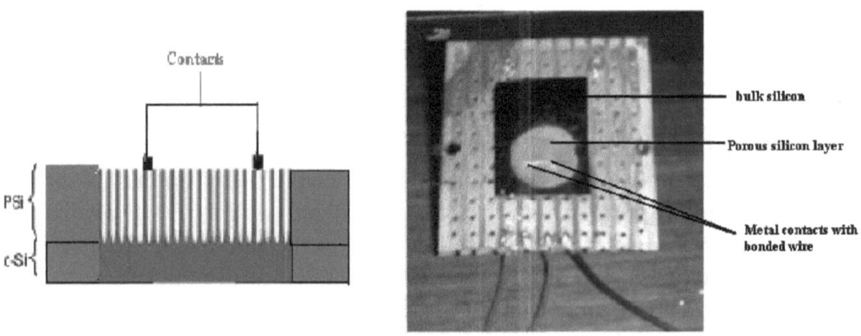

Fig. 2 Schematic and picture of the sensor

2.2 Measurement Setup

The equivalent capacitance and resistance between the two contacts is measured by Agilent 4844 CV bridge for different frequencies and applied voltage. An applied voltage of around 500 mV amplitude is chosen for measurement. All the experiments are performed at room temperature of around 22°C.

The solutions used for measurement are 5%, 10% and 25% glucose solutions commercially available from dextrose anhydrous (Claris). The 15% and 20% solutions have been prepared in the laboratory from 25% solution. The measurements have been taken upto 25% glucose since it is the highest concentration of commercially available monosacharride which is isomolar to human plasma. Measurements have also been taken with potassium chloride solution available in the form of Potchlor solution (Claris). Each ml of the solution contains 150 mg of potassium chloride. Five sets of solutions have been prepared by dilution with deionised water of 18 MΩcm resistivity. Similar solutions have been used for sodium chloride (Claris) with a lower initial concentration (1 ml of the solution containing only 9 mg of sodium chloride). To estimate the dehydration status, control solutions are prepared by mixing sodium chloride, potassium chloride, glucose and urea in approximate proportion to their contribution to body fluid osmolality, which is $Na^+ = 92\%$, $K^+ = 3\%$, glucose = 2% and urea = 3% dissolved in water [14]. One stock control solution is made having osmolality of 1400. Thereafter, serial dilution by addition of appropriate quantity of deionised water is done to produce solutions of desired osmolality. Sodium, potassium, glucose and urea are quantified in each and every sample of control solution by uranyl acetate method, tertaphenyl boron method, glucose oxidase and peroxidase method and modified Berthelot method respectively.

Solutions are added onto the sample in controlled quantity through a micropipette (100 μl). The cylinder is placed on the sample through which the solutions are applied using a pipette in controlled amount. After the reading is noted down, the

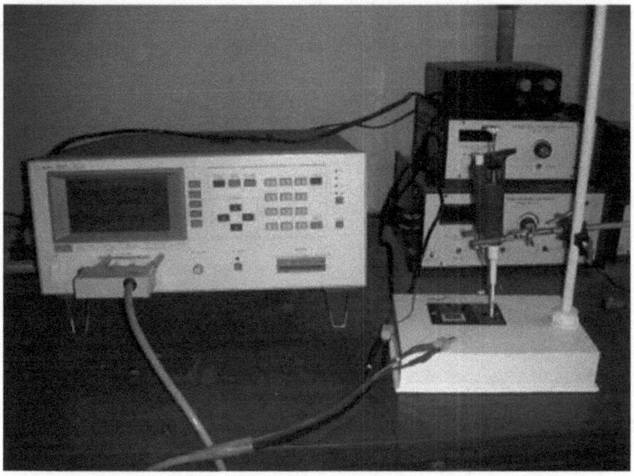

Fig. 3 Picture of the setup

solutions are removed and the sample is dried by using nitrogen jet and then kept in vacuum for sometime before the next measurement. The picture of the measurement set up is shown in Fig. 3.

3 Results and Discussions

3.1 C-V and I-V Characteristics

3.1.1 Contacts with Metal Paste

C-V characteristics of the sensor have been measured as shown in Figs. 4 and 5 respectively. Figure 4 shows the variation of capacitance of the sensor with frequency for thermally and chemically oxidized sample. It is observed that the value of the capacitance decreases with increase in frequency from 100 Hz to 1 MHz. The measured capacitance is the reactive part of the effective impedance between the contacts. This impedance is a function of frequency, pore capacitance, crystallite resistance and geometry of the metal contacts. The value of the capacitance is found to increase exponentially with low frequency which can be explained from the model in the next section. The differences in the capacitance value for the thermally oxidized and chemically oxidized sample can be explained from the equivalent model. The stability of the measurements is ensured by the conformal deposition of the metal paste by screen-printing method.

Figure 5 shows the variation of capacitance with applied voltage and is found to be insignificant for both thermally oxidized and H_2O_2 oxidized samples. This may be attributed to the fact that the formation of space charge region at the metal-porous silicon junction is not significant.

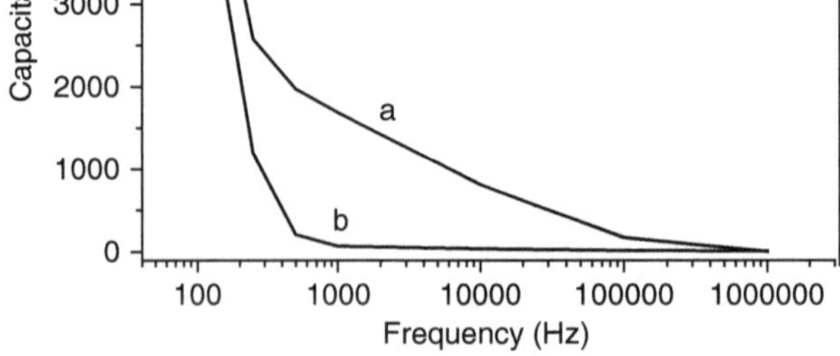

Fig. 4 Variation of capacitance with frequency for the same geometry of the metal contacts

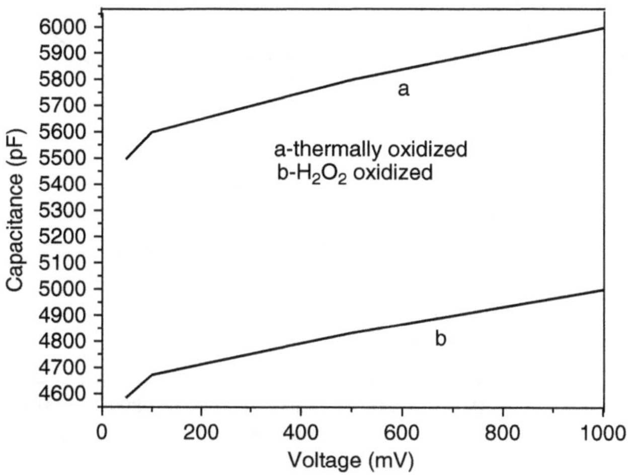

Fig. 5 Variation of capacitance with applied voltage for a frequency of 100 Hz

The value of the resistance measured is around 150 Ω and is independent of applied frequency and voltage. This may be attributed to the formation of fairly ohmic contacts between the metal and the porous silicon. Thus the resistance is primarily the resistance of the underlying bulk silicon which is independent of frequency and applied voltage.

3.1.2 Contacts with Vacuum Evaporation of Aluminium

C-V and I-V characteristics of the sensor with vacuum evaporated metal contacts is shown in Figs. 6 and 7 respectively. Figure 6a shows that the value of the capacitance decreases with increasing frequency like the previous case but the value is almost

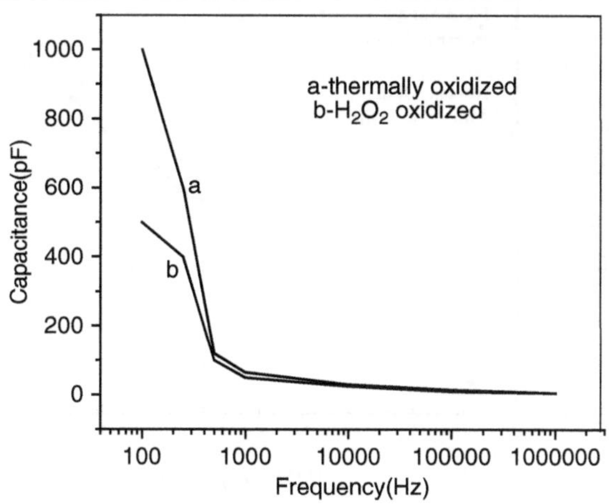

Fig. 6a Variation of capacitance with frequency for the same geometry of the metal contacts

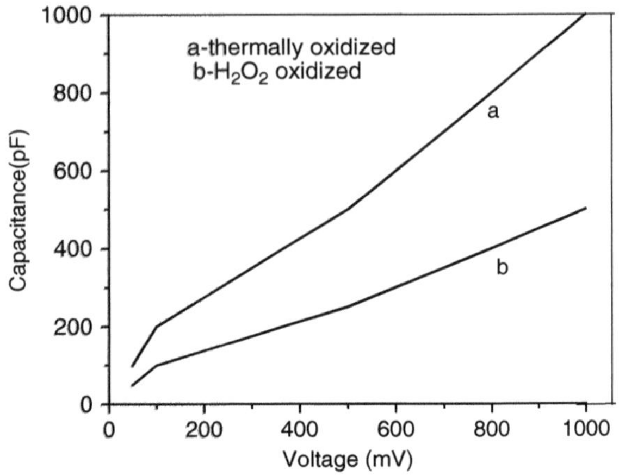

Fig. 6b Variation of capacitance with applied voltage for a frequency of 100 Hz

6 times lower compared to that of the previous case. This may be attributed to the less conformal deposition of metal which lead to lower surface area thus lowering the capacitance value. Also the variation of capacitance with applied voltage is quite significant in this case as can be observed from Fig. 6b. This may be attributed to the presence of a space charge region at the metal-porous silicon interface.

The I-V characteristics has been measured with a Keithley meter and is shown in Fig. 7. It is observed that the value of resistance is of the order of 2 kΩ which is almost 10 times greater than the previous case. Also there is a cutoff voltage in both the directions of the voltage polarity which further confirms the presence of space charge region. However, the identical characteristics in both the directions is primarily due to the presence of both forward and reverse biased junctions.

The decrease in the value of capacitance with frequency, the presence of a non-ohmic IV characteristics in case of vacuum evaporated metal contacts and the

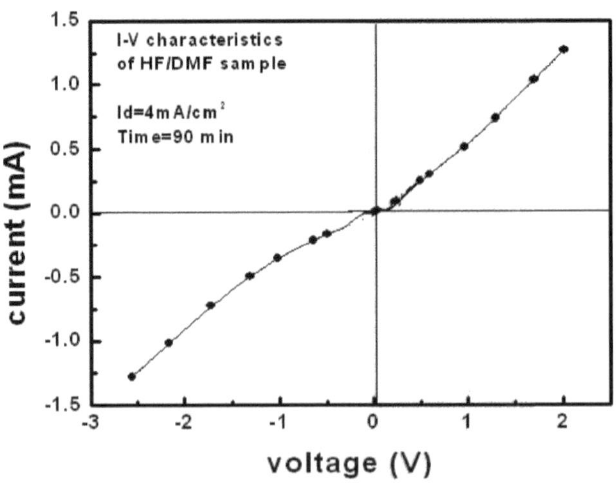

Fig. 7 I-V characteristics

increased value of capacitance with conformal deposition of metal can be explained with an equivalent model discussed in Sect. 3.2.

3.2 Equivalent Model

The equivalent capacitance and resistance between the two lateral contacts is dependent on the crystallite and pore dimensions, space charge region between the contact and the porous silicon and the contact geometry. For a particular contact geometry, the equivalent RC network comprises of the series combination of two parallel RC networks – one of them is contributed by the silicon crystallite and silicon dioxide junction and the other is due to the pore, two RC junction capacitances(R_j and C_j) due to the interface between the contact and the macroporous silicon and R_b is the resistance of the bulk silicon lying underneath as shown in Fig. 8a. At low frequencies only one of the capacitive impedance dominates. As frequency increases both the capacitive impedances play an important role resulting in a decrease in capacitance with frequency. For thermally oxidized samples, the thickness of the oxide is more, resulting in a decrease of the pore diameter and an increase in the pore capacitance. This can lead to an increased capacitance value at low frequency for thermally oxidized samples. The difference in the decrease of capacitance with frequency in Figs. 4 and 6 can be attributed to the different space charge region impedance.

In case of vacuum evaporated metal contacts, the junction impedance ($R_j//C_j$) plays an important role and thus the I-V characteristics is non-ohmic. In case of metal paste, the equivalent resistance is primarily R_b, which is the bulk resistance and hence is almost independent with applied voltage and frequency.

The increase in capacitance due to conformal deposition of metal can be explained by the following analysis. Let us consider two metal contacts of area $A = lw$, separated by a distance d, h is the thickness of the porous silicon layer, a is the dimension of the crystallite, s is the dimension of the pore, $N_1 N_2$ are the number of crystallites under ohmic contact as shown in Fig. 8b.

Total number of crystallites under ohmic contact $= N_1 N_2 = lw/(a+s)^2$

Total surface area of the crystallites under ohmic and conformal contact = $4ahN_1N_2$

(a)

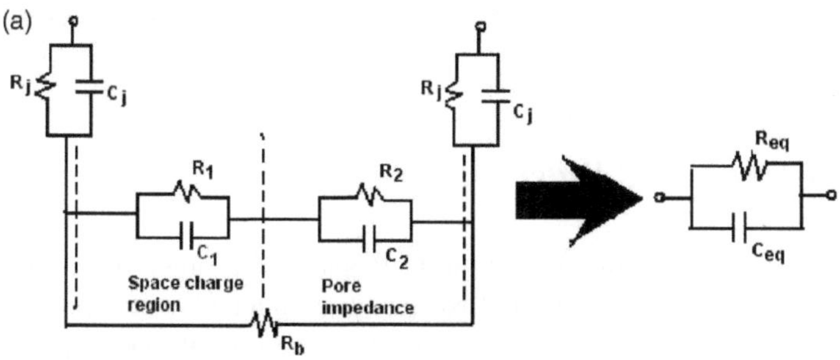

Fig. 8a Equivalent impedance between the lateral contacts

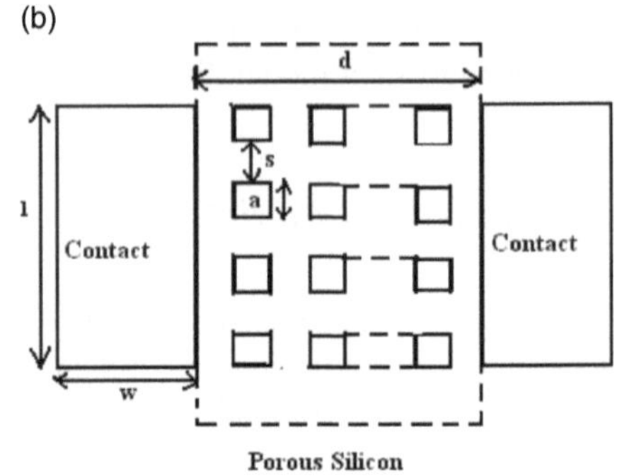

Fig. 8b Top view of the porous silicon layer between the contacts

For non-conformal contact, h decreases.

Total number of crystallites in the gap between the contacts $= N_1^/ N_2^/ = ld/(a+s)^2$

The total capacitance of the gap between the contacts can be obtained as:

$C_{total} = \varepsilon_0 lh/d(s/a{+}1/k)$ where k is the relative permittivity of silicon. Normally the value of C_{total} between two contacts for typical contact geometries can be evaluated upto the order of femtofarad.

Now, if V is the voltage maintained between the contacts, q is the charge needed to be present at the ohmic contact surfaces facing each other, then the surface charge density on crystallites under ohmic contact is given by:

$$\rho_s = q(a+s)/ahl$$

Thus the total charge drawn from the battery is $Q{=}\rho_s N_1 N_2$

Therefore effective capacitance as seen by external circuit is

$$C_{effective} = 4w/(a+s)^*(\varepsilon_0 lh/d(s/a{+}1/k))$$

From the above equation we can observe that if w is much larger than a and s which is practically the case, then the effective capacitance increases by many orders(upto nF) from the capacitance of the gap. Figure 9 shows the variation in capacitance with $w/(a+s)$ which is also verified by ANSYS v8.0 using the capacitance extraction feature.

It is observed from Fig. 9 that the capacitance increases linearly with w/(a+s) and for a typical dimension of $w = 2\,\text{mm}$ and $a = s = 1\,\mu\text{m}$, and conformal deposition of metal, the capacitance is around 4nF which matches closely with the experimental results.

3.3 Response to Glucose and Water

From the previous sections, we observe that the C-V and I-V characteristics for contacts with metal paste is more stable and reduces the effect of metal porous silicon junction. Thus the biochemical solutions have been tested with metal paste based contacts.

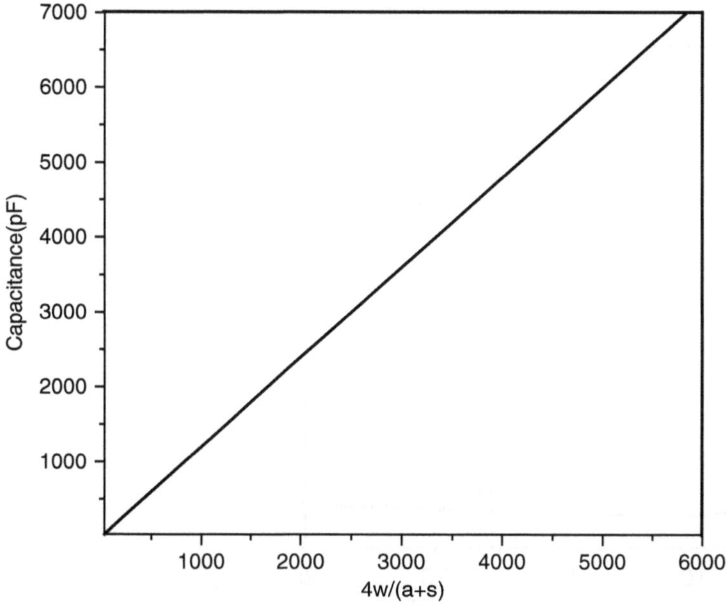

Fig. 9 Variation of capacitance with 4w/(a+s)

The response of the sensor to glucose and water is shown in Fig. 10. Figure 10a shows the response of the H_2O_2 oxidized and thermally oxidized samples with different concentrations of glucose at a frequency of 100 Hz. It is observed that the relative change in capacitance is more significant and linear for thermally oxidized sample in comparison to H_2O_2 oxidized sample. The change in capacitance in H_2O_2 oxidized sample can be probably attributed to the series effect of dipole capacitance originating from the double layer capacitance and the dielectric constant based capacitance. When the double layer capacitance stabilises with increased concentration, the effective change in capacitance also becomes constant since the dipole capacitance is usually lower than the dielectric capacitance. The change in the capacitance of the porous silicon sensor follows the trend of the electrolyte capacitance measured at the same frequency with the same instrument. In thermally grown oxide, the change in capacitance is primarily dominated by the dielectric capacitance. The effect of the dipole capacitance is mostly screened by the thick thermally grown oxide.

Figure 10b shows the relative change in capacitance with frequency for the thermally oxidized sample. It is observed that the relative change is highly dependent on the applied frequency and it decreases with increasing frequency. This may be due to the fact that at low frequencies the change in the dielectric constant dominates the modulation of the space charge region by the presence of electrolyte. For frequencies between 100 kHz and 500 kHz, the relative change is negative for all the different concentration of glucose. This can be probably attributed to the increase in

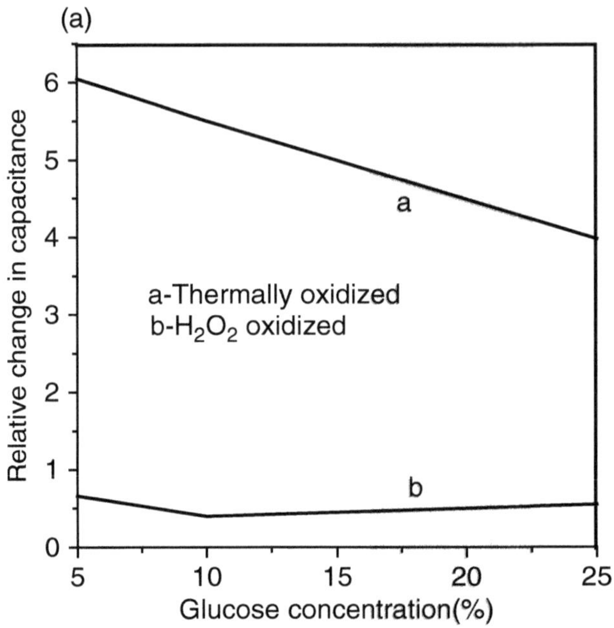

Fig. 10a Variation of capacitance with water (0% glucose) and glucose concentration at a frequency of 100 Hz

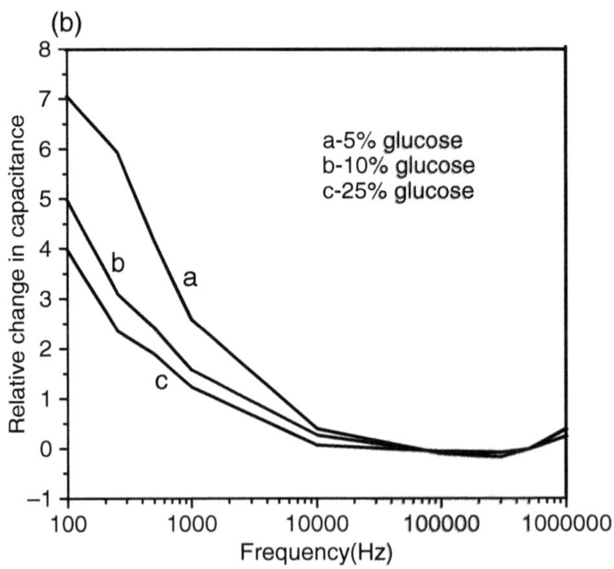

Fig. 10b Variation of change in capacitance with frequency for different glucose concentration

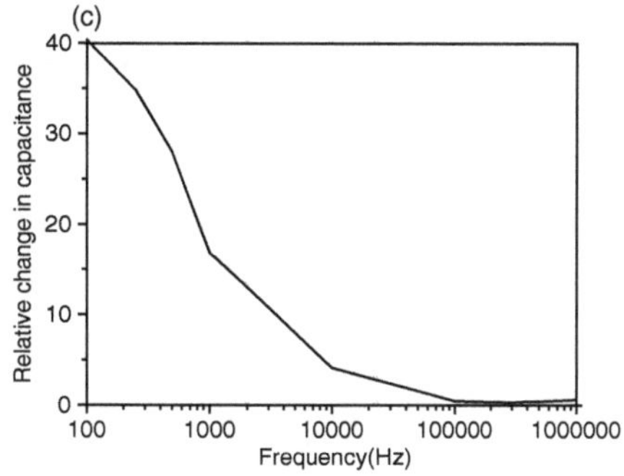

Fig. 10c Variation of relative change in capacitance with water for thermally oxidized sample

impedance, which might occur at the molecular resonant frequency. The response and recovery time of the sensor is significantly low.

Figure 10c shows the response of the sensor with thermal oxidation in presence of deionised water. It has been demonstrated that water increases the capacitance of the porous silicon layer [3]. It is observed that the capacitance initially decreases with frequency up to a value of 400 kHz after which it starts increasing slightly. A change in dielectric constant, dipole moment and possible adsorption on the surface of porous silicon can be used to explain its behaviour.

3.4 Response to Potassium and Sodium Chloride

The response of the sensor with sodium and potassium chloride are shown in Figs. 11a and b respectively. It is observed that there is a significant relative change in the capacitance of around 1800 at a frequency of 100 Hz when exposed to a

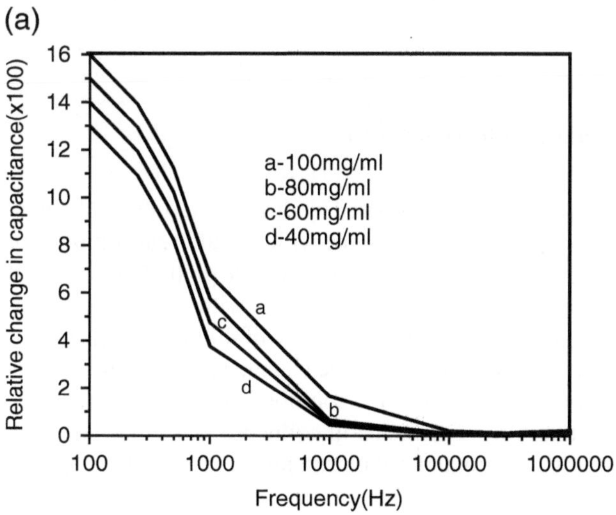

Fig. 11a Variation of relative change capacitance with frequency for different KCl concentration

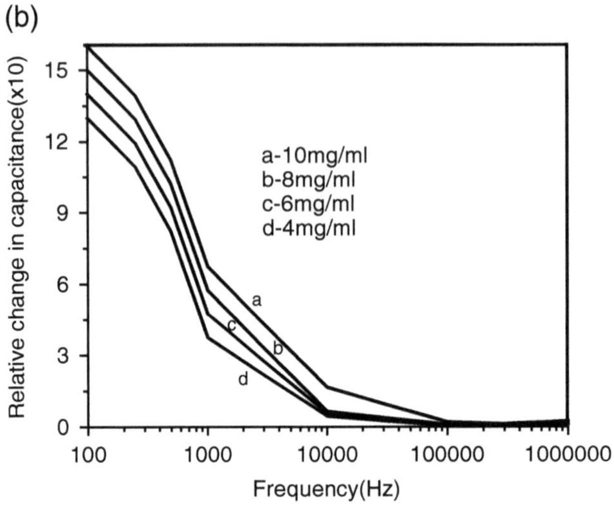

Fig. 11b Variation of relative change capacitance with frequency for different NaCl concentration

solution of potassium chloride containing 100 mg in 1 ml. The change in the capacitance decreases almost linearly in both the cases with decrease in concentration. The relative change of capacitance as obtained with macroporous silicon is higher compared to the relative change in impedance obtained from nanoscale interdigitated electrode arrays as reported in [15]. The nature of the change can be attributed to the relatively high mobility of the potassium and sodium ions, which penetrates to a large extent within the pores affecting the space charge region at the silicon-silicon oxide interface. However the selectivity of the sodium and potassium ions can be ensured through the use of some selective polymeric materials.

3.5 Response to Solutions of Dehydration

The mimickery solutions of dehydration are prepared by different sodium and potassium ion concentration. The change in the capacitance with the different solutions of osmolality representing corresponding states of dehydration are plotted in Fig. 12. It is observed that the relative change in capacitance increases with increasing osmolality of the solution and is around 1200 for an osmolality of 400 at a frequency of 100 Hz which is quite significant. However the relative change in capacitance in this case is lower than that of the pure sodium chloride solution because of the fact that the dehydration solutions contain other chemicals like glucose and urea. The high sensitivity of the sensor towards sodium and potassium ion has been advantageous for sensing dehydration.

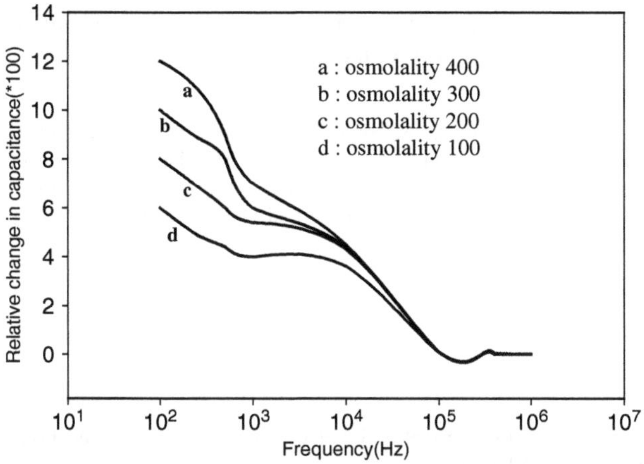

Fig. 12 Variation of relative change capacitance with frequency for different solutions of dehydration

4 Conclusion

In this paper, electrical sensing of biochemical solution using macroporous silicon has been presented. The fabricated porous silicon layers are around 30 μm thick with both crystallites dimensions and pore diameter around 1 μm and has been electrically characterised over a wide frequency range. The value of the capacitance can be varied over a wide range from pF to nF by tailoring the dimensions of the electrical contacts over and above the values determined by its morphology. The C-V and I-V characteristics of macroporous silicon has been studied with two types of contacts—vacuum evaporated aluminium and aluminium paste on macroprous silicon to maximize the sensitivity. The value of the capacitance of the sensor is found to be significantly large in case of metal paste. An electrical model of the porous silicon along with the contact geometry has been used to explain the basic device characteristics. The decrease in the value of capacitance with frequency, the presence of a non-ohmic IV characteristics in case of vacuum evaporated metal contacts and the increased value of capacitance with conformal deposition of metal can be explained with an equivalent model. The response of the porous silicon layer has been studied after oxidation both by hydrogen peroxide and by thermal method with different concentrations of glucose, potassium and sodium chloride solutions due to the physiological importance of the solutions and solutions of different osmolality mimicking the different stages of dehydration The porous silicon layer with thermally grown oxide shows a higher relative change in capacitance (of the order of 6) compared to a value of less than 1 for a frequency of 100 Hz with 5% glucose solution. The relative change in capacitance with the varying concentrations of the potassium chloride and sodium chloride is found to decrease with frequency. The nature of the response is attributed to the combined effect of dielectric constant, dipole moment and molecular dimension. This work establishes the potential of macroporous silicon for biosensing application with electrical detection.

Acknowledgements The authors would like to acknowledge AICTE, Govt. of India and DST, Govt. of India for funding the project. Thanks are also due to Prof. A. Mondal, Department of Chemistry, Bengal Engineering and Science University, Shibpur for his support and Dr. U. Gangopadhyay of IC Center, Jadavpur University for his cooperation and Dr. Dipankar Chakraborty for supplying the solutions of different osmolality.

References

1. L.T. Canham, "Silicon quantum wire array fabrication by electrochemical and chemical dissolution of wafers", *Appl. Phys. Lett.*, 1990, p. 57
2. H. Ouyang, L.A. DeLouise, M. Christophersen, B.L. Miller, P.M. Fauchet, *Proc. of SPIE* 5511, 2004, p. 71
3. M. Archer, M. Christophersen, P.M. Fauchet, "Macroporous silicon electrical sensor for DNA hybridization detection", *Biomedi. Microdevices,* 2004, p. 203
4. L.A. Balagurov, D.G. Yarkin, E.A. Petrova, "Electronic transport in porous silicon of low porosity made on a p+ substrate", *Mater. Sci. Engg.*, 2000, p. 127
5. M. Archer, M. Christophersen, P.M. Fauchet, "Electrical porous silicon chemical sensor for detection of organic solvents", *Sens. Actuators B,* 2005, p. 347
6. A. Motohashi, M. Ruike, M. Kawakami, H. Aoyagi, A. Kinoshita, A. Satou, "Identification of water molecules in low humidity and possibility of quantitative gas analysis using porous silicon gas sensor", *Jpn. J. Appl. Phys.*, 1996, p. 4253
7. A. Motohashi, M. Kawakami, H. Aoyagi, A. Kinoshita, A. Satou, "Gas identification by a single gas sensor using porous silicon as the sensitive material", *Jpn. J. Appl. Phys.*, 1995, p. 5840
8. K. Watanabe, T. Okada, I. Choe, Y. Sato, "Organic vapour sensitivity in a porous silicon device", *Sens. Actuat. B*, 1996, p. 194
9. S.J. Kim, S.H. Lee, C.J. Lee, "Organic vapour sensing by current response of porous silicon layer", *J. Phys. D: Appl. Phys.*, 2002, p. 3505
10. C. RoyChaudhuri, J. Kanungo, S.K. Dutta, S. RoyChaudhuri, S. Majhi, H. Saha, "Macroporous silicon as an electrical platform for biosensing applications", 2nd International Conference on Sensing Technology, November 26–28, 2007, Palmerston North, New Zealand, pp. 493–498
11. S.N. Cheuvront, M.N. Sawka, FACSM; Sports Science Exchange 97, Vol. 18, No. 2, 2005 Hydration Assessment of Athletes
12. L.A. Popowski, R.A. Oppliger, G.P. Lambert, R.F. Johnson, A.K. Johnson, C.V. Gisolfi, *Med. Sci. Sports Exerc.*, Vol. 33, No. 5, 2001, pp. 747–753
13. N.P. Walsh, S.J. Laing, S.J. Oliver, J.C. Montague, R. Walters, L.J. Bilzon, *Med. Sci. Sports Exerc.*, vol. 36, no. 9, 2004, pp. 1535–1542
14. C. RoyChaudhuri, D. Chakraborti, J. Kanungo, S.K. Dutta, U. Gangopadhyay, H. Saha, "Porous Silicon Based Microelectrode Array For Detection and Quantification Of Dehydration", National Seminar on Physics and technology of sensors, Mar. 7–9 , Mumbai, 2007
15. P.V. Gerwen, W. Laureyn, W. Laureys, G. Huyberechts, M.O. De Beeck, K. Baert, J. Suls, W. Sansen, P. Jacobs, L. Hermans, R. Mertens, "Nanoscaled interdigitated electrode arrays for biochemical sensors", *Sens. Actuators B*, vol. 49, 1998, pp. 73–80

Part III
SAW Sensors

Combined Infrared and Acoustic Beacon Tracker and its Application on an Autonomous Following Vehicle

Graham M. Brooker, Craig Lobsey and Kate McWilliams

Abstract This paper outlines the design and implementation of a combined IR and acoustic beacon tracker and its evaluation as the guidance sensor on a radio controlled car. The implementation is unique insofar as it combines the broad acquisition angle of IR sensors (nearly 180°) with the precision angle measurement capability of an acoustic monopulse tracker. Additionally it uses the time difference of arrival of the IR and acoustic signals to measure the range to the beacon. A control loop was designed using the sensor to allow the car to follow the beacon at a fixed distance. Tracker and control performance on the car were good with the vehicle capable of following a fast moving person with a turn radius of less than 3 m.

Keywords Beacon tracker · infrared · acoustic monopulse · vehicle control

1 Introduction

This paper outlines the development and implementation of a low-cost system that allows a number of small robotic vehicles to find and then accurately follow their human "owners". It is an expanded version of a conference paper [3]. RFID and GPS based options were considered initially, however they were rejected as unsuitable because the measurement of relative angles and small range differences accurately is not cost effective using these techniques. More conventional time-of-flight techniques were deemed to be more suitable. Range measurement could either be undertaken using standard pulsed ultrasound, or more effectively using a beacon that transmitted an electromagnetic and an acoustic signal simultaneously, and then

Graham M. Brooker
Australian Centre for Field Robotics, University of Sydney, Sydney, Australia,
e-mail: gbrooker@acfr.usyd.edu.au

Craig Lobsey
Australian Centre for Field Robotics, University of Sydney, Sydney, Australia

Kate McWilliams
University of Cape Town, South Africa

using a receiver to measure the difference in the arrival times, a technique known as time difference of arrival (TDOA).

The main advantage of this method is that it allows the fast electromagnetic signal to carry a unique ID code which would identify a specific beacon to a specific receiver. TDOA systems have been developed for indoor localisation over the past decade and include the well known MIT cricket [10] as well as various other pulsed systems [5, 12]. Issues with multipath and acoustic interference have resulted in the use of spread-spectrum techniques [2, 9]. Some researchers opted to use a sequential approach to overcome the interference problems [4].

2 Operational Principles of Prototype

Initial experiments were conducted using RF modules designed for keyless entry. However, it was found that those tesred have a substantial (>100 ms) and variable period between the transmission of the encrypted signal and its acknowledgement making them unsuitable for the measurement of arrival time differences of less than 15 ms required for a 5 m maximum range.

An alternative technique using standard infrared remote control hardware, shown conceptually in Fig. 1, was implemented to produce a range measurement system which works as follows:

- The beacon transmits a 600 μs burst of IR light with a wavelength of 960 nm on-off modulated at 38 kHz.
- It then transmits a 1 ms burst of ultrasound at a frequency of 40 kHz.
- The receiver detects the IR pulse, which, to all intents is instantaneous, followed shortly afterwards by the slower acoustic pulse.
- Because the speed of sound is reasonably constant (about 340 m/s) and slow, it is easy to measure the time difference, ΔT, from which the range can be calculated.

The relative angle of the beacon with respect to the receiver is determined using a variation of the conventional amplitude comparison monopulse technique

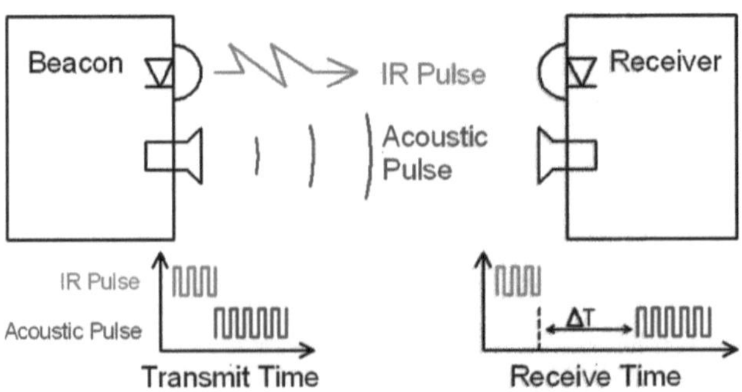

Fig. 1 Conceptual diagram of the system showing the range measurement technique

developed during WWII for tracking aircraft using radar [7, 11]. Though this process is usually applied using electromagnetic signals, a number of acoustic applications have been made [1, 6], where its implementation as an aid to the blind has been quite successful.

One of the novel aspects of the sensor described in this paper is that both IR and acoustic implementations of the angle measurement are made. In the IR implementation, two receivers are placed between an opaque septum as shown in Fig. 2, so that as the angle of arrival varies, first the one receiver is illuminated, and then both, and finally the other. This produces four states: No signal (if the beacon is outside the angle limits of both receivers), angle left (if one receiver is illuminated), angle straight on (if both are illuminated) and angle right (if the other receiver is illuminated). This can be used to drive a bang-bang controller to adjust the coarse steering of the robot.

The size of the septum that divides to two receivers can be adjusted to control the angle over which both receivers are illuminated, and so defines the "dead" band of the controller.

Because of the finite size of the receive aperture lens and diffraction around the septum, there will be a fuzzy region of transition between the receiver seeing the beacon and not. This will depend on the intensity of the beacon, and the ambient light level. It has been estimated that a dead band of much less than 10° could not be reliably obtained by this method.

In the case of the acoustic system, because the amplitude of the received signal is determined by the receiver beam pattern, a far more subtle measure of the angle is possible. Two receiver transducers are mounted with a slight outward squint so that if the acoustic signal arrives at an angle, the amplitude received by one of the

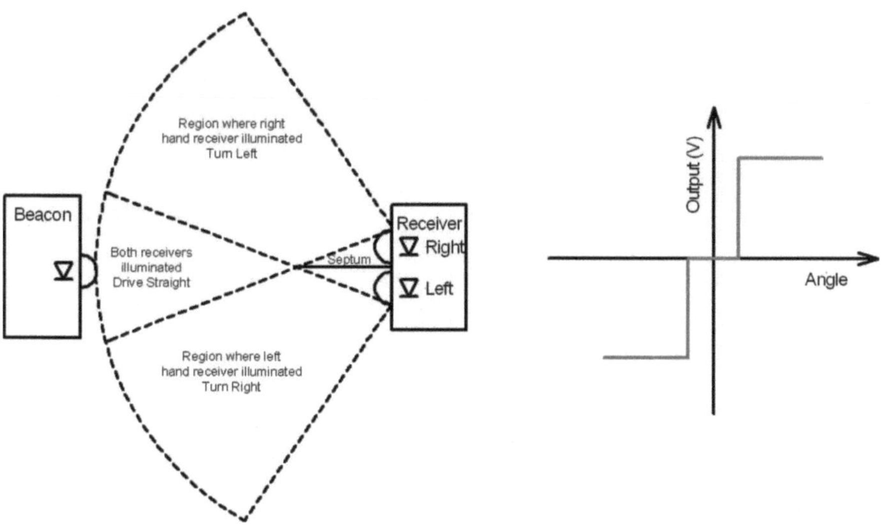

Fig. 2 Conceptual diagram showing digital angle measurement and transfer function

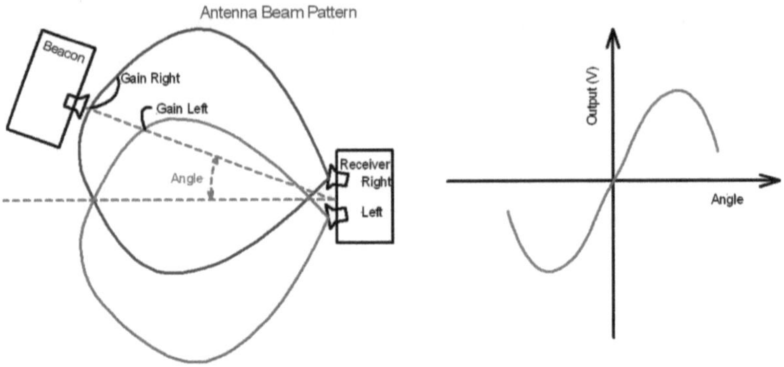

Fig. 3 Conceptual diagram showing analog angle measurement and the transfer function

receivers will be higher than that received by the other as illustrated in Fig. 3. When the illumination is symmetrical the two signal amplitudes are equal and there is no dead band region because the sensor becomes a null tracker.

To visualise the effect of adjusting some of the parameters of these transducers, a simulation model was built using the ultrasonic transducer normalised gain pattern supplied by the manufacturer and reproduced as a polar plot in Fig. 4.

The difference between the signal voltages received from the left and the right transducer can be used to produce the receive antenna transfer function shown in Fig. 5a. It can be seen that the peak magnitude of the error signal is dependent on the squint angle and that there is a region of about $+/-20°$ over which the relationship between the voltage and the beacon angle is almost linear. The error slope at the origin is a good indication of the tracking accuracy, and it can be seen that this

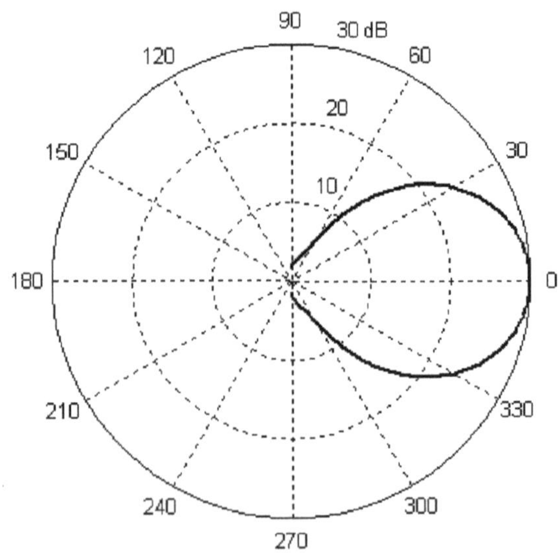

Fig. 4 Polar pattern of ultrasound sensor normalised gain

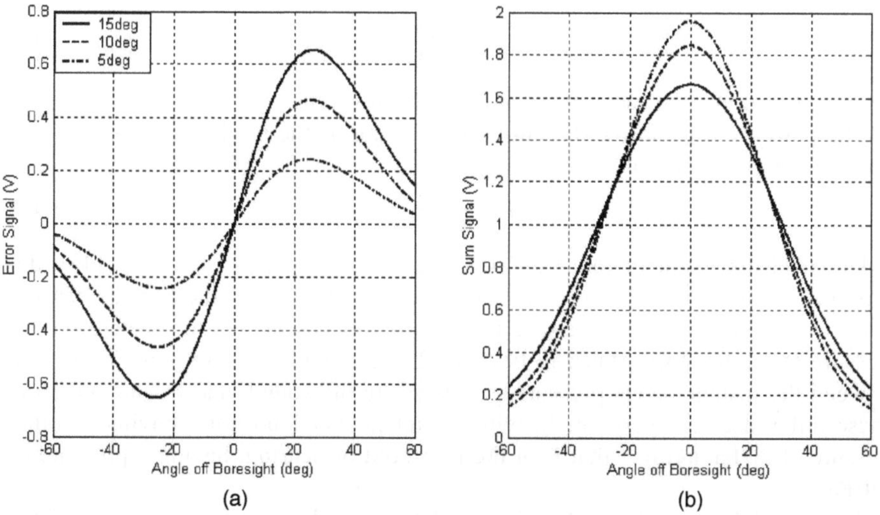

Fig. 5 Antenna transfer function (**a**) error voltage signal and (**b**) sum voltage signal

increases as the squint angle is increased. The downside is that as the squint angle is increased the on-boresight signal amplitude decreases, as shown in Fig. 5b, and at a half squint of more than 15° it becomes too low and the SNR is compromised.

In addition, the amplitude of each of the received signals decreases considerably as the beacon angle increases off the beam axis in the vertical direction. For a symmetrical beam pattern, at an elevation offset of 30° the gain is reduced by 10 dB, and in the end a threshold will be reached below which the SNR will be too low for reliable measurements to be made. Because the signal level is also dependent on the range to the beacon, the angle limit threshold will also be range dependent. These factors will result in a complex operational volume envelope around the receiver.

2.1 Theoretical Performance

In acoustic systems, the acoustic power density and sound pressure levels are determined at a reference distance from the transmitter. In air this is usually 30 cm and in underwater sonar systems it is 1 m. Therefore, for an acoustic transmitter radiating a power, P_t, omnidirectionally, the power density, I_{iso} (W/m^2), at a range of 30 cm will be the total radiated power divided by the surface area of a sphere with a radius of 30 cm.

$$I_{iso} = \frac{P_t}{4\pi 0.3^2} \tag{1}$$

The antenna gain, or Directivity Index (DI) is defined as the ratio solid angle subtended by the surface of the sphere (4π) to the area of the beam.

$$G \approx \frac{4\pi}{\theta\phi}, \tag{2}$$

where θ, ϕ (rad) are the two orthogonal 6 dB beamwidths.

The power density in the direction of the beam, is the product of this gain and the isotropic power density

$$I = \frac{P_t G}{4\pi 0.3^2}. \tag{3}$$

Unlike electromagnetic radar, the performance of acoustic systems is seldom determined in terms of the power density but rather in terms of the Sound Pressure Level (SPL). Atmospheric pressure is measured in Pascals (N/m^2) where the air pressure at sea level is defined to be 101,325 Pa. Sound pressure is the difference between the instantaneous pressure generated by the sound, and the air pressure. These differences are very small, with an unbearably loud noise having a sound pressure of 20 Pa and one that is at the threshold of hearing, having a pressure of 20 μPa.

The sound pressure level, L_p (dB), is defined as the square of the ratio of the sound pressure, P (Pa), to $P_{\text{ref}} = 20\,\mu\text{Pa}$ threshold level. It is typically represented in dB form as

$$L_p = 10\log_{10}\left(\frac{P}{P_{\text{ref}}}\right)^2. \tag{4}$$

The relationship between the sound pressure, P (Pa), and the transmitted power density, I (W/m^2) is

$$P^2 = IZ_{\text{air}}, \tag{5}$$

where Z_{air} is the acoustic impedance of air.

This can be determined by taking the product of the air density, ρ_{air} (kg/m^2) and the speed of sound, c (m/s).

$$Z_{air} = \rho_{air} c = 1.229 \times 340 = 418\Omega \tag{6}$$

Rewriting (4) in terms of (5) and (6) gives the SPL, L_p (dB), in terms of the acoustic power density, I (W/m^2), at a given range.

$$\begin{aligned} L_p &= 10\log_{10}\frac{IZ_{air}}{P_{ref}^2} \\ &= 10\log_{10} I + 10\log_{10} Z_{air} - 20\log_{10} P_{ref} \\ &= 10\log_{10} I + 26.21 + 93.9 \\ L_p &= 10\log_{10} I + 120 \end{aligned} \tag{7}$$

As the sound propagates out to a target, the sound pressure level drops with the sum of the square of the range plus an attenuation component. This transmission loss, H (dB), is given by

$$H = 20\log_{10}\frac{R}{R_{ref}} + \alpha R, \tag{8}$$

where R is the range to the target (m) and R_{ref} is the range at which the SPL is defined (0.3 m in this case) and α is the attenuation in dB/m.

The SPL at the receiver, L_r (dB) can be calculated from these two equations

$$L_r = L_p - H. \tag{9}$$

In this application the ultrasonic beacon and receiver are muRata piezoelectric transducers with the following specifications [8]:

- MA40S4S: SPL 120 dB at 40 kHz; 0 dB relative to 0.0002 μbar per 10 Vpp at a range of 30 cm
- MA40B8R: Sensitivity –63 dB at 40 kHz; 0 dB relative to 1 Vrms/μbar
- Beamwidth 55° (6 dB)

The acoustic power density 30 cm from the transmitter driven by 20 Vpp can be determined easily if it is realised that 1 μbar = 0.1 Pa, making 0.0002 μbar = 20 μPa which is the standard reference level for SPL.

The SPL increases with the square of the applied voltage, therefore

$$L_p = 120 + 20\log_{10}\frac{20}{10} = 126\text{dB}$$

The transmission loss at a range of 3 m is then calculated using the attenuation, α, for sound through air which is approximately 1.2 dB/m

$$H = 20\log_{10}\frac{3}{0.3} + 1.2\times 3 = 23.6\text{dB}$$

The received sound pressure level is

$$L_r = L_p - H = 126 - 23.6 = 102.4\text{dB}$$

This excites the receive transducer to produce an electrical signal. For the specified sensitivity of –63 dB, the output voltage is

$$20\log_{10} V_{out} = 102.4 - 63$$

Therefore

$$V_{out} = 10^{(102.4-63)/20} = 93\mu\text{V}$$

This voltage is more than adequate to ensure a good SNR when followed by a high gain preamplifier

3 Tracker Implementation

3.1 Beacon

The beacon is made as light and unobtrusive as possible so that it can easily be attached to a persons belt. From Fig. 6a it can be seen that it consists of a PIC microcontroller (PIC16F62X) which drives two IR LEDs with a peak current of 80 mA for

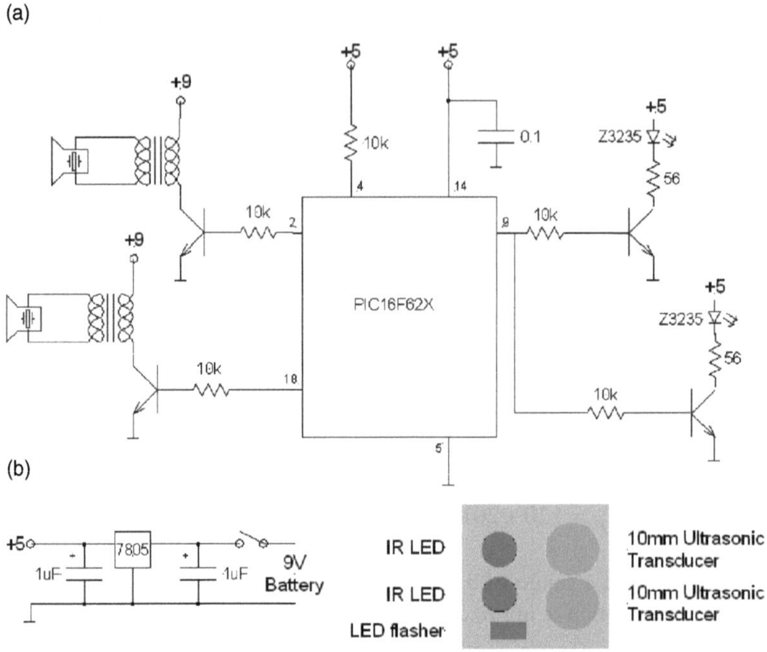

Fig. 6 Details of the beacon showing (**a**) the schematic diagram and (**b**) the transducer layout on the face

a 600 μs pulse with a modulation frequency of 38 kHz. The pulse repetition interval is 50 ms making the duty cycle $0.012 \times 0.5 = 0.6\%$ and reducing the average current consumption to 0.48 mA per LED.

The micro-controller also drives two ultrasonic transducers at 40 kHz through switching transistors and step-up transformers to provide each with 20 Vpp which is their maximum rated drive level. Two transducers are used to widen the azimuth coverage so that the beacon will be seen by the receiver even if it is at an angle.

Figure 6b shows the layout for the beacon transducers. It is important that the IR LEDs and the Ultrasonic Transducers be squinted in the horizontal plane to improve angular coverage. It is also important that the two Ultrasonic Transducers be mounted one above the other, as close together as possible to reduce the magnitude of the elevation lobing.

A visible LED, not shown in the schematic, can be included if necessary and strobed for a brief period once every second to indicate that the beacon is switched on and working.

In this prototype the 9 V supplied by an alkaline battery was regulated down to 5 V using a linear regulator. This is inefficient and it is proposed that a small DC/DC converter should be used which would double the battery life to about 12 hours for the alkaline battery. A higher energy density rechargeable would improve the time still further.

3.2 Receiver

The receiver comprises a pair of muRata piezoelectric ultrasonic transducers amplified by a conventional stereo pre-amplifier as shown in Fig. 7. These transducers, and the amplifiers should be matched to ensure that the output voltage is balanced or a measurement bias will result. It is suggested that the amplifiers are replaced by a pair of gain-controlled analog receiver ICs manufactured by SenseTech if better range performance is required.

The piezoelectric transducers receive the 1ms long ultrasound pulse and convert it to an electrical signal. Because of the narrow bandwidth of both the transmit and receive transducers, the pulse shape is not rectangular, but has a finite rise and fall time as can be clearly seen in Fig. 8. These received signals are amplified are then detected and filtered to produce a rectified envelope of the pulse as shown.

From the schematic diagram it can be seen that the sum of the envelope voltages from the left and right channels is then converted to a digital pulse using a comparator configured with a Schmidt trigger input and with an adjustable threshold level. The leading edge of this pulse then triggers a monostable with a duration of 0.5 ms as shown in Fig. 9. The reason for this is to ensure that the envelopes are sampled at their peak, and that occurs after about 0.5 ms as can be seen in the measured data. The summing amplifier inverts the two detected signals to produce an inverted pulse with a peak amplitude of just over 200 mV in this case.

The comparator trigger threshold (labelled trigger threshold) is set to –140 mV in this example, and it can be seen from the figure that the threshold decreases by about 40 mV during the period that the comparator is triggered. This hysteresis ensures that noise induced multiple triggering does not occur as the pulse transitions through the threshold level.

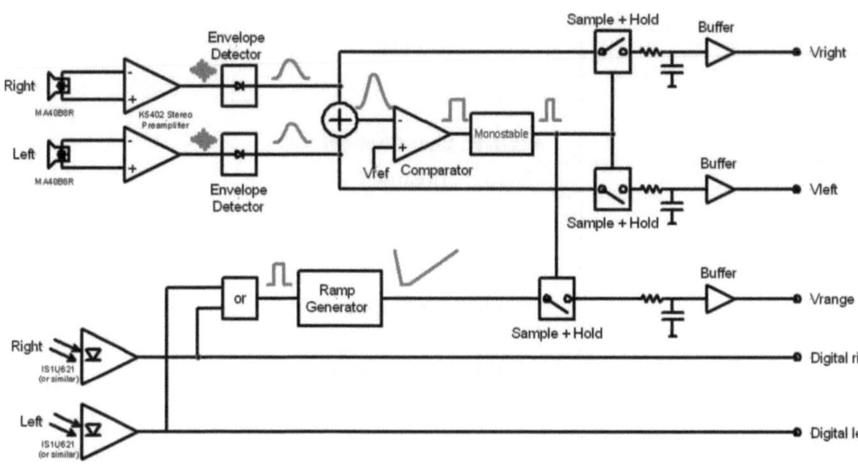

Fig. 7 Receiver schematic block diagram

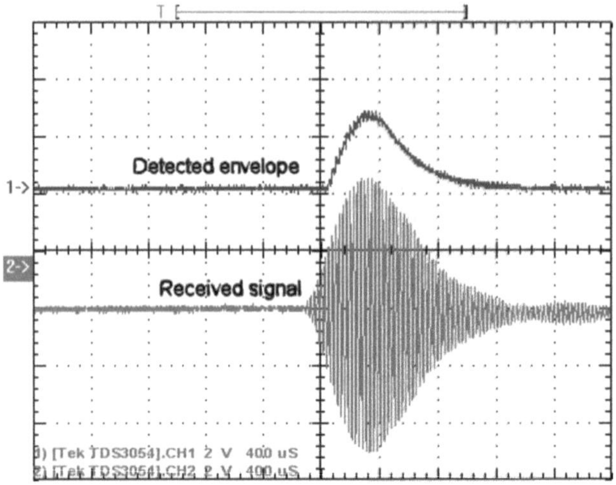

Fig. 8 Received acoustic signal and the detected envelope from one of the channels

The comparator output is a TTL signal which remains high for the period that the pulse envelope has crossed the detection threshold. Its duration is about 1 ms, but because the pulse is not rectangular, will depend on the size of the detected pulse and the threshold setting.

Because the comparator output can vary, it is used only to trigger the monostable which is then used to perform a number of functions. In the first instance it officially marks the receipt of the acoustic pulse for the range measurement circuitry which is explained later. Secondly, it triggers a sample and hold which samples the detected envelope for both the receivers. These values are then held for the 50 ms inter pulse period.

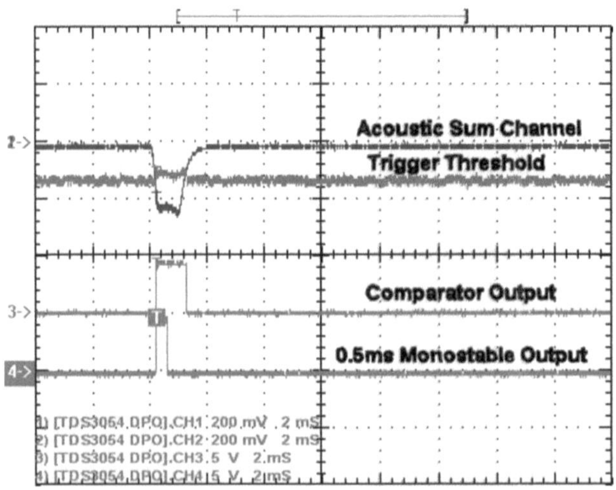

Fig. 9 Receiver timing

The sample and hold circuitry consists of an analog switch followed by a resistor and a capacitor, the output of which is buffered by a non-inverting follower. The RC time constant of this circuit is designed to smooth the sampled voltage slightly (otherwise just a capacitor could be used). During the 0.5 ms that the analog switch is closed, the capacitor charges and then tracks the envelope voltage. This is known as the sample period or sample aperture. When the switch is opened, the high input impedance of the buffer and the high off resistance of the analog switch ensure that the charge remains on the capacitor for the complete inter pulse period with minimal droop. This is known as the hold period, and it can be seen from Fig. 10 that the voltages, V_{left} and V_{right}, remain constant for the inter-pulse period.

It can also be seen that the voltages are equal on-boresight and that, as the beacon moves to the right, V_{right} becomes larger than V_{left}, and the reverse happens when it moves the other way.

The IR circuitry consists of a pair of buffered IR receivers that produce a TTL low on receipt of an IR signal modulated at 38 kHz. These signals are read into a PIC micro-controller (not shown in the schematic) for further processing.

The pulsed inputs are latched into the PIC on every cycle and are used to determine in which quadrant the beacon is located, from whether the left or the right or both receivers have been triggered. These outputs can be used to control a search algorithm in the absence of an acoustic signal.

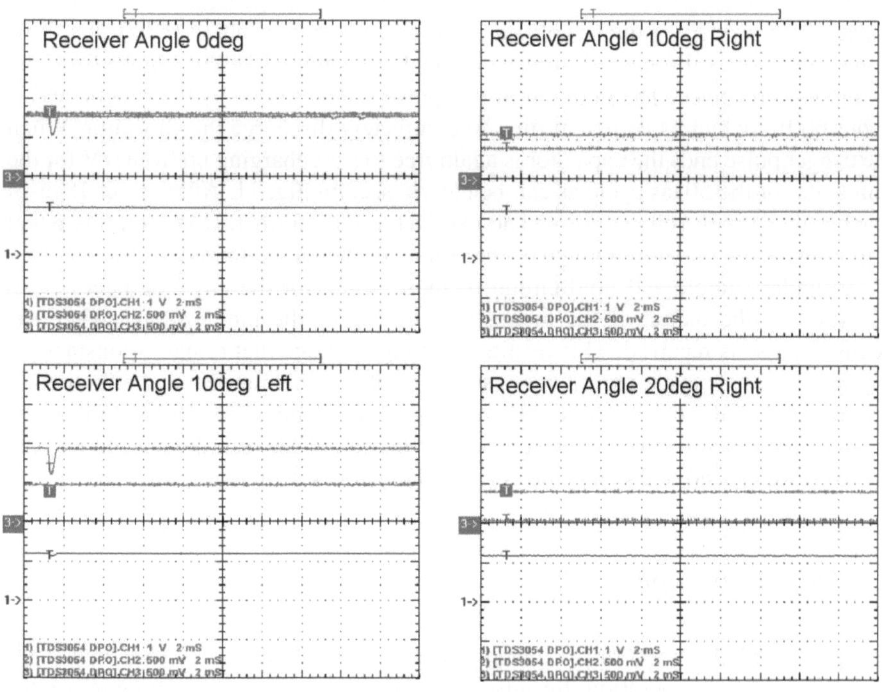

Fig. 10 Left and right analog outputs as a function of receiver angle with respect to the beacon position

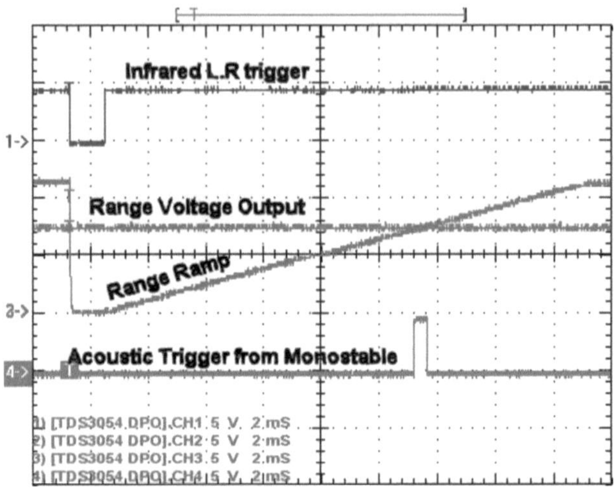

Fig. 11 Analog range signals showing the infrared trigger that starts the timing ramp and the acoustic trigger that is used to sample the ramp voltage

As shown in Fig. 11, an output from the PIC, which is the logical OR of the two IR receiver inputs serves as a low-going trigger pulse for the analog range voltage generator. The range circuit comprises a constant current generator that charges a capacitor to produce a linear voltage ramp with a slope of about 0.75 V/ms. This ramp capacitor is discharged through an analog switch when the trigger from the IR inputs is received. The duration of this pulse is determined by the beacon and is about 600 μs which is sufficient time to completely discharge the capacitor. When the trigger pulse ends the capacitor is again free to start charging up from 0 V for the remainder of the 50 ms cycle as can be clearly seen in Fig. 11. After about 18 ms it reaches the supply voltage (12 V) and can increase no further. However, this is not important as the maximum range of interest is reached after 15 ms.

A similar sample and hold circuit to those wired for the angle signals is also triggered by the output of the monostable to sample the ramp voltage when the acoustic pulse is received. This results in a range voltage that remains constant for a complete 50 ms cycle until a new sample is taken. This range voltage is directly proportional to the time between the receipt of the IR pulse and the receipt of the acoustic pulse and hence is representative of the range between the beacon and the receiver. At a speed of 340 m/s the conversion factor is 2.94 V/m.

3.3 DC Power Supply

A pair of DC/DC converters convert the +12 V input into +/ − 15 V and +5 V required for circuit operation. Because these modules generate switching transients at both their input and outputs, they are filtered using the appropriate LC networks. In addition, the +/ − 15 V rails are regulated to +/ − 12 V to reduce this switching noise still further.

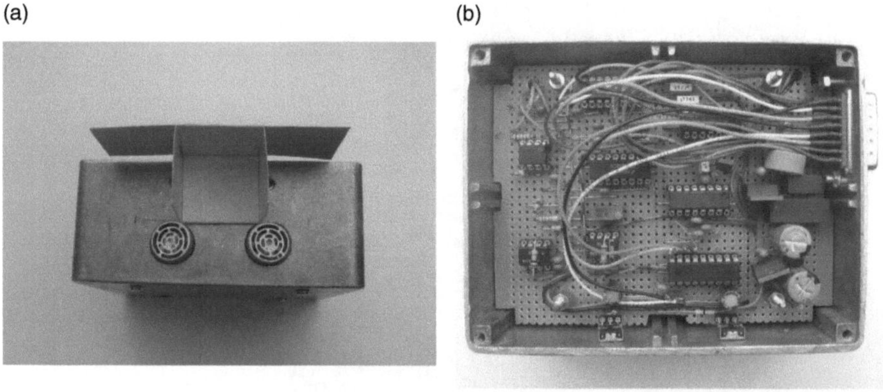

Fig. 12 Photographs of receiver box showing (**a**) both ultrasound and IR receivers and a construction to adjust the viewing angle of the IR receivers and (**b**) the prototype electronics built on veroboard

3.4 Construction

Construction of the prototype was straightforward. The ultrasonic transducers were soldered, at the correct squint angle, to the input posts on the stereo preamplifier. This board was mounted in a diecast box with the transducers protruding through large holes drilled in the one side, as can be seen in Fig. 12. The remainder of the electronics was assembled onto a piece of veroboard and mounted above the preamplifier using standoffs. Additional holes were drilled in the wall of the box to allow the infrared transducers an unrestricted view.

Finally, a cardboard septum was constructed to restrict the view from the infrared transducers, as can be seen in Fig. 12a, so that only about 10° of overlap occurred.

4 The Vehicle

To demonstrate the capabilities of the beacon tracking receiver, it was mounted onto the chassis of an off-road radio-controlled truck (Tamiya Overlander). This vehicle was selected as it has a simple pulse-width modulated servo input for both the steering and the throttle that could easily be controlled by a microcontroller.

Power supplies were developed to convert the vehicle battery supply (nominally 6.5 V) to the appropriate voltages for both the receiver and the processor.

4.1 Sensor Mounting Position

The receiver is mounted about 5 cm off the ground over the back wheels, with the rest of the prototype electronics mounted forward on the chassis as shown in Fig. 13. The reason for this configuration was to ensure that the relatively large mass of

Fig. 13 Radio controlled car chassis with sensor and prototype control electronics

the sensor was over the drive wheels and that the distance from the beacon to the receiver was as large as possible.

Such a low mounting position suffers from a number of drawbacks:

- The unit will have to look up toward the beacon if it is mounted on the owner's belt (about 1 m off the ground)
- The angle to the beacon will vary with distance and will require a fairly wide elevation beamwidth to compensate (about 30° for a ground range between 1 m and 5 m)
- There is a chance that the unit will look directly into the sun which might have a detrimental effect on any IR sensors used.

Figure 14 shows the normalised elevation gain for the ultrasound transducer as a function of beacon range for a number of tracker tilt angles. It can be seen that at short range, where the relative angle is large, the gain is low and as the range increases, the gain increases rapidly before reaching a plateau. This variation is acceptable, because the received signal power is proportional to the square of the range, so the increased gain compensates for the R^2 loss.

A reasonable compromise is to use a tilt angle of 20° as that maximises the gain at a ground range of 2 m without having it roll-off too quickly at longer ranges where the signal level will be lower. This angle could be optimised further by modelling the actual received signal level as a function of range instead of just using the antenna gain, but this wasn't done.

4.2 Vibration

Because the receiver comprises two sensitive ultrasonic transducers followed by a pair of high gain amplifiers, it is very sensitive to loud noises in the vicinity,

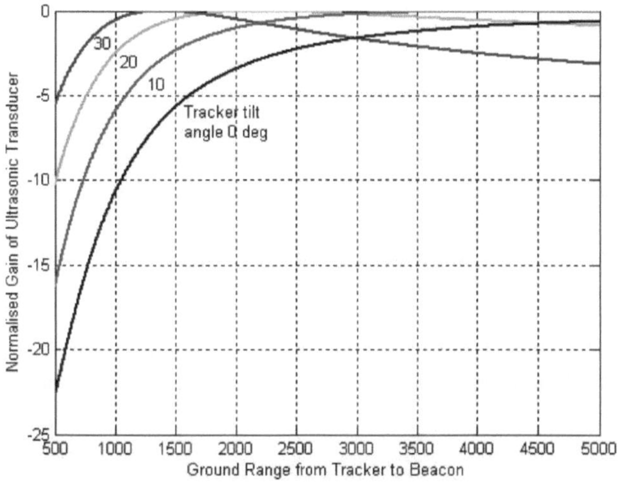

Fig. 14 Effect of beacon range on ultrasonic receiver gain

and in this prototype, vibration of the housing. To address the vibration issue, the transducers should be isolated from the housing, and as an added precaution, the housing should be mounted on the platform using shock mounts. This problem is illustrated in Fig. 15 which shows the effect of tapping the receiver box.

The issue with extraneous noise may not be a serious problem as the tests conducted, involved clapping loudly within 10 cm of the transducers. If it does prove to be an issue then it may be addressable by using other less sensitive transducers, or by further narrow-band filtering.

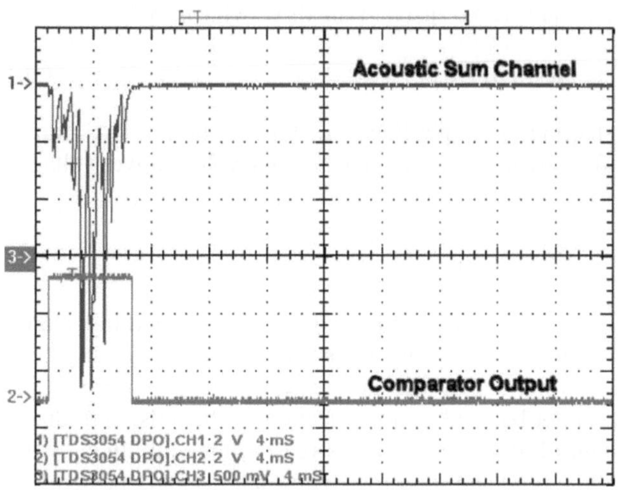

Fig. 15 Effect on acoustic detection of conducted vibration induced by tapping the housing

4.3 Beacon Mounting Effects

There are two restrictions to the bacon mounting position. Firstly because of the limited angle of view of its transducers, it must be mounted so that it faces the robot while the owner is walking, and secondly, because neither the IR nor the acoustic signals will penetrate clothing effectively, it must be mounted in a position where shirts and jackets cannot obscure its view.

The obvious mounting point is that the beacon should be clipped to the tail of any jacket or if the owner is in his shirt sleeves, attached to the belt.

Alternative mounting points which may be feasible are to attach it to the back of a cap or hat or to the heel of a shoe. However, in either case this will require that the elevation angle be adjusted, and may have an adverse effect on the short range performance.

5 Control Algorithms

The GumStix Atmel Atmega-128 8-bit microcontroller runs the software which performs the tracking, throttle and steering functions. In order for the vehicle to properly track the target, the presence of the acoustic signal and at least one infrared signal is required. An additional output from the sensor indicates this presence and notifies the controller that the incoming sensor data is valid. With this valid data, the vehicle can track the target using proportional control. The microcontroller digitises the range and the two angle voltages as well as sampling the digital outputs of the sensor.

Figure 16 shows a flowchart of the tracking algorithm executed by the microcontroller. The digitised range voltage, V_{range}, is subtracted from a set-point, V_{set}, to produce the difference which controls the speed of the drive-wheels, $\dot{R}$

$$\dot{R} = K_1(V_{\text{range}} - V_{\text{set}}). \tag{10}$$

To control the vehicle steering, a simple linear tracking algorithm was implemented by estimating the offset error, ΔAz, using the left and right receiver voltages (V_{left} and V_{right})

$$\Delta Az = K_2 \frac{V_{\text{left}} - V_{\text{right}}}{V_{\text{left}} + V_{\text{right}}}. \tag{11}$$

The value is normalised to full scale and used to set the steering angle. Normalisation of the error signal is necessary to provide accurate steering control independent of signal amplitude. Unfortunately, at low signal to noise ratios, this normalisation process cannot be used as amplification of the noise introduces unwanted inputs into the control algorithm. For this reason, the amplitude of the received acoustic signal is monitored and if it drops below a threshold, the acoustic data is ignored.

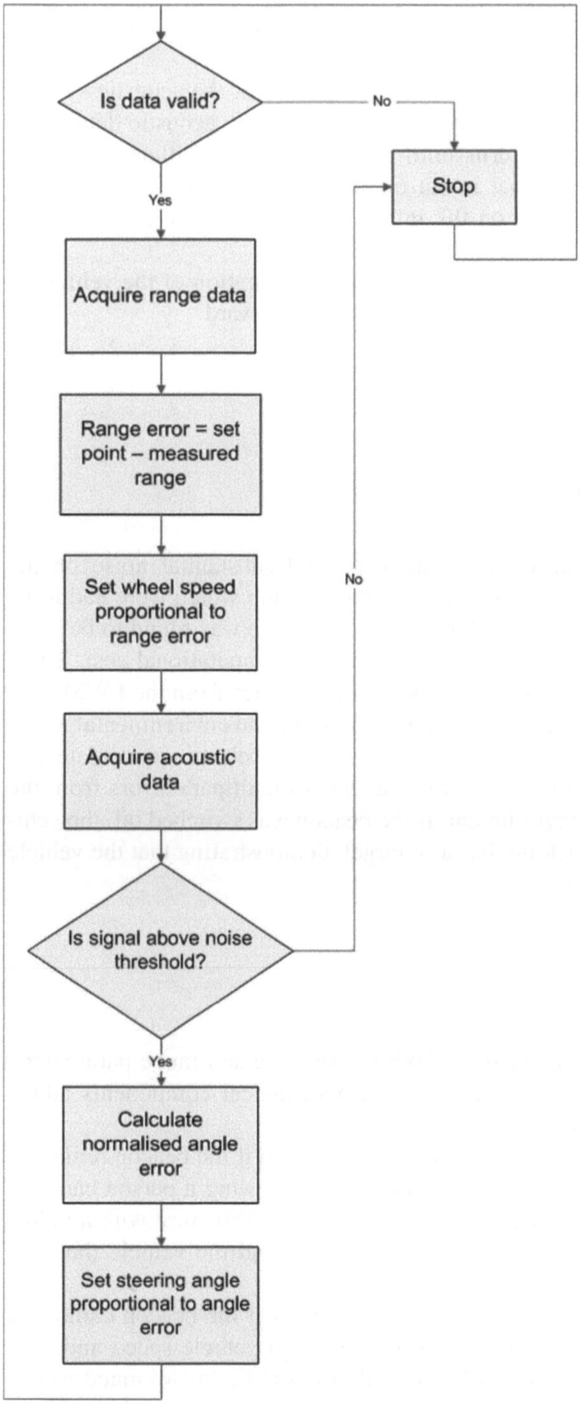

Fig. 16 Flowchart showing the tracking algorithm implemented on the Atmega-128 microcontroller

The algorithm also includes a small dead band to prevent oscillation around the centre point.

Because the infrared detection range is longer than that of the acoustic system, it is desirable that the vehicle track the beacon even when the acoustic data is not present. In this mode, the vehicle turns until the infrared signal illuminates both optical sensors. If the acoustic signal is out of range, the vehicle moves forward using a bang-bang controller based on the infrared tracking signal only, until the acoustic signal is acquired.

It must also be noted that due to the mechanical configuration of the vehicle, it can only turn towards the beacon when it is also moving forward.

6 Preliminary Results

6.1 Noise and Multipath

Initial tests of the complete prototype system showed substantial noise on the acoustic signals. The noise led to instabilities in the steering which were had to be reduced using a digital low-pass filter. A ten-sample average was found to be sufficient to create smooth steering behaviour with minimal computational cost. There are many possible sources of this noise, including interference from the PWM controlled servo motors, digital noise from the sensor circuitry and environmental noise.

Although the vehicle could detect a loss of signal and behave appropriately, it would occasionally track a false target. This was due to multipath errors from the acoustic signals in a cluttered environment. If the beacon was switched off, the vehicle would immediately stop tracking the false target, demonstrating that the vehicle was indeed tracking a reflection.

6.2 Control Algorithm

Proportional control was implemented for both the steering and range parameters, which was sufficient as the slow response of the mechanical components added considerable damping to system.

The beacon following capability of the system was good if the beacon remained within the receiver beam. The vehicle was capable of following a person carrying the beacon at a fast walk, and was capable of performing a 180° turn with a radius of less than 3 m. Figure 17 shows a plot of the beacon and the vehicle that was extracted from a video taken of one of the indoor trials.

At maximum vehicle speed, there was some overshoot if the beacon came to a sudden stop. Therefore, for applications that require high vehicle speed and sudden changes in control input, a PI or PID controller could be incorporated to improve the response. This, however, would increase the processing requirements of the controller.

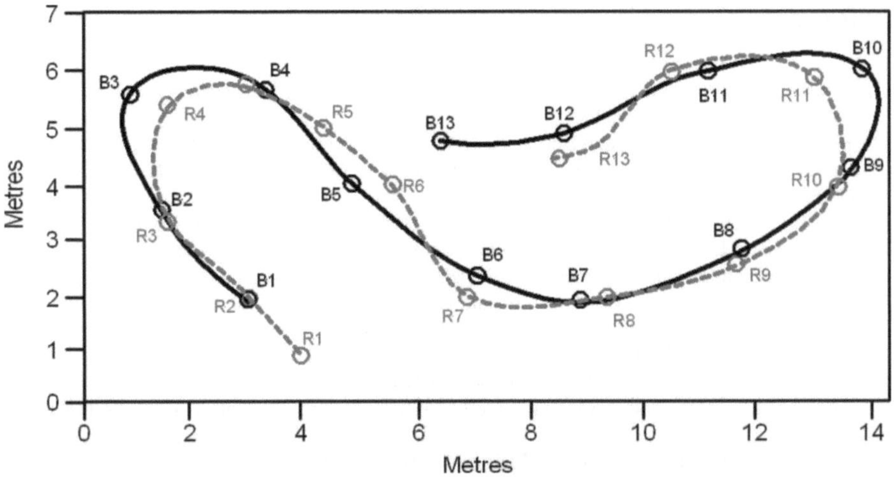

Fig. 17 Plot of the vehicle receiver R_x and the beacon B_x positions

6.3 Beacon Height

The sensor was found to be particularly sensitive to the position and angle of the beacon relative to the receiver. This sensitivity is due to the beam pattern of the acoustic transducers and could be improved by using transducers with a larger beamwidth. This characteristic highlighted the advantage of including the bang-bang controller using the infrared sensors which would allow the vehicle to continue to follow the beacon and return to a position where the acoustic signal was again usable if it had been lost.

7 Conclusions

This paper has discussed the design and implementation of a simple dual IR and acoustic tracker and its evaluation on a radio controlled car chassis. The implementation is unique insofar as it combines the broad acquisition angle of IR sensors (nearly 180°) with the precision angle measurement capability of an acoustic monopulse tracker. Additionally it uses the difference in the time of flight of the IR and acoustic signals to measure the range to the beacon as well.

Preliminary results of the tracker performance were promising with the vehicle capable of following a fast moving person and to execute reasonably tight turns.

Acknowledgements We would like to thank the ACFR (University of Sydney), the University of Cape Town and Itech Corporation for supporting this project.

References

1. Aguilar, R. and Meijer, G. (2002, November 29). Low-Cost Ultrasonic Fusion Sensor for Angular Position. SeSense 2002.
2. Bortolotto, G., Masson, F. and Bernal, S. (2001, 12–14 September). USRPS - Ultrasonic Short Range Positioning System. IX Reunion de Trabajo en Prosesamiento de la Informacion y Control (RPIC 2001), Santa Fe, Argentina.
3. Brooker, G., Lobsey, C. and McWilliams, K. (2007, 26–28 November). Combined infrared and Acoustic Beacon Tracker Implementation on an Autonomous Vehicle. 2nd International Conference on Sensing Technology, ICST'07, Palmerston North, New Zealand.
4. Fukuju, Y., Minami, M., Morikawa, H. and Aoyama, T. (2003). DOLPHIN: An Autonomous Indoor Positioning System in Ubiquitous Computing Environment. IEEE Workshop on Software Technologies for Future Embedded Systems (WSTFES'03).
5. Girod, L. and Estrin, D. (2001, 29 October–3 November). Robust Range Estimation Using Acoustic and Multimodal Sensing. Intelligent Robots and Systems (IROS 2001), Maui, Hawaii.
6. Kuc, R. (2002). Binaural Sonar Electronic Travel Aid Provides Vibrotactile Cues for Landmark, Reflector Motion and Surface Texture Classification. IEEE Trans. on Biomedical Engineering 49(10): 1173-1180.
7. Leonov, A. and Fomichev, K. (1986). Monopulse Radar, Artech House.
8. muRata. (2008). Piezoelectric Ceramic Sensors (PIEZOTITE). Retrieved January, 2008.
9. Palmer, R. (2002). A Spread Spectrum Acoustic Ranging System – An Overview. IEEE Canadian Conference on Electrical & Computer Engineering.
10. Priyantha, N. (2005). The Cricket Indoor Location System. PhD, MIT,
11. Skolnik, M. (1990). Radar Handbook, McGraw Hill.
12. Ward, A., Jones, A. and Hopper, A. (1997). A New Location Technique for the Active Office. IEEE Personal Communications: 42–47.

Electrostatic Microactuator Design Using Surface Acoustic Wave Devices

Don W. Dissanayake, Said F. Al-Sarawi and Derek Abbott

Abstract An integration of low power operated Surface Acoustic Wave (SAW) devices with the electrostatic microactuators for microfluidic and similar applications is presented in this chapter. Passive, low power, and small area devices can be interrogated wirelessly using SAW devices, which can respond to a uniquely coded signal for a secure and reliable operation. The novel approach relies on converting the interrogating coded signal to surface acoustic wave that is then correlated with an embedded code. A theoretical analysis of how the actuation mechanism operates is carried out and simulation results of the new microactuator are discussed. At the initial analytical stage, for simplicity, a basic SAW delay line structure is used to generate an electrostatic field between output interdigital transducer (IDT) of the SAW device and a thin conductive plate (actuator), which is placed on top of the output IDT. The static and transient displacement analysis of the actuator is carried out using ANSYS simulation tools. A comparison between the static displacements obtained from ANSYS based simulations and Rayleigh-Ritz based analysis is also presented and discussed.

Keywords SAW device · wireless · microactuator · electrostatic · IDT · ANSYS

Don W. Dissanayake
Centre for High Performance Integrated Technologies and Systems (CHiPTec), School of Electrical and Electronic Engineering, University of Adelaide, Australia,
e-mail: don@eleceng.adelaide.edu.au

Said F. Al-Sarawi
Centre for High Performance Integrated Technologies and Systems (CHiPTec), School of Electrical and Electronic Engineering, University of Adelaide, Australia,
e-mail: alsarawi@eleceng.adelaide.edu.au

Derek Abbott
Centre for Biomedical Engineering (CBME), School of Electrical and Electronic Engineering, University of Adelaide, Australia,
e-mail: dabbott@eleceng.adelaide.edu.au

1 Introduction

Micro Electro Mechanical Systems (MEMS) based microvalves and micropumps are of greater interest when it comes to nanolitre drug delivery and similar biomedical applications. Actuators are one of the important components in microvalves and micropumps for fluid manipulation.The design of a suitable actuator device to modulate the fluid at the microscale, for accurate operation, is of great importance. Many types of microactuators have been developed to match different requirements for various applications [1, 2]. With miniaturization, physical scaling laws inherently favour some technologies and phenomena over others. In some cases, technologies that can be made by micromachining work well at the microscopic scale, but have no analogy or usefulness in the macroscopic domain. Moreover most of these actuators are too complicated to fabricate within a micropump structure.

Notably SAW devices are used to develop micromachines such as ultrasonic micromotors and fluid transfer methodologies such as flexural micropumps [3, 4]. Currently available microfabrication technologies such as photolithography and X-ray lithography with a combination of other processes have enabled the use of SAW devices for a variety of self-contained MEMS applications, which have advanced functionality and performance. The key benefits of these micromachines are in their small size, ease of production, and low-cost.The use of SAW devices for micro actuation applications provides the great benefit of controlling and interrogating the devices remotely without direct user intervention [1, 2, 5].

In this chapter, SAW based electrostatic actuator modeling is discussed. Consequently such a microactuator can be embedded in a microfluidic device to modulate the fluid flow using less power. In Sect. 2, the use of SAW devices for micro actuation is discussed. Section 3 explains the operation of the SAW device based microactuator. An underlying theoretical model is then elaborated in Sect. 4 and followed by a method to derive the electric potential wave at the output IDT in Sect. 5. Section 6 shows the boundary condition analysis for the proposed model. Section 7 discusses the simulation and results and is followed by a conclusion.

2 SAW Based Microactuator

SAW device related technology has been utilised to design and develop MEMS based microaccelerometers and gyroscopes for military and similar applications [2, 6]. The technology used in those applications is similar to the capacitor effect generated by programmable tapped delay lines, which use the principle of air gap coupling [7] between the SAW substrate and a silicon superstrate. These capacitors are then used to control the amount of Radio Frequency (RF) coupling from the input IDT on the SAW substrate to the output terminal on the silicon chip [6]. It is a well known method to use a sandwich structure of semiconductor on piezoelectric substrate to form so called space-charge coupled SAW devices and SAW convolvers [7].

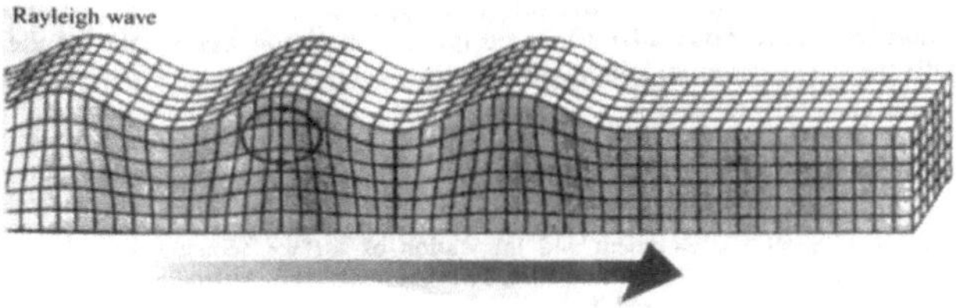

Fig. 1 Rayleigh Wave Mode which shows elliptical particle displacements [8]

Such an approach can be utilised in the design of a SAW based microactuator. The proposed approach for the actuator design is converse to the method used in [6] for the microaccelerometer design. Being an elastic deformation wave on a piezoelectric substrate, the SAW induces charge separation. Thus it carries an electric field with it, which exists both inside and outside the piezoelectric substrate and decays according to Laplace's equation. In this SAW device based actuator, a thin conductive plate is placed on top of the output IDT, which is separated by an air-gap. The conductive plate does not alter the mechanical boundary conditions of the SAW substrate, but causes the surface to be equipotential and the propagating electric potential to be zero at the surface of the conductive plate. As a result an electrostatic force is generated between the conductive plate and the output IDT in the SAW device causing micro deformations in the conductive plate.

It is highly important to choose a suitable SAW mode for the SAW device especially when it is designed for microfluidic applications. Therefore, different acoustic modes need to be analysed for example, one would expect horizontal shear SAW mode to be more useful than Rayleigh SAW mode. This is because when a SAW device is in vitro, Rayleigh SAW mode is rapidly damped out, due to the pressure generated by the surrounding fluid [5]. In contrast to the general approach of using the SAW device surface to transfer the fluid, the proposed approach uses an actuator (conductive plate),which is on top of the SAW substrate. As a result the fluid flow is isolated from the SAW substrate and hence it prohibits the interaction of the fluid with the SAW mode. Moreover, Rayleigh SAW mode is best suited for space-charge related applications as most of the energy in this mode is concentrated within one wavelength of the substrate [7], as shown in Fig. 1. Hence, considering the measures taken to isolate the SAW substrate from the fluid in this proposed approach, Rayleigh SAW mode is chosen.

3 Proposed Microactuator Operation

A SAW substrate made out of 128-YX-Lithium Niobate and a conductive plate made out of silicon are used to generate an air gap coupled SAW based electrostatic actuator. The device operation is as follows. The input IDT generate Rayleigh waves using inverse piezoelectric effect based on the RF signal that is being fed to the SAW device through the microstrip antenna. The Output IDT regenerates the electric signal using the piezoelectric effect of the SAW device. As it was explained in Sect. 2, the generated electrostatic field between this propagating electric potential wave and the conductive plate on top of the output IDT creates a compulsive and repulsive force between the two. Since the conductive plate is a thin flexural plate, it bends as a function of the applied electrostatic field enabling its use as a microactuator.

4 Theoretical Model

In piezoelectric materials, the frequency of the induced electric field wave is sufficiently small enough so that the electromagnetic coupling effects can be safely discarded. This means local perturbations are felt almost instantaneously through out the substrate, hence the electric field can be treated as quasi-static [9, 10, 11]. In the piezoelectric analysis this is achieved by setting the permeability to zero, corresponding to an infinite speed of the electromagnetic wave. Using Maxwell's equations, electromagnetic quasi static approximation can be written as

$$\mathbf{E}_i = -\frac{\partial \Phi}{\partial x_i}, \tag{1}$$

$$\nabla \cdot \mathbf{D} = 0, \tag{2}$$

where $\mathbf{E}$ is the electric field, Φ is the electric potential, $\mathbf{D}$ is the electric displacement and $i = 1,2,3$ shows the direction of interest as shown in Fig. 2. From the mechanical equations of motion,

$$\rho \frac{\partial^2 \mathbf{u}_i}{\partial t^2} = \sum_j \frac{\partial \mathbf{T}_{ij}}{\partial x_i}, \tag{3}$$

where ρ is the density of the piezoelectric substrate, $\mathbf{u}$ is the mechanical displacement and $\mathbf{T}$ is the mechanical stress.

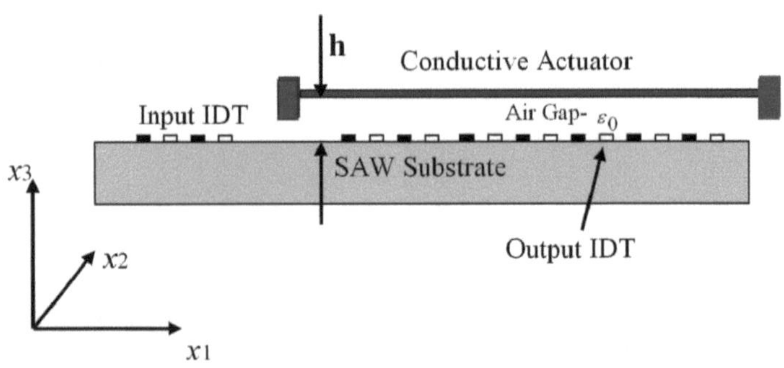

Fig. 2 SAW Actuator model and boundary conditions. The SAW substrate is the rotated crystal cut to make sure the wave propagation in x_1 direction and h is the height of the air gap

The relation between the mechanical strain **S** and mechanical displacement can be written as

$$\mathbf{S}_{ij} = \frac{1}{2}\left[\frac{\partial \mathbf{u}_i}{\partial x_j} + \frac{\partial \mathbf{u}_j}{\partial x_i}\right]. \tag{4}$$

Constitutive equations for piezoelectric material give the relationship between **D**, **E**, **T** and **S** in a SAW substrate as

$$\mathbf{T}_{ij} = \sum_k \sum_l \mathbf{c}^E_{ijkl}\mathbf{S}_{kl} - \sum_k \mathbf{e}_{kij}\mathbf{E}_k, \tag{5}$$

$$\mathbf{D}_i = \sum_j \sum_k \mathbf{e}_{ijk}\mathbf{S}_{jk} + \sum_j \varepsilon^S_{ij}\mathbf{E}_j, \tag{6}$$

where **c** is the stiffness tensor for constant electric field, **e** is the piezoelectric coupling tensor, ε is the permittivity tensor for constant strain and i, j, k, l go from 1 to 3. Equations (1–6) leads to a system of four coupled equations, which are represented by (7) and (8) below.

$$\sum_j \sum_k \left[\mathbf{e}_{kij}\frac{\partial^2 \Phi}{\partial x_j \partial x_k} + \sum_l \mathbf{c}^E_{ijkl}\frac{\partial^2 \mathbf{u}_k}{\partial x_j \partial x_l}\right] = \rho\frac{\partial^2 \mathbf{u}_i}{\partial t^2}, \tag{7}$$

$$\sum_i \sum_j \left[\varepsilon^S_{ij}\frac{\partial^2 \Phi}{\partial x_i \partial x_j} - \sum_k \mathbf{e}_{ijk}\frac{\partial^2 \mathbf{u}_j}{\partial x_i \partial x_k}\right] = 0. \tag{8}$$

This relationship between mechanical displacement and electric potential is considered in a semi-infinite, isotropic and homogenous linear elastic space as shown in Fig. 1. The crystal cut ensures the Rayleigh wave propagate in the x_1-x_3 sagittal plane as in Fig. 2. Sagittal plane is the plane defined by the real wave vector (wave number) k and the unit vector normal to the surface of the SAW substrate (x_3 direction). The method of partial waves can be used to solve the problem of wave propagation on anisotropic substrates such as piezoelectrics [9, 12]. As a result, plane wave solutions of the form in equations (9) and (10) are considered for the mathematical modeling of this device,

$$\mathbf{u}^m_j(x_1, x_3, t) = \alpha^m_j e^{\mathrm{i}kb^m x_3}.e^{\mathrm{i}k(x_1 - vt)}, \tag{9}$$

$$\mathbf{\Phi}^m(x_1, x_3, t) = \alpha^m_4 e^{\mathrm{i}kb^m x_3}.e^{\mathrm{i}k(x_1 - vt)}, \tag{10}$$

where the α_j values are linear coefficients that depend on the decaying constant b, v is the SAW velocity in the substrate, $m = 1, 2, 3, 4$ and $j = 1, 2, 3$.

By substituting the plane wave solutions given in (9) and (10) into the coupling wave equations (7) and (8), a linear system for the coefficients α_j can be obtained and solved. A similar approach has been adopted in microaccelerometer modelling in [6]. Following the approach discussed in [6], a general solution is obtained and the resulting equations can be seen from (11) and (12). The weighting coefficients

C of these plane waves are chosen to satisfy the mechanical and electrical boundary conditions at the surface of the piezoelectric substrate specific to this SAW based actuator model is presented in Sect. 6,

$$\mathbf{U}_j(x_1,x_3,t) = \left[\sum_m \mathrm{C}_m \alpha_j^m e^{ikb^m x_3}\right].e^{ik(x_1-vt)}, \tag{11}$$

$$\mathbf{\Phi}(x_1,x_3,t) = \left[\sum_m \mathrm{C}_m \alpha_4^m e^{ikb^m x_3}\right].e^{ik(x_1-vt)}. \tag{12}$$

5 Electric Potential at Output IDT

Once the plane wave equation is evaluated for the electric potential wave in the SAW devices, the following analysis can be applied to find out the SAW electric potential at the output IDT. It is assumed that the crystal cut is best set so that the SAW velocity is directed to the x_1 direction to allow an orthogonal interaction between the SAW and the output IDT. Moreover to simplify the analysis, the acoustic wave is assumed to pass the output IDT once, and interact with no reflections caused by impedance mismatches [8, 13]. Since the behavior of the propagating waves and the placement of the IDT being periodic, the analysis is carried out for a single period as shown in Fig. 3.

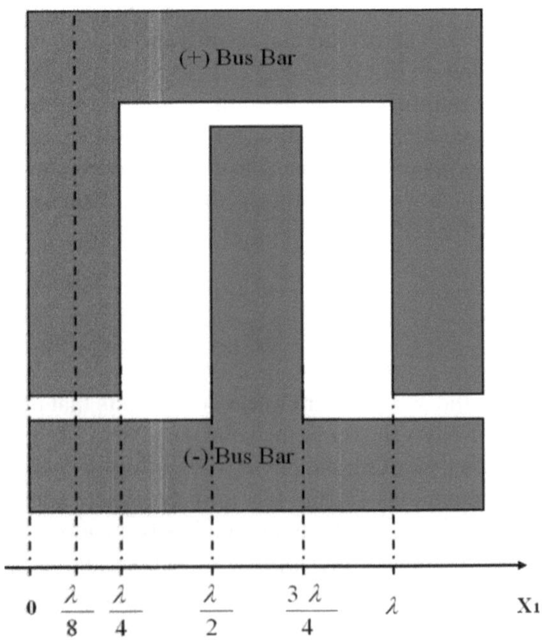

Fig. 3 IDT finger representation for one wavelength with a metalisation ratio of 0.5. Here λ is the wavelength of the SAW

Output IDT fingers are represented as square waves with the period defining wave length of the SAW and the duty cycle defining the finger width [14]. This width is $\lambda/4$ for a metallisation ratio of 0.5. For the SAW based electrostatic actuator model, initially a basic model of the IDT configuration was used where each finger is connected to one of two IDT bus bars alternatively as shown in Fig. 3. Since one bus bar is assigned a positive value and the other bus bar is assigned a negative value, each finger will be assigned a positive or negative value determined by the finger's connection to a particular bus bar. At the output IDT, the IDT is considered as a combination of square waves as it was explained above and the SAW is defined as a plane wave as it was shown in Sect. 4. In order to obtain the time response of the surface acoustic wave interaction with the output IDT, a cross-correlation is applied to the SAW and the IDT signals as shown in Fig. 4. As a result the resultant electric potential wave at the output IDT region is a combination of both the electric potential at the IDT and the electric potential at the IDT finger gaps as expressed in (13).

$$\boldsymbol{\Phi}(x_1,x_3,t)=\begin{cases}\Psi, & \text{for } 0\le x_1\le \frac{\lambda}{4}\\ \Omega, & \text{for } \frac{\lambda}{4}<x_1<\frac{\lambda}{2}\\ -\Psi, & \text{for } \frac{\lambda}{2}\le x_1\le\frac{3\lambda}{4}\\ -\Omega, & \text{for } \frac{3\lambda}{4}<x_1<\lambda\end{cases}, \tag{13}$$

where $\Psi=\boldsymbol{\Phi}(\frac{\lambda}{8},x_3,t)$ and $\Omega=\boldsymbol{\Phi}(x_1,x_3,t)$

$$\text{for } \boldsymbol{\Phi}(x_1,x_3,t)=\left[\sum_m \mathrm{C}_m\alpha_4^m e^{ikb^m x_3}\right].e^{ik(x_1-vt)}.$$

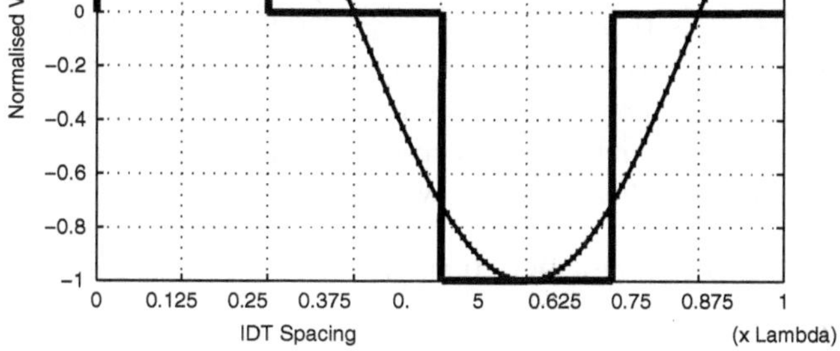

Fig. 4 Correlation between SAW electric potential and the output IDT

6 Boundary Condition Analysis

Weighting coefficients in (11) and (12) need to be determined based on the electric and mechanical boundary conditions applicable for the SAW device based actuator model.

Mechanical boundary conditions: For simplicity, the IDT mass is taken to be negligible so that the mechanical force acting on the SAW substrate can be discarded. Hence the surface is considered be mechanically free. This can be mathematically expressed as

$$\sum_j \mathbf{T}_{3j} = 0. \tag{14}$$

Electrical boundary conditions: The surface ($x_3 = 0$) where the output IDT is present and the gap between the output IDT fingers were considered separately for electric boundary conditions. As explained in Sect. 5 the electric potential wave at the IDT fingers act as an equipotential time varying wave throughout the positive and negative IDT fingers (does not vary with x_1). Where as the electric potential wave between the fingers still a time varying and a moving wave (varies with x_1). Since the conductive plate is placed at a height h above the output IDT, the electric potential goes to zero at $x_3 = h$. It should be noted that the electric potential above the surface satisfies Laplace's equation. Also the electric potential and the electric displacement in the x_3 direction are continuous at the surface ($x_3 = 0$) [6, 11]. These conditions can be mathematically expressed as

$$\begin{aligned} \mathbf{E}(x_3 = 0^-) &= \mathbf{E}(x_3 = 0) = \mathbf{E}(x_3 = 0^+), \\ \mathbf{D}(x_3 = 0^-) &= \mathbf{D}(x_3 = 0) = \mathbf{D}(x_3 = 0^+), \end{aligned} \tag{15}$$

and Laplace's equation can be written as

$$\mathbf{\Phi}_L(x_1, x_3, t) = \left[Ae^{kx_3} + Be^{-kx_3}\right].e^{ik(x_1 - vt)}. \tag{16}$$

The electric potential at $x_3 = h$ is zero because of the conductive plate. Using this condition, B can be eliminated from (16) and can be rewritten as

$$\mathbf{\Phi}_L(x_1, x_3, t) = A\left[e^{kx_3} - e^{2kh-kx_3}\right].e^{ik(x_1 - vt)}. \tag{17}$$

It should be noted that the analysis has to be carried out in two parts, considering the space above the output IDT ($0 \le x_1 \le \frac{\lambda}{4} \cup \frac{\lambda}{2} \le x_1 \le \frac{3\lambda}{4}$) and the space above the output IDT gap ($\frac{\lambda}{4} < x_1 < \frac{\lambda}{2} \cup \frac{3\lambda}{4} < x_1 < \lambda$). But the procedure for derivation of A is the same for both the cases. Therefore the approach is shown only for one case.

Considering the plane wave equation (13) for electric potential and (17), and the continuity equation (15), A can be evaluated as follows,

$$\mathbf{\Phi}_L(x_1, 0, t) = \mathbf{\Phi}(x_1, 0, t),$$

therefore,

$$A\left[1 - e^{2kh}\right].e^{ik(x_1 - vt)} = \left[\sum_m \mathrm{C}_m \alpha_4^m\right].e^{ik(x_1 - vt)},$$

hence,

$$A = \frac{\sum_m \mathrm{C}_m \alpha_4^m}{\left[1 - e^{2kh}\right]}.$$

Considering the relationship between the electric displacement and the electric field $\mathbf{D} = \varepsilon_0 \mathbf{E}$, and (1), a relationship between the electric displacement and the electric potential in the gap can be written as

$$\mathbf{D}_3(x_1, x_3, t) = -\varepsilon_0 \frac{\partial \mathbf{\Phi}_L(x_1, x_3, t)}{\partial x_3}. \tag{18}$$

Using the above, the electric displacement at $x_3 = 0$ can be written as

$$\mathbf{D}_3(x_1, 0, t) = -k\varepsilon_0 \frac{\sum_m \mathrm{C}_m \alpha_4^m \left[1 + e^{2kh}\right]}{\left[1 - e^{2kh}\right]} . e^{ik(x_1 - vt)}. \tag{19}$$

Another expression for the electric displacement at $x_3 = 0$ is obtained from the equation (6).

$$\mathbf{D}_3(x_1, 0, t) = \sum_j \sum_k \mathbf{e}_{3jk} \mathbf{S}_{jk} + \sum_j \varepsilon_{3j}^S \mathbf{E}_j. \tag{20}$$

By equating (19) and (20), the relevant electric boundary condition can be obtained and specific values for the weighting coefficients **C** can be calculated by solving an eigenvalue problem.

7 Simulations and Results

7.1 Static Analysis

The performance of the SAW device based electrostatic actuator behavior is simulated using the ANSYS simulation tools [15]. Initially the static analysis is carried out to determine the displacement of the actuator. In order to mimic the effect of the electric potential wave generated at the output IDT of the SAW device, a set of interleaved electrodes are used and every alternative electrode is coupled, so that one set of electrodes act as the positive bus bar and the other as the negative bus bar. Hence the whole SAW device in the microfluidic device is replaced at simulation level. Material properties of silicon are used for the doubly-clamped conductive plate, which in turn acts as the microactuator. The conductive plate dimensions are chosen to be $1000\,\mu\text{m} \times 2\,\mu\text{m} \times 10\,\mu\text{m}$ ($L \times H \times W$). The gap between the electrodes and the conductive plate h is taken to be $10\,\mu\text{m}$ and considered to be filled with air. AC sinusoidal wave with a frequency of 500 MHz and a peak voltage value of 10 volts is used to emulate the electric potential wave at the output IDT (as given in (13)). The conductive plate is connected to ground so that the plate acts as an equipotential surface.

To verify the simulation results, commonly used Rayleigh-Ritz iterative method was used as well. For comparison purposes, displacement versus voltage results are plotted in MATLAB and are shown in Fig. 5.

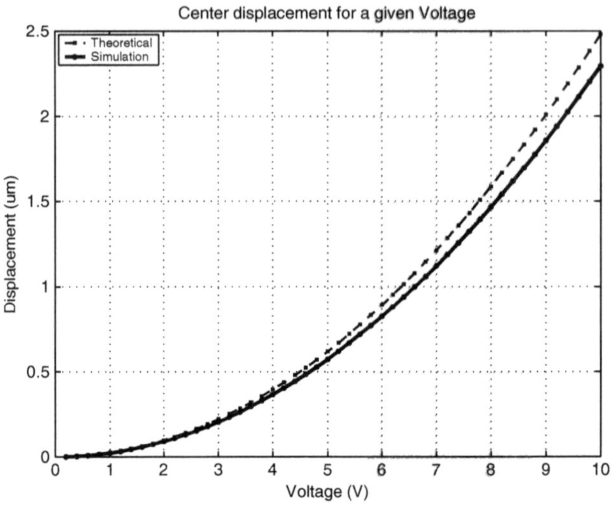

Fig. 5 Comparison of simulated and theoretical results for the SAW actuator. Displacement vs Voltage plot for the mid-beam displacement in the conductive plate actuator above the SAW device

A contour plot obtained from ANSYS simulations which depicts the bending of the actuator, is shown in Fig. 6. The actuator displacement can be optimised by reducing the gap between the conductive plate and the output IDT, reducing the thickness of the conductive plate, and reducing the stress level applied at the actuator by optimising the clamping mechanism.

7.2 Transient Analysis

It should be noted that when a conductive beam is subject to a dynamically changing electrostatic field, the displacement behaviour needs to be calculated analytically using an iterative process. At the initial stage, ANSYS based Finite Element Analysis (FEA) method greatly helps to analyse such a dynamic behaviour as the ANSYS solvers are equipped with analytical algorithms. This section depicts the transient simulation results carried out for a conductive plate with the same dimensions mentioned in the static analysis above. Moreover the input electrical signal is similar to the one that was used for the static analysis.

As can be seen from these simulations, the micro displacements are successfully obtained using this method. Figure 7 shows the mid-beam and the quarter-beam displacement variation over the time (250-time-periods). Based on authors' previous work, it was shown that beam displacements up to $2\,\mu$m can be achieved using SAW device based actuation [16]. After carrying out a few different transient analysis simulations with different end times and comparing the obtained transient displacements with the static displacements achieved, it is proven even

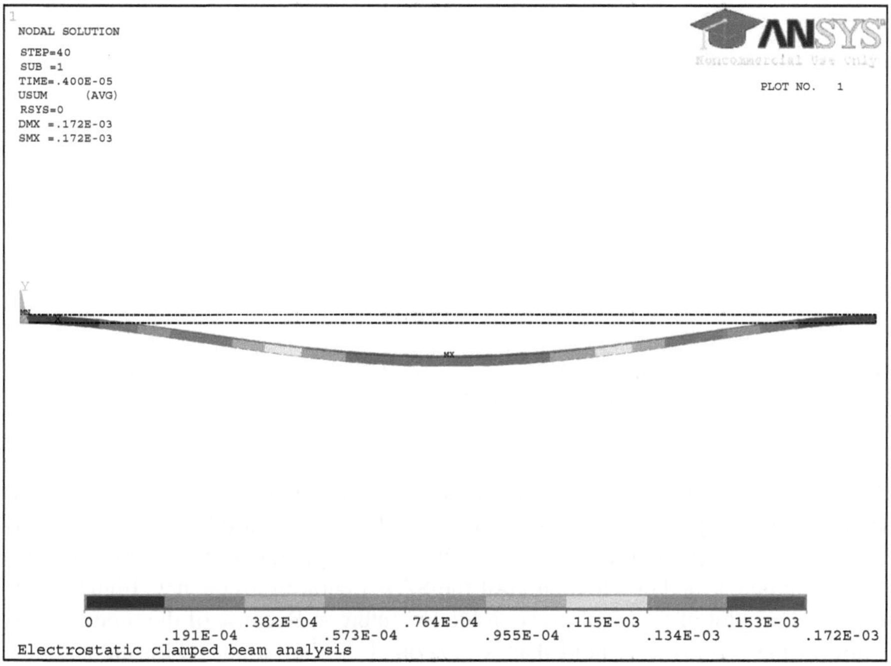

Fig. 6 A side view of the displaced actuator. A thinner actuator (2 μm) is analysed to achieve higher displacements [17]

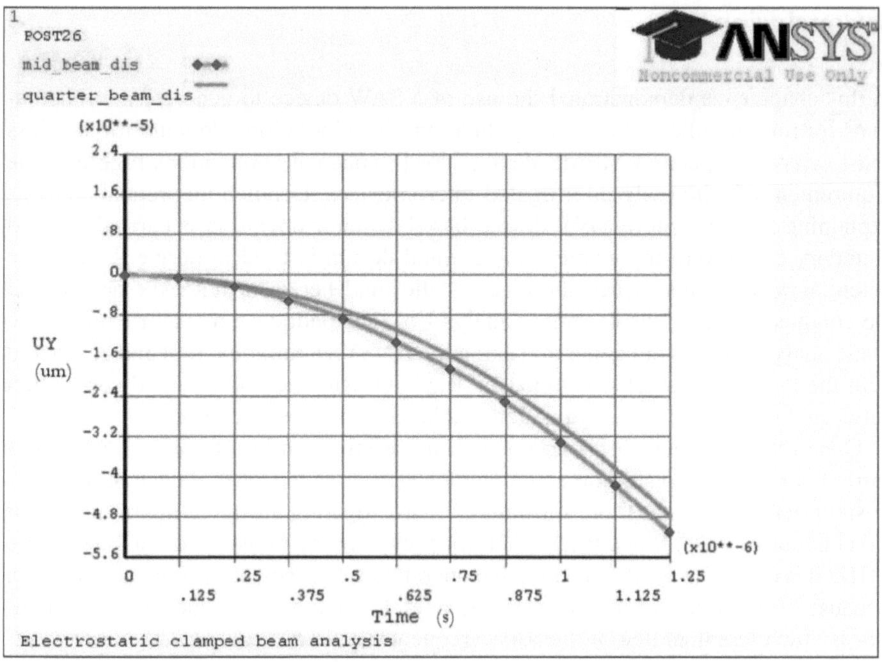

Fig. 7 Displacement vs Time plot of the mid-beam of the conductive plate. Analysis carried out for 250 × T, where T is the time period of the SAW signal [17]

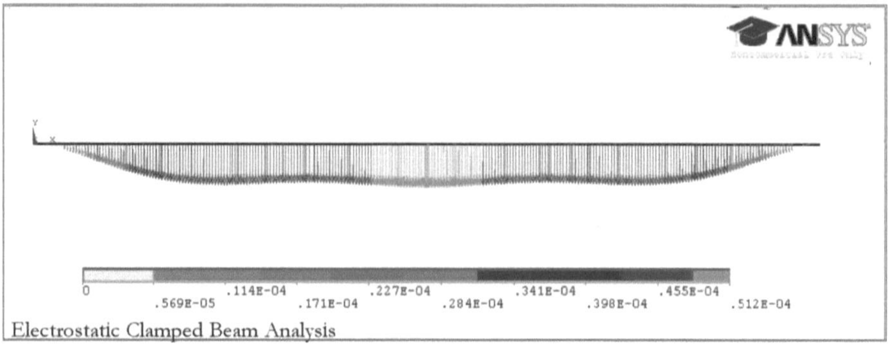

Fig. 8 A Vector plot of the displaced conductive plate obtained after the transient analysis. Vector shows nodal displacements of the conductive plate [17]

after 250-time-periods still the dynamic displacement is not showing any periodic nature but in the process of gaining more displacement. Based on the results obtained so far, it is evident that the actual vibration frequency of the conductive plate is a very much scaled down version of the SAW operating frequency. Figure 8 depicts a vector plot of the displaced conductive actuator. Because of the time varying nature of the electrostatic field that is generated by the output IDTs, the actuator consists of flexural displacement component as it can be seen from the vector plot.

8 Conclusion

In this chapter we demonstrated the use of a SAW device to generate microactuations for microfluidic and similar applications. The increasing demand for wireless and low power operated MEMS devices for biomedical applications highlight the requirement of wirelessly interrogated micro devices. Detailed theoretical analysis explaining how the entire SAW device based actuator operation is carried out and boundary conditions applicable for presented design is used to derive the electric potential wave forms, hence the electrostatic field between the SAW device and the conductive plate. Displacement analysis of the conductive actuator is obtained. Static analysis results are generated using the ANSYS simulation tool and compared with the theoretical results obtained by Rayleigh-Ritz method. A good correlation between the theoretical and simulated displacement curves are observed.

Once the static analysis is completed, more importantly a transient analysis is carried out in order to study the dynamic behaviour of the SAW device based electrostatic actuator. This is more substantial in investigating the operating frequency of the conductive plate. Since the SAW frequency is in the range between 100 MHz – 1 GHz it is crucial to verify the effective operating frequency of the conductive plate. Because of the time varying electrostatic field, the operating frequency of the actuator is much less than that of the SAW frequency.

Acknowledgements The authors would like to thank the Australian Research Council (ARC) and the School of Electrical and Electronic Engineering (University of Adelaide) for the funding and the support provided for this research.

References

1. Dissanayake D, Tikka A, Al-Sarawi S, Abbott D (2006) Radio frequency controlled microvalve for biomedical applications. In: Proc. SPIE–Smart Materials IV, 6413, Australia, 64130, pp 1–13
2. Varadan V K, Varadan V V (2000) Microsensors, micromechanical systems (MEMS), and electronics for smart structures and systems. Smart Materials and Structures, 9:953–972
3. Wixforth A (2003) Acoustically driven planar microfluidics. Superlattices and Microstructures, 6:389–396
4. Strobl C J, Guttenberg Z V, Wixforth A (2004) A Nano–and pico-dispensing of fluids on planar substrates using SAW. IEEE Transactions on Ultrasonics, Ferroelectrics, and Frequency Control, 51(11):1432–1436
5. Jones I, Ricciardi L, Hall L, Hansen H, Varadan V, Bertram C, Maddocks S, Enderling S, Saint D, Al-Sarawi S, Abbott D (2007) Wireless RF communication in biomedical applications. Smart Materials and Structures, 17 015050 (10pp)
6. Subramanian H, Varadan V K, Varadan V V, Vellekoopz M J (1997) Design and fabrication of wireless remotely readable MEMS based microaccelerometers. Smart Materials and Structures, 6(6):730–738
7. Milstein L B, Das P (1979) Surface acoustic wave devices. IEEE Communications Magazine, 17:25–33
8. Gardner J W, Varadan V K, Awadelkarim O O (2001) Microsensors, MEMS, and Smart Devices. Tsinghua University Press, Beijing
9. Gantner A, Hoppe R H W, Köster D, Siebert K G, Wixforth A (2005) Numerical simulation of piezoelectrically agitated surface acoustic waves on microfluidic biochips. http://www.opus-bayern.de/uni-augsburg/volltexte/2005/142/, visited on 14/02/2008
10. Kannan T (2006) Finite element analysis of surface acoustic wave resonators. MA thesis, University of Saskatchewan
11. Maugin G A (1985) Nonlinear electromechanical effects and applications. World Scientific Publishing Co Pty Ltd, Singapore
12. Adler E L (2000) Bulk and surface acoustic waves in anisotropic solids. International Journal of High Speed Electronics and Systems, 10(3):653–684
13. Ippolito S J, Kalantar-zadeh K, Wlodarski W, Powell D A (2002) Finite-element analysis for simulation of layered SAW devices with XY LiNbO3 substrate Proc. In: SPIE–Smart Structures, Devices, and Systems, 4935, Australia, pp 120–131
14. Skinner J L, Cardinale G F, Talin A A, Brocato R W (2006) Effect of critical dimension variation on SAW correlator energy. IEEE Transactions on Ultrasonics, Ferroelectrics and Frequency Control, 53(2):497–501
15. ANSYS Incorporation. Home page, http://www.ansys.com/, visited on 14/02/2008
16. Dissanayake D, Al-Sarawi S, Abbott D (2007) Surface Acoustic Wave Device Based Electrostatic Actuator for Microfluidic Applications. In: Proc. 2nd International Conference on Sensing Technology, Palmerston North, New Zealand, pp 381–386
17. Dissanayake D W, Al-Sarawi S, Abbott D (2007) Surface Acoustic Wave Device based wireless passive microvalve for microfluidic applications. In: Proc. SPIE–BioMEMS and Nanotechnology III, 6799, 67990H pp 1–10

Wide Band Linearization of a Millimetre-Wave, Linear Frequency Modulated Radar Employing a Surface Acoustic Wave, Delay Line Discriminator

David G. Johnson and Graham M. Brooker

Abstract Phase-Locked Loops (PLLs) are critical components of most modern communication and radar systems, commonly using Direct Digital Synthesis to form the reference signal. An alternative linearisation method employing a Delay Line Discriminator within a PLL is put forward, with experimental results showing better than 10cm range resolution between a pair of closely-located targets over a linearised bandwidth of ~4GHz.

Keywords Linearisation · MMW · FMCW · radar

1 Introduction

Whilst Phase-Locked Loops (PLLs) are far from new [1, 2, 3, 4], they continue to be the workhorse behind modern communications systems, and are a critical component of many radar systems. The continual development of PLLs is therefore as important today as it ever has been. The trend towards Direct Digital Synthesizer (DDS) chips, produced by companies such as Analog Devices [5, 6], has become the norm for large bandwidth RF designs when accompanied by high-M,N frequency divider feedback networks. However, alternative methods such as the Delay Line Discriminator (DLD) technique discussed in this paper remain worthy of consideration.

This paper describes the basic functionality of a PLL circuit capable of linearising signals with 30dB fluctuation in power with 300MHz bandwidth, which when passed through a highly-linear 13x frequency multiplier provides 4GHz of linearised signal output centred at 94 GHz.

David G. Johnson
Australian Centre for Field Robotics, University of Sydney, Sydney, Australia,
e-mail: d.johnson@acfr.usyd.edu.au

Graham M. Brooker
Australian Centre for Field Robotics, University of Sydney, Sydney, Australia,
e-mail: gbrooker@acfr.usyd.edu.au

S.C. Mukhopadhyay, G.S. Gupta (eds.), *Smart Sensors and Sensing Technology*,

The paper is organised as follows: Sect. 2 outlines the need for the linearisation of radar transmitters. Section 3 provides an overview of the radar's front-end. Section 4 discusses the specifics of the loop-linearisation process. Section 5 provides a performance analysis of the linearization process and discusses results for two closely separated targets.

2 The Need for Linearisation

One of the simplest and most reliable methods of radar operation utilises a Frequency Modulated Continuous Wave (FMCW) waveform, or 'Chirp', whereby targets at a constant range, R, are detected at a unique beat frequency, f_b, described by

$$f_b = \frac{2R}{c} . \frac{\Delta f}{T} \tag{1}$$

where Δf is the extent of the chirp that is generated over time, T, and c is the speed of propagation.

For an ideal linear chirp, the range resolution, δR, is related to the resolution of the beat frequency, $\delta f_b, (\delta f_b \approx 1/T)$. Rearranging (1) therefore gives:

$$\delta R = \frac{c}{2\Delta f} \tag{2}$$

The beat frequency can be easily determined by applying the Fast Fourier Transform (FFT) to data digitised at the receiver input. Two factors influence the resolution with which the beat frequency, and hence range can be measured:

- The spectral purity of the transmitted signal at any given moment in time.
- The non-linearity of the frequency chirp, as seen in Figure 1.

Noise caused by spectrally impure signals and/or linearity errors broadens the signal-peaks in the subsequent FFT and hence prevent targets closely located in range, and hence frequency, from being resolved. The true range resolution, δR, is then a function of both chirp bandwidth and chirp linearity, given by (3) in the case of a quadratic nonlinearity [7].

$$\delta R = \sqrt{\left(\frac{c}{2\Delta f}\right)^2 + (R.Lin)^2} \tag{3}$$

where the linearity, Lin, is defined as the change in chirp slope, $S = \Delta f/T$, normalised by the minimum slope,

$$Lin = \frac{S_{\max} - S_{\min}}{S_{\min}} \tag{4}$$

Sensor performance can therefore be improved by minimising the value of Lin. Whilst the correction of non-linear signals can, to some extent, be performed in

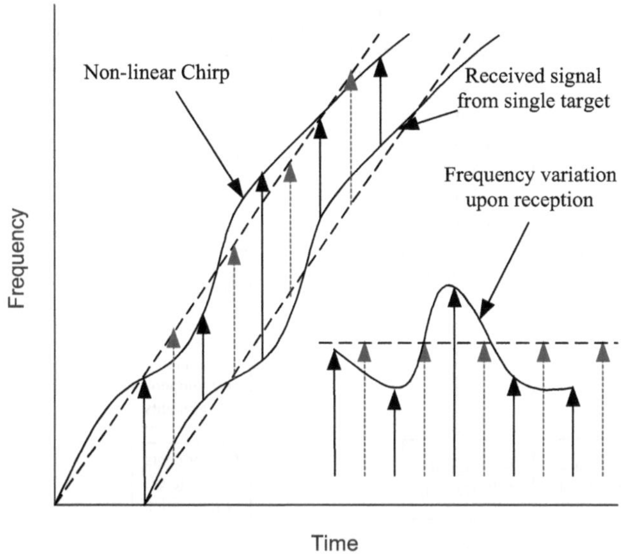

Fig. 1 Un-Linearised Frequency Chirp

post-processing [8, 9], these methods require additional doppler compensation techniques to be applied to each detected target, and in any case it is intuitively more sensible to ensure the transmitted signal is as pure and as linear as possible. This can be achieved partially by the choice of transmitter components, but also by the specific linearisation measures to be discussed in Sect. 4.

3 Radar Front-End Overview

Over the past 5 years a number of 94GHz FMCW radars have been developed at the ACFR (e.g. [10]).

Whilst the antenna, drive circuitry and signal processing have been continually developed over time, a common front-end, sourced from Elva-1 [11] has been employed and forms the backbone for this radar, as seen in Fig. 2. The transmit chain is based on a low frequency VCO followed by an Impatt Diode Active Frequency Multiplier which is inherently more linear than the alternative: Gunn oscillators; but is nevertheless limited to a linear bandwidth of ~500 MHz at MMW frequencies when operated in an 'open-loop' arrangement. In order to improve the linearity at the transmitting antenna, the oscillator must be phase & frequency locked to a stable reference source.

The linearity of the VCO in open-loop is shown in Fig. 3, where it can be seen that an IF bandwidth of 600 MHz is achievable, which when multiplied to MMW frequencies gives a theoretical bandwidth of 7.8 GHz. However, the useful

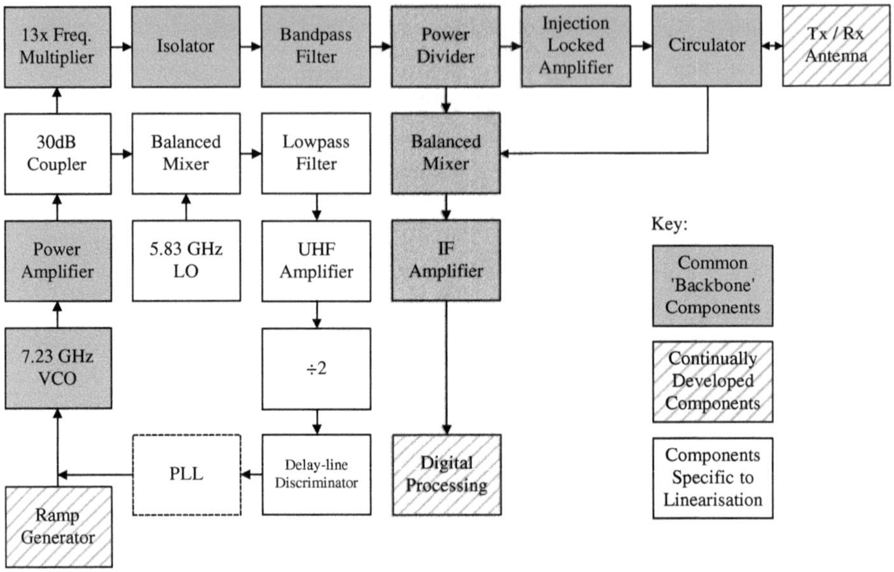

Fig. 2 Schematic of MMW Radar Front-End

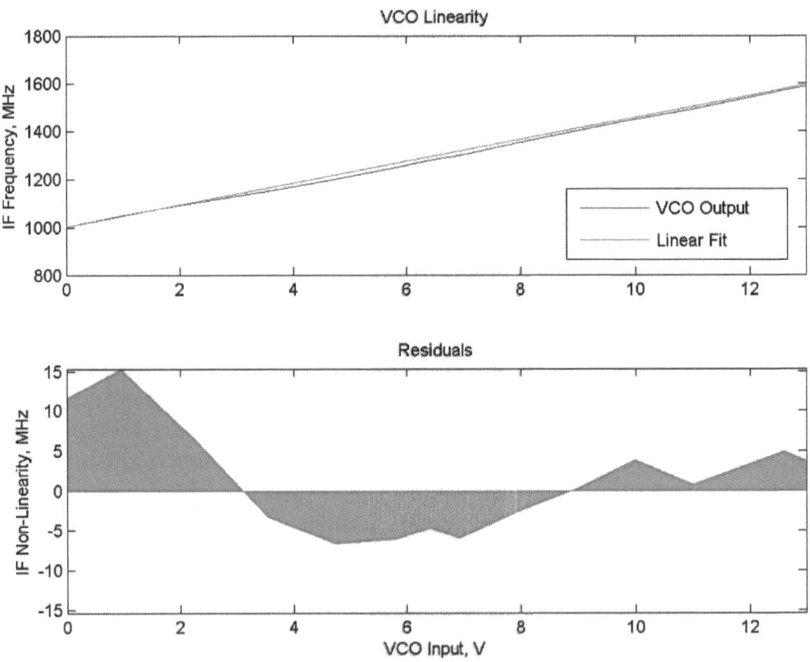

Fig. 3 VCO Linearity in Open-Loop

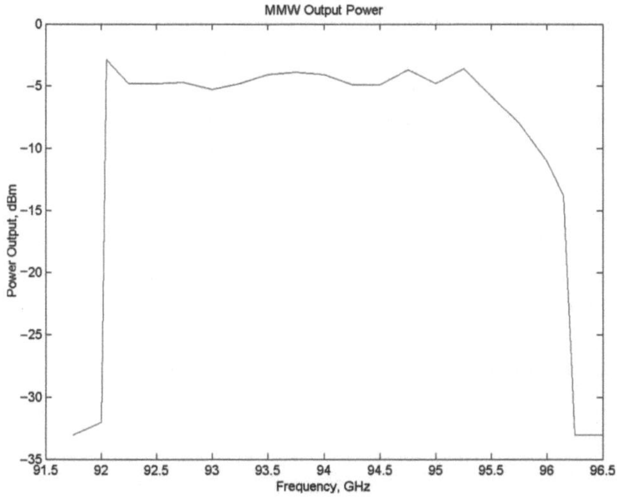

Fig. 4 MMW Output Power

bandwidth in the MMW region is limited by that of the MMW components to ~4GHz as seen in Fig. 4. For this reason, linearisation efforts will be restricted to a 300MHz bandwidth in the IF region.

4 Closing the Loop

4.1 The Generally Accepted Method

The accepted method for implementing a Frequency & Phase Locked Loop for FMCW radar is via Direct Up-Conversion as depicted in Fig. 5. A high-speed DDS

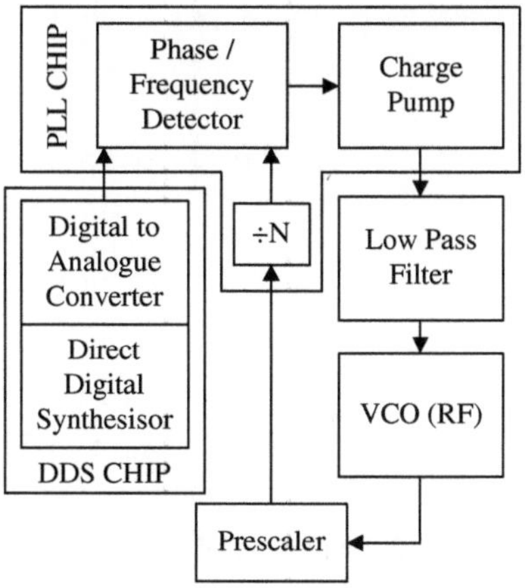

Fig. 5 Typical Direct Up-Conversion PLL

chip, pre-programmed with the required frequency slope parameters and containing a high dynamic range Digital to Analogue Converter (DAC), outputs a series of incremental frequency reference signals to the Phase/Frequency Detector (PFD) within the PLL chip. The reference signals are then compared to a sub-sampled version of the VCO output within the PFD, which in combination with a charge pump and a Low Pass Filter outputs a corrected voltage to the VCO, resulting in a frequency closer to that desired.

The obvious drawback of the DDS approach in terms of the linearity of the frequency slope is the fact that the frequency is stepped rather than swept linearly. However for modern, fast DDS chips, the duration of each frequency step is typically less than a cycle, hence providing phase transients between frequency-steps are well catered for [12], the overall sweep linearity can be quite reasonable.

4.2 Our Method

In our method, shown in Fig. 6, a linear voltage ramp is passed to a VCO, combined with a PLL correction voltage. Further linearisation is then achieved by tapping off

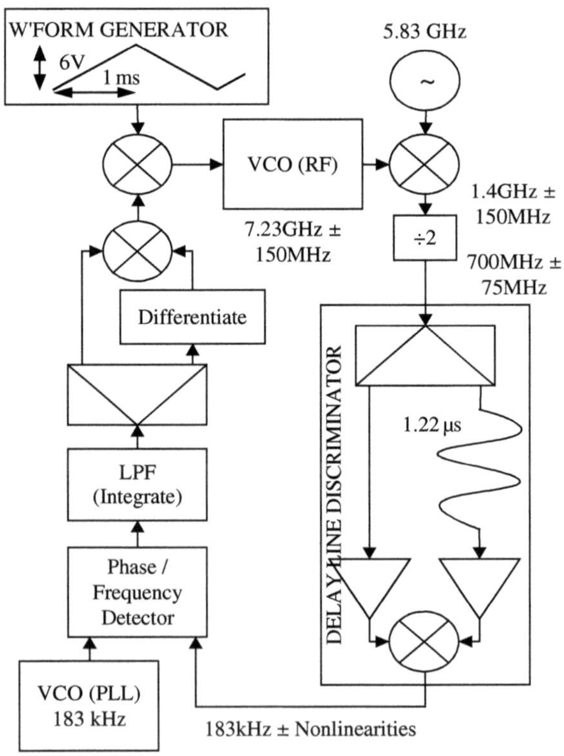

Fig. 6 PLL Employing DLD

an Intermediate Frequency (IF) signal prior to multiplication of the VCO output to MMW, which is then fed into a Surface Acoustic Wave (SAW) Delay Line Discriminator (DLD).

Extensive work has been written on the subject of SAW devices (e.g. [13]) to which the reader is referred, although DLD devices are not commonly used. In brief, surface waves are piezoelectric microwaves which can be propagated across the surface of certain crystals, such as quartz, at a velocity of ~3500 m/s (or 3 μs per cm), hence a 1 μs coaxial-cable delay-line >200 m in length can be replaced with a surface-mount device a few mm long. If extremely stable operation is required, then the delay line must be temperature stabilised as changes in temperature effect the length, and hence delay, through the device.

The 1.22 μs SAW delay line employed here has a bandwidth of ~300 MHz with a centre frequency of 700 MHz. The IF is therefore passed through a frequency divider to bring it within the pass band of the delay line.

Mixing a delayed portion of the VCO signal with itself produces a false-target at a fixed range, hence when the output of the VCO is a constant frequency slope; the output from the DLD is a constant beat frequency. By passing the measured beat frequency to a PLL chip and comparing its phase to that of the fixed frequency expected from the false-target for a constant slope, the voltage produced by the PLL loop-filter should correct exactly for any non-linearities in the original frequency slope.

A triangular waveform was desired in order to allow the determination of a target's Doppler frequency – from the difference between a target's beat frequencies on up and down sweeps. The correction signal must therefore be inverted along with the slope.

One limitation of the current implementation of the loop filter is that whilst the DLD frequency remains locked, the upper and lower bounds of the chirp may drift over time. This may be corrected using a second frequency-locked-loop in conjunction with the first, to lock a relative position of the chirp in place over time.

5 Results

These results were originally presented in [14].

5.1 Slope Linearization

The various stages of the linearization process are depicted in Fig. 7.

Due to the fact that the DLD technique locks to a signal that is a function of the frequency slope rather than to a fixed input frequency, measurement of the closed-loop linearity, comparable to that of Fig. 3, relied on the post processing of a full up-down sweep, as seen in Fig. 7a. The triangular ramp generator waveform was varied by ~6 V over the region of the VCO previously determined to be the most

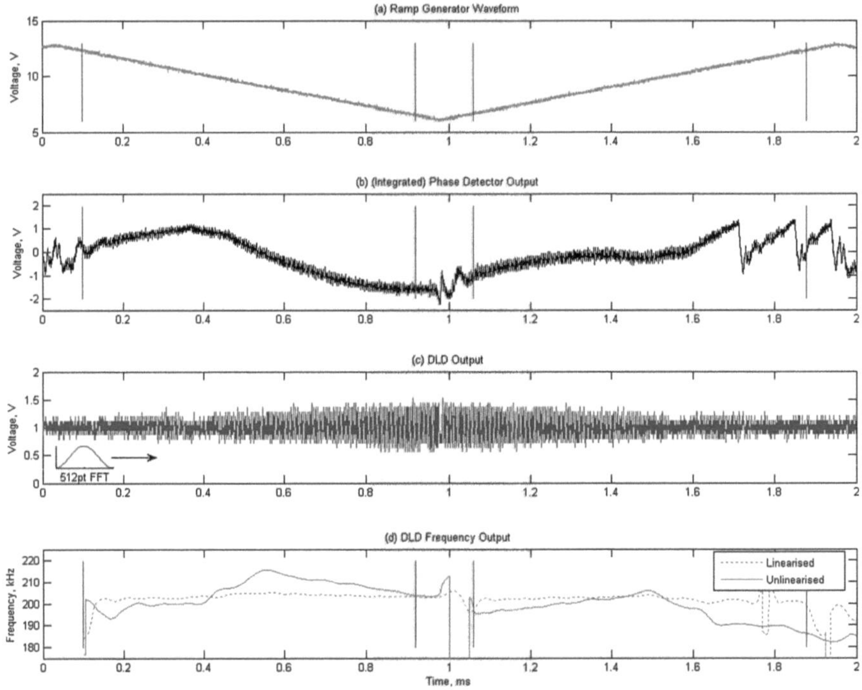

Fig. 7 Operation and performance of the Sweep Linearization Process

linear, as can be seen from Fig. 3. A subset of this region, corresponding to a MMW range of 92.5–96 GHz, was then chosen such that transient effects within the PLL loop filter generated by changes in slope polarity were minimised. By sampling at a rate of 5 MHz, a 4 k FFT could then by applied to the IF signal within this region for later range profile generation.

Figure 7b displays the output of the PLL loop filter Integrator on both up and down sweeps. Employing a positive edge-triggered phase comparator with a transfer characteristic given by (5), the deviation in frequency slope from the ideal is essentially given by the magnitude of this curve when the phase difference between the two input signals ($\phi_{DLD}-\phi_{REF}$) is truly less than ±180°.

$$V_{PC} = \frac{V_{CC}}{2\pi}(\phi_{DLD} - \phi_{REF}) \tag{5}$$

It can be seen that the linearization begins to break down towards the end of the up-sweep. This is due to a small difference (~5%) in the sweep time on the up-sweep causing the change in phase comparator output to exceed the relatively narrow bandwidth of the loop filter (which is still attempting to lock to a reference signal fixed at 203 kHz), allowing the relative phase to pass over the 2π phase-boundary.

The final two parts of Fig. 7 display the DLD output in both the temporal and frequency domain. Figure 7d was formed by sliding a Hanning window and 512 pt FFT over the raw DLD signal and applying a peak detection and interpolation algorithm to the result. The mirroring between the results from up- and down-sweeps is made slightly less obvious by the change in up-sweep duration previously mentioned causing a step change in DLD frequency output around 1 ms and a slight scale-reduction of up-sweep data in the un-linearised case. However, whilst lock is maintained on the first part of the up-sweep the sweep rate is almost identical to that of the down-sweep. It should be noted that the sharp deviation in linearity observed at around 1.7 ms is more an artefact of the peak detection algorithm rather than a kink in the frequency slope associated with the phase transition seen in Fig. 7b. For the better performing down-sweep an improvement in linearity from 0.12 to <0.02 can be observed by applying (4).

5.2 Two-Target Discrimination

As stated previously, a radar's true range resolution is best measured by its ability to discriminate between two closely separated targets. Range measurements were therefore made against a pair of closely located targets in both linearised and open-loop modes. The trihedral corner-reflector targets (~3 m^2 at 94 GHz), were mounted on a tripod (7 m ± 5 cm) down-range from the radar within the 1-degree beamwidth of the MMW antenna. The tripod was then placed ~2 m in front of a brick wall. Figure 8 depicts the apparatus setup:

It was later realised that the 1.22 μs delay line corresponded coincidentally to a target at 7.1 m for the bandwidth and sweep-time employed. Fortunately this did not impact on results due to the range offset caused by waveguide lengths within the radar.

A 4 k FFT was then applied to the IF signal on the down sweep as described in Sect. 5.1. According to the measurements of DLD frequency deviation made previously, a spectral peak bandwidth of ~20 kHz was expected for the un-linearised case and this can indeed be seen in the FFT of both the DLD and IF signals in Fig. 9.

Comparing the upper graphs in Figs. 9 and 10 the narrowing of the spectral peak of the false-target generated by the DLD can be easily observed.

Examining the lower graphs in Figs. 9 and 10 it would appear at first glance that the two targets can be discerned just as easily without additional linearization.

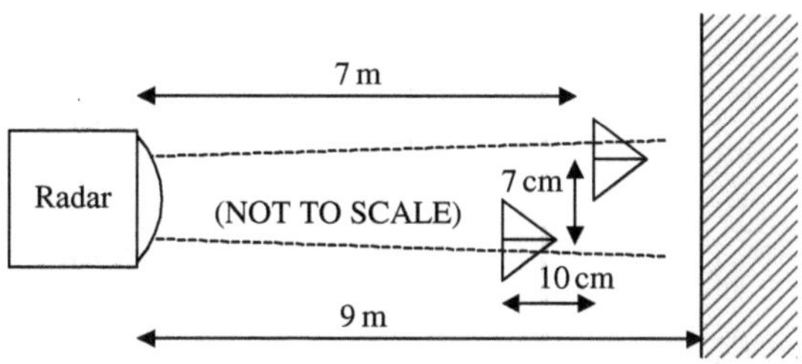

Fig. 8 Experimental setup

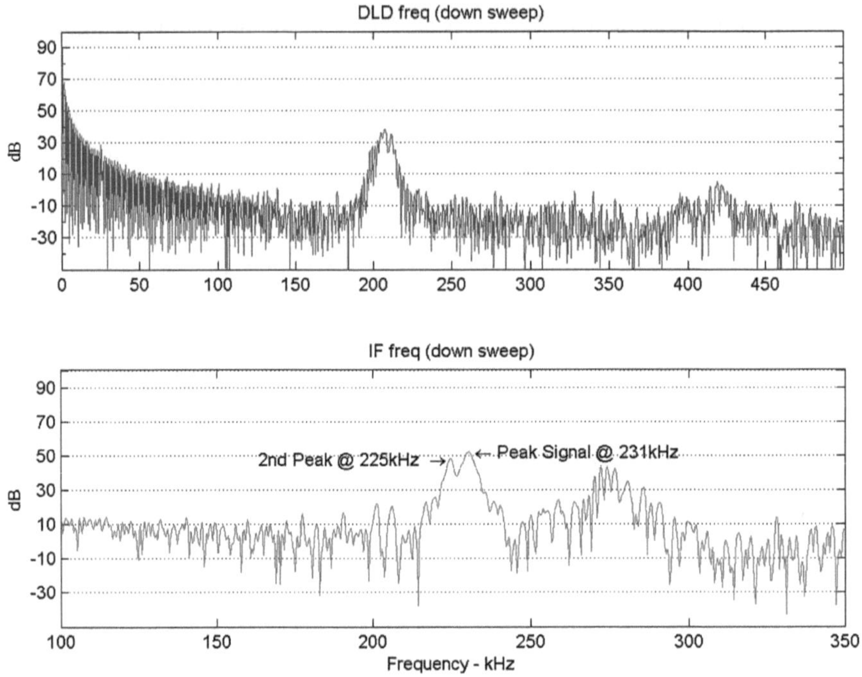

Fig. 9 FFT Results for 2-Targets (Un-linearised)

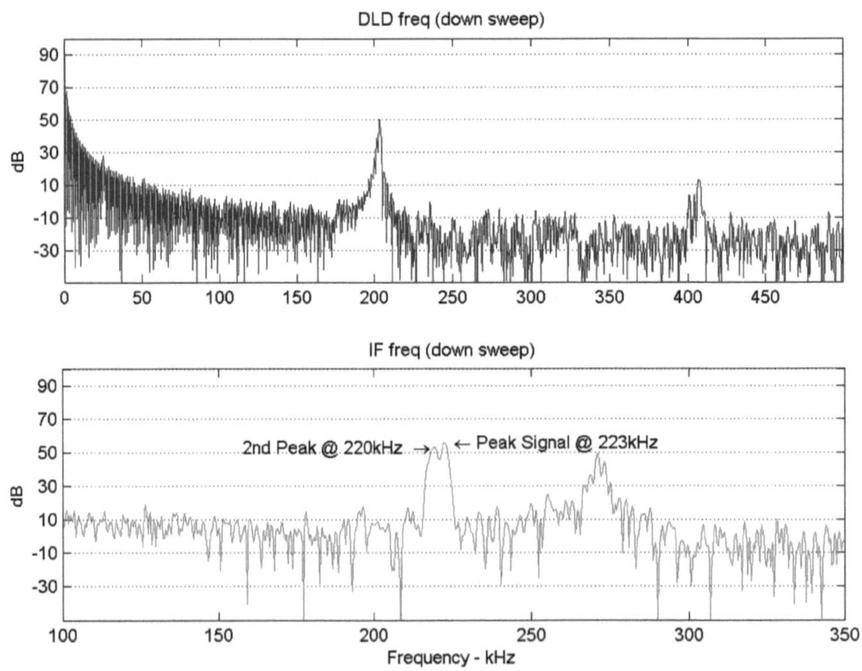

Fig. 10 FFT Results for 2-Targets (Linearised)

However, applying (1) to the highlighted peaks in the un-linearised case, with the appropriate bandwidth/sweep-duration ratio, places the targets 19 cm apart, compared to 10 cm for the linearised case, as required. The wall is also more clearly discernable in the linearised case.

6 Conclusion

A method for linearising an FMCW radar signal over a large bandwidth, without the need for a DDS, has been developed and tested experimentally. Results show a better than 6 fold improvement over the already highly-linear open-loop configuration, with a range resolution of <10 cm readily achievable.

One remaining drawback of this method is that whilst the frequency-slope gradient may be kept constant over a chirp, the upper and lower frequency bounds tend to drift over time. This will be fixed in the next iteration of the PLL loop filter by means of a secondary frequency locked loop.

References

1. S.C. Gupta, "Phase-Locked Loops", Proceedings of the IEEE, 63 (2), pp 291–306 (1975)
2. M. Curtin, P. O'Brien, "Phase-Locked Loops for High Frequency Receivers and Transmitters-Part 1", Analog Devices Inc., Analog Dialogue 33–3 (1999)
3. M. Curtin, P. O'Brien, "Phase-Locked Loops for High Frequency Receivers and Transmitters-Part 2", Analog Devices Inc., Analog Dialogue 33–5 (1999)
4. M. Curtin, P. O'Brien, "Phase-Locked Loops for High Frequency Receivers and Transmitters-Part 3", Analog Devices Inc., Analog Dialogue 33–7 (1999)
5. AD9954-400 MSPS, 14-Bit, 1.8V CMOS, Direct Digital Synthesizer, Analog Devices, Norwood, MA, Datasheet, Rev. A, Jan 2007 [Online]. Available: http://www.analog.com
6. AD9956-2.7 GHz DDS-based AgileRF™ Synthesizer, Analogue Devices, Norwood, MA, Datasheet, Rev. A, Sep 2004 [Online]. Available: http://www.analog.com
7. G.M. Brooker, "Understanding Millimetre Wave FMCW Radars", Proceedings International Conference on Sensing Technology, pp 152–157 (2005)
8. S. Scheiblhofer, S. Schuster, A. Stelzer, "Signal Model and Linearization for Nonlinear Chirps in FMCW Radar SAW-ID Tag Request", IEEE Transactions on Microwave Theory and Techniques, 57 (4), pp 1477–1483 (2006)
9. M. Vossiek, T. v. Kerssenbrock, P. Heide, "Novel Nonlinear FMCW Radar for Precise Distance and Velocity Measurements", Microwave Symposium Digest, 1998 IEEE MTT-S International, pp 511–514, vol 2 (1998)
10. G.M. Brooker, S. Scheding, M.V. Bishop, R.C. Hennessy, "Development and Application of Millimeter Wave Radar Sensors for Underground Mining", IEEE Sensors Journal, 5 (6), pp 1270–1280 (2005)
11. FMCW Radar Front-End Module, Datasheet, [Online]. Available: http://www.elva-1.com
12. M. Pichler, A. Stelzer, P. Gulden, C. Seisenberger, M. Vossiek, "Frequency-Sweep Linearization for FMCW Sensors with High Measurement Rate", Microwave Symposium Digest, 2005 IEEE MTT-S International, pp 1693–1696 (2005)

13. M. Feldmann, J. Hénaff, Surface Acoustic Waves for Signal Processing, Artech House, Boston (1989)
14. D.G. Johnson, G.M. Brooker, "4GHz Bandwidth Closed Loop Linearisation of a Millimetre-Wave, Linear Frequency Modulated Radar, Proceedings 2nd International Conference on Sensing Technology, Palmerston North, New Zealand, pp 534–539 (2007)

Part IV
Fibre Bragg Grating Sensors

Fiber Bragg Grating Sensors and Piezoelectric Actuators in Co-Located Configuration for Active Vibration Control Applications

C. Ambrosino, G. Diodati, A. Laudati, G. Breglio, M. Giordano, A. Cutolo and A. Cusano

Abstract An active vibration control hybrid system using Fiber Bragg Grating (FBG) sensors and piezoelectric (PZT) actuators for vibration suppression is presented. An aluminum test-structure have been provided with FBGs sensors bonded below the same number of PZT actuators in a co-located configuration. A novel bonding procedure was used to realize the co-located sensors/actuators cells combining epoxy based glue filled with glass micro-balloons matching the size of optical fibers. Closed loop operation was implemented by using a Proportional-Derivative controller and proper designed power amplifiers to drive the actuators based on the FBGs response. Positioning criteria based on sensitivity considerations have been implemented to define the positioning of the sensors/actuators cells. The system was first validated through open loop tests comparing experimental results with numerical predictions. Finally, "closed loop" configuration tests have been carried out demonstrating of vibration-noise reduction up to 17 dB.

C. Ambrosino
Engineering Department, University of Sannio, C.so Garibaldi 107 82100 Benevento, Italy,
e-mail: carmen.ambrosino@unisannio.it

G. Diodati
C.I.R.A. (Italian Aerospace Research Center), via Maiorise 81043 Capua CE, Italy

A. Laudati
Optosmart s.r.l., via Pontano 61, 80121 Napoli, Italy, e-mail: info@optosmart.com

G. Breglio
University "Federico II" Department of Electronic and Telecommunication Engineering via Claudio 21, Naples, Italy

M. Giordano
National Research Council, Institute for Composite and Biomedical Materials, Naples, Italy

A. Cutolo
Engineering Department, University of Sannio, C.so Garibaldi 107 82100 Benevento, Italy,
e-mail: carmen.ambrosino@unisannio.it

A. Cusano
Engineering Department, University of Sannio, C.so Garibaldi 107 82100 Benevento, Italy,
e-mail: carmen.ambrosino@unisannio.it

Keywords Fiber bragg gratings · piezoelectric actuators · active vibration control · co-located configuration · smart systems

1 Introduction

A smart structure has the capability to respond in real time if the external environment is changing in order to maintain the mission requirements [1]. Space systems, automotive, civil structures, machine tools and medical systems are some of the numerous applications of this technology [2, 3]. In the design of various smart system the suppression of vibration constitutes a critical requirement. In fact, vibrations are undesirable because they cause reduced system effectiveness, unpleasant noise, human discomfort and malfunction in systems. In the last two decades, due to the numerous applications in which they could be successfully adopted, Active Vibration Control (AVC) methods, focused on reducing the sound radiation of light structures, have attracted increasing interest. In the aeronautic field, the need of such a method springs out especially in order to suppress aircraft and helicopter interior acoustic noise having frequencies up to 1 kHz. In fact within this range the traditional passive damping techniques are practically ineffective and the system redesign is often much costly [4, 5]. The basic idea of AVC systems is to operate directly on the structural vibrations causing the radiated sound. In active control, the effect of unwanted noise is cancelled by voluntary addition of controlled signals, opposite in sign but equal in magnitude [6]. To this aim, active materials for actuating or sensing include piezoelectric materials, shape memory alloys (SMAs), electrostrictive materials, electrorheological fluids, magnetostrictive materials and fiber optics [7]. As actuators, piezoceramic materials are the most used [8]. They have established themselves as actuators in a great variety of technical applications due to their short reaction time, high frequency response behaviour, mechanical simplicity, light weight and compact construction. In recent years, a great number of researchers have investigated surface mounted or embedded piezoelectric ceramic patches, such as lead zirconate titanate (PZT), receiving much attention in vibration control [9, 10, 11]. As sensors, on the other hand, optical fiber sensors have also been used in the control of smart structures and in active control applications and have been examined in a great number of scientific papers [12, 13]. Also, Fiber Bragg Gratings sensors present a large series of advantages to render them ideal physical optical gauges both for smart structures applications, structural health monitoring [14, 15, 16, 17, 18], impact and damage detection [19, 20]. This is due to their accuracy, resolution, sensitivity, immunity to electromagnetic interference and high multiplexing capability [14], exhibiting much more precise measurements than that of electrical strain gauges. Also in vibration measurements they have already been adopted [21]. In this work, piezoelectric actuators have been selected in combination with FBG sensors in order to prove the feasibility of this hybrid system to be a valid method in AVC applications. Although this approach has been already used in past works reported in literature [21, 22, 23, 24], in this work a co-located configuration, already presented in [25], was developed. To this aim, a test aluminum proof in fixed-fixed beam configuration has been equipped with four fiber Bragg gratings sensors. Each FBG device is bonded below a corresponding PZT element on the aluminum plate by using a suitable bonding procedure forming a co-located configuration. A proportional derivative controller has been used to drive the PZT elements based on the signals provided by the sensing elements. Experimental results of "open loop" characterization are here reported. Finally, preliminary results in "closed loop" configuration tests are reported demonstrating the great potentiality of this hybrid system to be employed as valid candidate for AVC applications.

2 Technologies and Test Proof

2.1 Sensing System

Fiber Bragg grating sensors have been widely used in the last decades for smart structures applications [14, 15, 16, 17, 18, 19, 20]. They can offer advantages over conventional electrical strain gauges since they are lightweight, no electromagnetic interference, can be integrated into structures [15]. They have the advantages of being linear in response, with high multiplexing capabilities simplicity and low-size [26]. The sensed information (temperature or strain) is encoded directly into wavelength. In fact, a Fiber Bragg grating is formed by generating a permanent periodic or semi- periodic perturbation in the fiber's core refractive index [15]. According to Bragg's law, the grating will only reflect a specific wavelength, Bragg wavelength (λ_B), which is given by [27]:

$$\lambda_B = 2n\Lambda \tag{1}$$

where n is the effective refractive index of the fundamental mode and Λ is the grating period. Any effect able to modify the guiding properties of the fiber (thermo-optic effect, pho toelastic effect) or geometrical quantities such as the grating period (strain and thermal extension or compression) will produce a shift in the reflected wavelength. Relative variations of λb can be obtained by differentiating the (1) respect to these two parameters, and substituting the relationships obtained applying the strain optic theory [15]. For bonded configuration and dynamic tests involving lower strain field amplitudes, it can be safely assumed that only the strain field parallel the fiber axis significantly affects the FBG response. This means that the transverse strain is uniform, and related to the axial component by the Poisson's ratio. In this way, Bragg's wavelength relative shift is proportional to axial strain according to the following expression:

$$\frac{\Delta\lambda_B}{\lambda_B} = \left\{1 - \frac{n_0^2}{2}\left[p_{12} - \nu(p_{11} + p_{12})\right]\right\}\varepsilon_1 = S_\varepsilon \varepsilon_1 \tag{2}$$

where n_0 is the effective refractive index of the unstrained optic fiber, p_{12} and p_{11} are the elements of the elasto – optic matrix [15], ν is the Poisson's ratio and

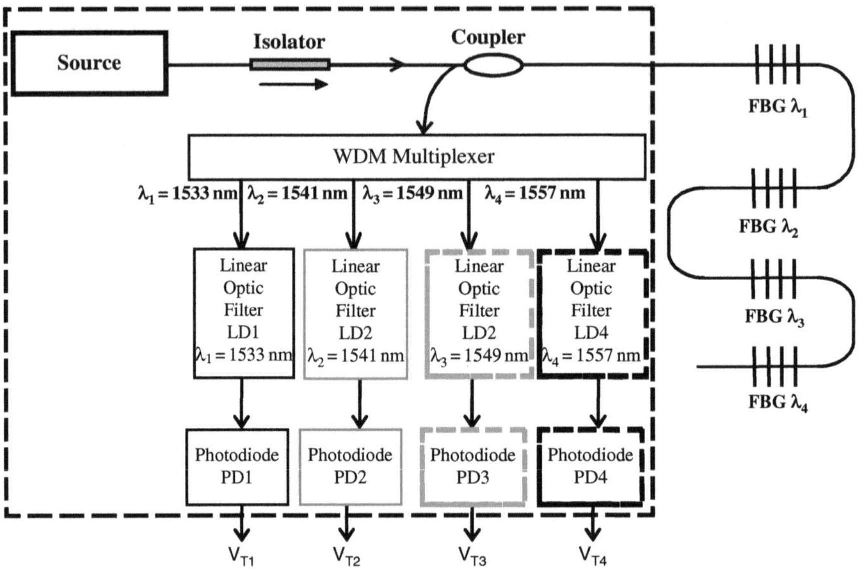

Fig. 1 Diagram of the interrogation unit

S_ε is FBG strain sensitivity. Typical values of strain – optic matrix elements and Poisson's ratio for standard optic fibers are: $p_{11} = 0.12$, $p_{12} = 0.27$ and $\nu = 0.17$, while $S_\varepsilon = 0.78 * 10^{-6}\,\mu\varepsilon^{-1}$ [28].

For these optic devices the interrogation system adopted relies on the use of a simple technique based on broadband interrogation and edge filtering described in [29, 30]. In order to interrogate simultaneously several FBGs (in this case four) at different Bragg wavelengths disposed on a single optical fiber with a bandwidth of 1 kHz, an integrated and modified system developed by Optosmart s.r.l. [31] has been used. The basic architecture is shown in Fig. 1. A WDM demultiplexer, in fact, has been inserted in the general optical chain in order to separate the signals reflected from the FBGs sensors in different channels. The source is represented by a superluminescent broadband light emitting diode (SLD), having an almost Gaussian shape transmitted spectrum. When each FBG is irradiated by the broadband source (SLD), it reflects a narrowband signal centered at its typical λ_B. Successively, the WDM separate the signals reflected from the FBGs sensors at the corresponding wavelengths. Once separated, each optical signal related to a given Bragg element will be processed by a linear filter with central wavelength matched with the Bragg wavelength corresponding to the selected sensor. The in-fiber edge filters are the key elements in this technique and are responsible for the wavelength-intensity conversions. They consist of an in fiber optical devices with a transmittance (reflectance) linearly varying with the optical wavelength realized by using the well assessed FBG technology [32]. A typical transmittance behaviour exhibited by these in fiber filters is provided in Fig. 2 where the case of linear filter operating at 1557 nm (central wavelength within the linear range) is shown. After the wavelength-amplitude

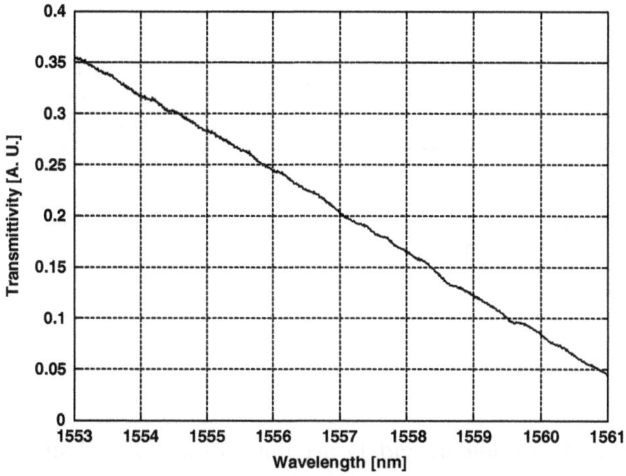

Fig. 2 Transmittance of a linear filter @1557 nm

conversion carried out by the optic filters, each optical signal is detected by a conditioned photodetector providing an output signal change ΔV_T, which in turn is a direct measure of the Bragg wavelength shift corresponding to a given sensing element. In fact, each photodetector signal is related to the convolution between the respective sensing grating spectrum and the transmission response of the optical filter through the relationship:

$$V_T = K_T \int_{FBG\,BAND} B_g(\lambda - \lambda_B) T(\lambda) S_{SD}(\lambda) d\lambda = K_T S_{SD}(\lambda_B) \int_{FBG\,BAND} B_g(\lambda - \lambda_B) T(\lambda) d\lambda \quad (3)$$

where K_T account for photodiode conversion, coupling coefficient, $S_{SD}(\lambda)$ is the spectral power density of the broadband source, $T(\lambda)$ represents the spectral response in transmission of the optical filter, $Bg(\lambda\text{-}\lambda b)$ represents the reflected spectrum from the sensing grating [33]. Additional functionalities such as optical amplification and intensity compensation not shown in Fig. 1 complete the interrogation unit.

In this work, as sensors, an array of four FBGs disposed on a single optical fiber has been used: each sensing element demonstrated a reflectivity >95%, a FWHM bandwidth of 0.8 nm and a physical length less than 3 mm. To match the central wavelengths of the optical filters, the four Bragg wavelengths were chosen to be $\lambda_1 = 1533$ nm, $\lambda_2 = 1541$ nm, $\lambda_3 = 1549$ nm and $\lambda_4 = 1557$ nm, respectively. In order to correctly operate, the WDM demultiplexer exhibited a four wavelength gates transmittance matched with the operating wavelengths of the FBG sensor array and the optical filters. In addition, in order to warrant the maximum dynamic range for each sensing element, the channel spacing of each wavelength gate was properly selected. All the optic filters demonstrated a nominal linearity range of 7 nm (in Fig. 1 a trasmittivity of the optic filter @1557 nm is shown). A superluminescent diode has been adopted as broadband source. Table 1 reports a summary of the system specifications.

Table 1 Systems specifications

CHANNEL	Central wavelenght for FBG	Central wavelenght for WDM gates	Gate width for WDM	Central wavelenght for optic filters	Nominal linearity range for optic filters
1	1531 nm	1531 nm	3.5 nm	1531 nm	7 nm
2	1541 nm	1541 nm	7 nm	1541 nm	7 nm
3	1549 nm	1549 nm	7 nm	1549 nm	7 nm
4	1557 nm	1557 nm	3.5 nm	1557 nm	7 nm

2.2 Actuation System

Piezoelectric materials are the most popular, widely used and well assessed smart materials [8, 9, 10, 11]. This is due to their lightweight, low-cost, higher actuating force, lower power consumption and easy-to-implement capability. When a piezoelectric material is subjected to an electrical charge or voltage, a mechanical force or strain is induced in the material and this enables the material to be used as an actuator. The induced strain is directly proportional to the applied electric field, and although very low, covers a wide frequency range. At low field amplitudes, the constitutive equations are relatively linear and when operated well below the resonant frequency, a piezo actuator behaves as a capacitor. In this work, the actuators selected are PPK11 patches (dim. $30 \times 30 \times 1$ mm), the specifications are reported in Table 2. A proper designed equipment comprising a four channels voltage amplifier specialized for capacitive loads and a Proportional Derivative PD controller to efficiently control the PZT actuators was developed at C.I.R.A.

2.3 Test-Proof

The test proof consists of a rectangular Aluminum plate (Type 7075T6, length 500 mm, width 200 mm, thickness 1.5 mm, Young Module $E = 7.2\ 10^{10}$N/mm^2). The plate is in fixed-fixed beam configuration (the 200 mm sides are fixed) and the constrains are in aluminum with properly designed pins in order to grip the structure. The lower spares of the constrains have been fixed to an optical table. The sensors/actuators locations were polished and treated with a proper catalyst before the bonding procedure. Later on, the sensing elements (FBGs) have been bonded to

Table 2 Piezo patch actuator specifications

Denomination	PPK-11 Stettner	Thickness [mm]	1
Length [mm]	30	Young Module E [10^{10}N/mm^2]	6.3
Width [mm]	30	Max Voltage [V]	500
Poisson Module	0.33	$\mathbf{d_{31}}$ Constant [10^{-12} C/N]	−350

Fig. 3 Equipped plate with four co-located couples PZTs/ FBGs

the upper part of the aluminum plate by using a 2-component fast curing adhesive, HBM X60, consisting of both a liquid and a powder. The fiber axis was chosen parallel to the long side of the aluminum plate. Glass micro-balloons with a diameter of approx. 125 μm have been added to the powder of the polymeric glue, increasing the volume and reducing the weight. An important advantage in their use is the significant improvement of the adhesive planarity providing a well supported adhesion between structure, sensing and actuator elements. Finally, the piezoelectric elements have been bonded upon the fiber surface. Schematic diagrams for the test proof and bonding configuration are reported in Figs. 3 and 4. A picture of the microballoons used to provide a supported adhesion of the sensor/actuator cell is shown in the inset of Fig. 4.

2.4 Sensor/Actuator Cells Positioning

In AVC applications, the selection of the sensors/actuators cells represent one of the key points in order to efficiently suppress vibrations over the target structure by implementing the minimum number of instrumented locations. Methods for optimal placement are relatively new [8, 34, 35] and further investigations are currently on going to really identify optimum criteria.

Here, in order to validate the hybrid architecture with co-located cells, a first tentative strategy was to select locations corresponding to the maximum sensitivity for the fiber optic strain sensors. To this aim, a numerical analysis of the structure have been carried out in NASTRAN® environment in the frequency range up to 1 KHz. The first three out-of-plane bending modes evaluated have resonances frequencies respectively of $f_1 = 30.3$ Hz, $f_2 = 59.3$ Hz, $f_3 = 81.9$ Hz. Here, for each vibration mode, the 2-D strain distribution has been evaluated and the regions where the maximum strain parallel to the fiber axis occurs were selected. Successively, in order to control the higher number of modes with the minimum number of sensor/actuator cells, a modal superposition technique was used: sensors/actuators locations were set by considering the condition of higher strain field along the optical fiber axis for the greater number of vibration modes simultaneously. According to this line of argument, the optimal position of the first sensor has been found to be

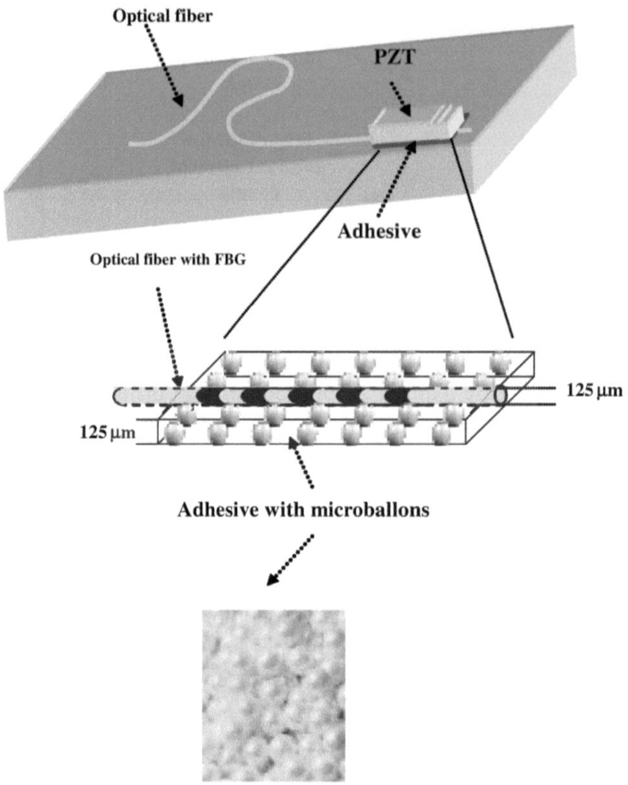

Fig. 4 Schematic bonding configuration for one of the four FBG and PZT pair

near the clamped edge at 250 mm along the × direction and 37.5 mm along the y direction starting from the center of the plate (see the sensor/actuator cell at Point 1 in Fig. 5). The positioning of the remaining three sensors (FBG@1549 nm located at Point 2, FBG@1533 nm located at Point 3 and FBG@1557 nm located at Point 4 in Fig. 5) has been retrieved based on symmetry considerations.

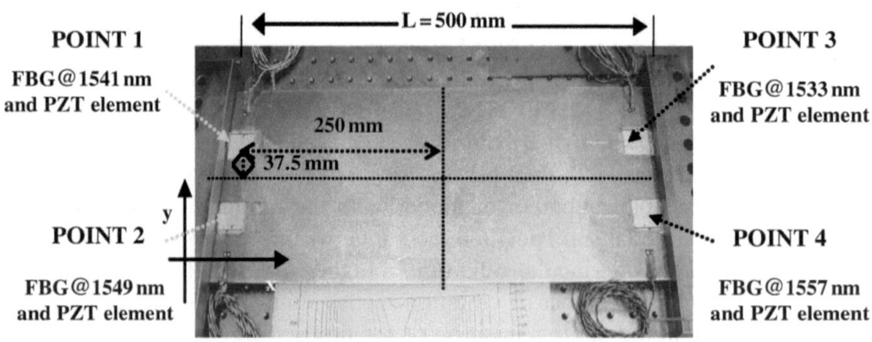

Fig. 5 Schematic for sensor/actuator placement

3 Results and Discussion

3.1 Open Loop Tests

In order to experimentally validate the system, open loop tests have been carried out. This step was fundamental also to retrieve the dynamic feature of the structure and at the same to confirm the numerical predictions and thus the positioning plan. Modal analysis methodology was adopted and in particular, in these tests, the Strain Frequency Response Function (SFRF) approach was considered [36]. As mechanical excitation, the piezoelectric bonded at point 4 in the sample (see Fig. 5) has been driven in order to excite the structure with a sweep signal from 1 to 400 Hz in 40 seconds. The approach based on the evaluation of the SFRF [36] was used in this study. To this aim, several sweeps have been performed and for each of them. the input signal (sweep signal) and sensor signals were simultaneously acquired. Successively, the SFRF for each sensor was obtained by evaluating the ration between the Fourier transforms of the sensor output and the excitation signal, respectively. Successively, for each complex SFRF term, real and imaginary parts were thus identified. In order to improve the SFRF analysis, properly designed filtering algorithms have been used [36]. Finally, final modal frequencies have been obtained as an average of the respective frequencies calculated in each sweep test. For sake of brevity, only results related to two of the four fiber Bragg gratings have been here reported (FBG@1549 nm and FBG@1557 nm bonded at Point 2 and Point 4, respectively, as it can be seen in Fig. 5) but similar results have been obtained also considering the other two sensors (with central wavelength @1541 nm and @1533 nm, located at Point 1 and Point 3 respectively, as it can be seen in Fig. 5). In Fig. 6a, real and imaginary parts of the SFRF related to the FBG@1549 nm for a single sweep has been shown. The SFRF has been represented in the frequency range [0–100] Hz to cover the first three vibration modes of the test-structure. The two curves are shown in the same figure with a suitable offset for a better comprehension. In Fig. 6b, real and imaginary parts of SFRF for the FBG@1557 nm are also reported. According to modal analysis, the resonant frequencies have been determined searching the zero crossing frequencies of the real part corresponding contemporaneously to a peak in the imaginary part related to the strain amplitude of natural modes in the given sensor location. Experimental resonant frequencies ($f_1 = 30$ Hz, $f_2 = 59$ Hz, $f_3 = 81$ Hz) were found demonstrating good agreement with numerical predictions. In order to confirm these experimental results, additional tests have been carried out by using a different excitation method. An impact hammer has been used as mechanical excitation revealing good agreement with results previously obtained. A resume of numerical and experimental resonant frequencies in the investigated frequency range is reported in Table 3.

3.2 Closed Loop Tests

In order to demonstrate the potentiality of the proposed approach for AVC applications, preliminary tests in closed loop configuration have been also carried out. In

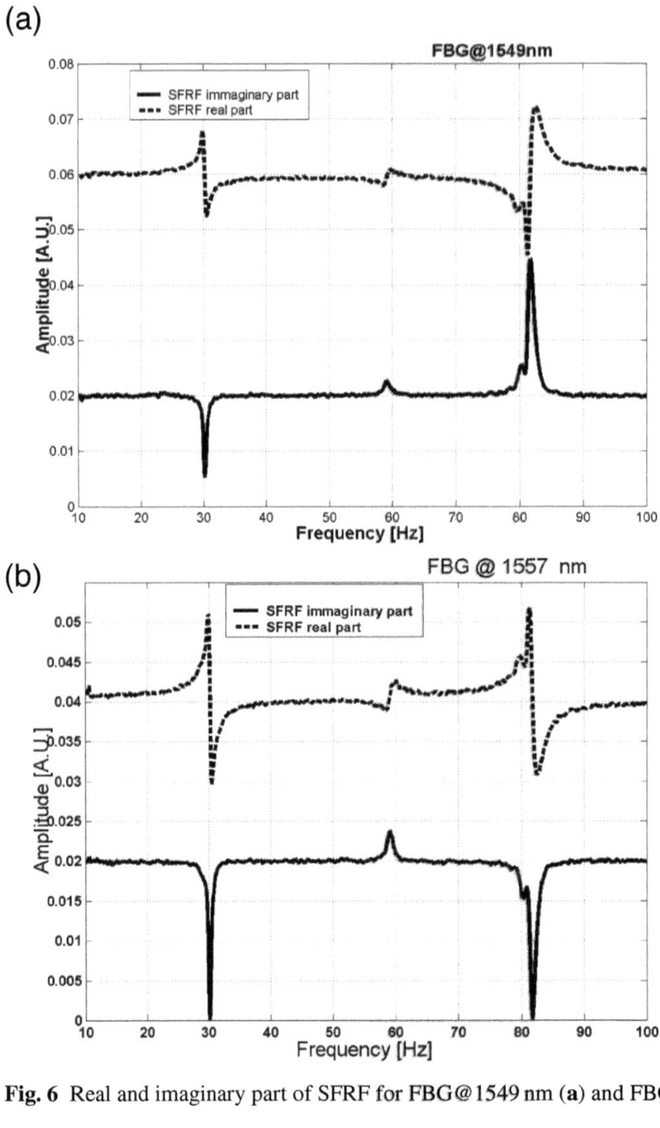

Fig. 6 Real and imaginary part of SFRF for FBG@1549 nm (**a**) and FBG@1557 nm (**b**)

Fig. 7 is shown the diagram of the experimental setup involved in the closed loop tests. As it can be noticed from figure, in order to excite the structure, the actuator at Point 3 has been used to excite the structure and in this way simulate the unwanted vibrations from external environment. The FBGs sensors located at Points 2 and

Table 3 Numerical and experimental characterization of the structure

Numerical natural frequencies	[Hz]	30.3	59.3	81.9
Experimental natural frequencies (sweep tests)	[Hz]	30	59	81
Experimental natural frequencies (hammer tests)	[Hz]	33	59	81

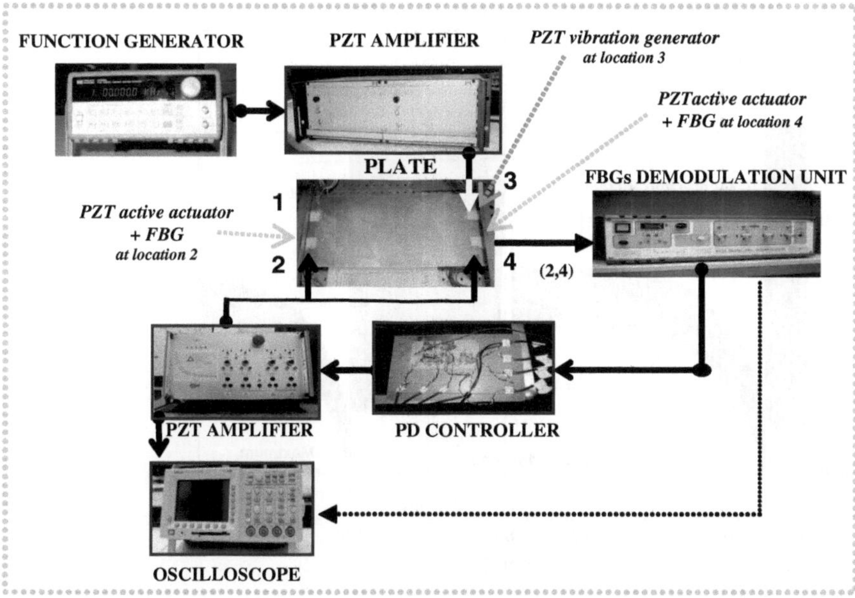

Fig. 7 Block diagram of closed loop configuration

4 (FBG@1549 nm and FBG@1557 nm) and the corresponding co-located actuators have been used in order to provide the control loop. The actuators have been both simultaneously or separately fed by the proper designed Proportional-Derivative (PD) controller. In these tests, in order to stimulate the structure, single frequency signals have been used to drive the PZT at point 3. For each frequency, the FBGs time responses at the active Points 2 and 4 have been continuously used as input signals for the controller determining the actual driving signals for the active actuators in the same locations. Contemporaneously, the time response of the FBG@1557 nm at point 4 was also used to monitor the vibration state of the structure while the closed loop is activated. In Fig. 8a and b, the FBG time responses at point 4 are shown for two excitation frequencies, 80 Hz and 30 Hz, corresponding to the more relevant resonant frequencies revealed in the open loop characterization tests (see Table 3). As it can be noticed from Fig. 8a, for an excitation frequency of 80 Hz and with both the PZT elements at point 2 and at point 4 active, a maximum vibration reduction of lightly less than 17 dB was observed. Moreover, with the single PZT at point 4 active a lower reduction of approx. 15 dB was obtained. This demonstrates an higher efficiency of the actuator element located at point 4 compared to the active PZT at point 2 to compensate for the unwanted vibrations. The location of the excitation point (point 3) in respect to the locations of the co-located sensors/actuators (points 2 and 4) involved in the closed loop operation can explain this evidence. In fact, due to the symmetry, the PZT at point 4 is able to almost completely compensate the vibration generated at excitation location. According to Fig 8b, also in the second case related to an excitation frequency of 30 Hz and with a control loop

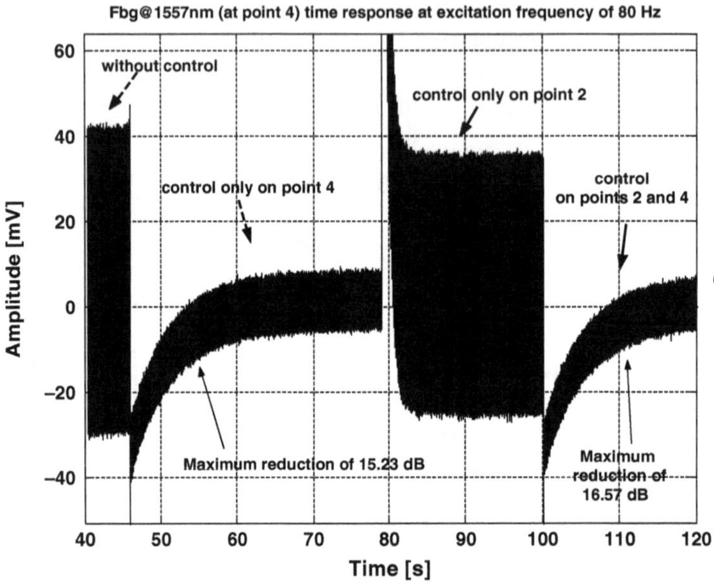

(a)

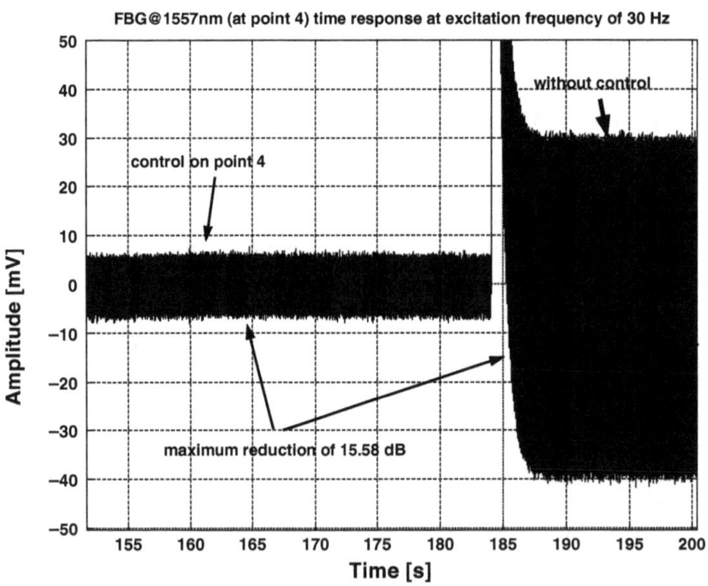

(b)

Fig. 8 FBG@1557nm responses under different control conditions at 80 Hz and 30 Hz

provided by the PZT at point 4, the effect of the PZT element at point 2 in the closed loop configuration was found to be strongly reduced compared with the efficiency of the PZT at point 4 confirming our conjecture. In this case a vibration reduction of 15.58 dB has been observed. Experimental results reveal the great potentiality of the proposed hybrid scheme to be used for active vibration control applications. A vibration reduction up to 17dB has been reached in the preliminary configuration. So, these preliminary results encourage the possibility to develop high performances integrated broadband active system if sensors/actuators positioning and control system architecture are properly optimized. In fact an important point to outline is that based on the experimental evidence it is obvious how the number and positioning criteria can not be based exclusively on sensitivity considerations. In fact, the influence region of each PZT actuators and power budget considerations should also be taken into account.

To quantitative characterize the active control system here proposed, an intense experimental work is actually running.

4 Conclusion

Based on the reported results, we can conclude that the hybrid active scheme proposed, consisting of the integration of FBG sensors and PZT actuators can be considered an effective alternative in active vibration control applications both in terms of performances and miniaturization. The novel bonding procedure adopted, comprising epoxy glue and glass micro-balloons, has been successfully tested, warranting higher planarity in the adhesion procedure. Through open loop tests the system has been validate also comparing experimental results with numerical predictions. Preliminary closed loop tests for the co-located configuration have provided encouraging results in terms of vibration suppression reaching a maximum reduction of lightly less than 17 dB in spite of these results are obtained in non optimal conditions. So, higher performances are thus expected as for the optimization of sensors/actuators and control system architecture has been undertaken in the experimental setup that is currently on going.

References

1. Chopra I., "Review of state of art of smart structures and integrated systems", AIAA Journal 40(11), (November 2002)
2. Song G., Sethi V., Li H.-N, "Vibration control of civil structures using piezoceramic smart materials: A review", Engineering Structures 28 (2006) 1513–1524
3. Crawley E., "Intelligent structures for aerospace: A technology overview and assessment", AIAA Journal 32(8), (August 1994)
4. C.R. Fuller, A.H. von Flotow, IEEE Control Systems Journal, 15, (December 1995) 9–16
5. Y. Yaman, T. Caliskan, V. Nalbantoglu, E. Prasad, D. Waechter "Active vibration control of a smart plate", ICAS2002, International Council of the Aeronautical Sciences, Paper 424, Toronto, Canada, September 8–13, 2002
6. S.P. Singh, H.S. Pruthi, V.P. Agarwal, "Efficient modal control strategies for active control of vibrations" Journal of Sound and Vibration 262 (2003) 563–575
7. R. Jha, J. Rower, "Experimental investigation of active vibration control using neural networks and piezoelectric actuators" Smart Materials and Structures 11 (2002) 115–121

8. E.F. Crawley, J. de Luis, "Use of piezoelectric actuators as elements of intelligent structures" AIAA Journal 25 1373–85 (1987)
9. T. Bailey, J.E. Hubbart, Distributed "Piezoelectric-polymer active vibration control of a cantilever beam". Journal of Guidance, Control and Dynamics 8(5), (1985) 605–611
10. E.F. Crawley, J. de Luis Use of piezoelectric actuators as elements of intelligent structures AIAA Journal 25 (1987) 1373–85
11. I. Chopra "Review of current status of smart structures and integrated systems". Proc SPIE 2717 (1996) 20–62
12. D.E. Cox, D.K. Lindner "Active control of vibration suppression in a flexible beam using a modal domain optical fiber sensor," Journal Vibration and Acoustics, 113 (1991) 369–382
13. P.A. Juang, M.N. Wu, "Active control using a fiber optic interferometric sensor," Smart Materials and Structures 4 (1995) 370–372
14. B. Culshaw, J. Dakin, Optical fiber sensors: Applications, analysis, and future trends, Artech House inc., Norwood (1997)
15. R. Measures, Structural monitoring with fiber optic Technology, Academic Press, London, 2001
16. F. Ansari, Sensing issues in civil structural health monitoring, ed. Springer, Dordrecht, 2005
17. Cusano, A., Capoluongo, P., Campopiano, S., Cutolo, A., Giordano, M., Felli, F., Paolozzi, A., Caponero, M. "Experimental modal analysis of an aircraft model wing by embedded fiber Bragg grating sensors", Sensors Journal, IEEE 6(1) (2006) 67–77
18. A. Cusano, P. Capoluongo, S. Campopiano, A. Cutolo, M. Giordano, M. Caponero, F. Felli, A. Paolozzi, "Dynamic measurements on a star tracker prototype of AMS using fiber optic sensors", Smart Materials and Structures 15 (2006) 441–450
19. A. Takeda, Y. Okabe, R. Tsuji, S. Takeda "Application of chirped fibre Bragg grating sensors for damage identification in composites", Proceedings of SPIE 4694 (2002) 106–17
20. P. Capoluongo, C. Ambrosino, S. Campopiano, A. Cutolo, M. Giordano, I. Bovio, L. Lecce, A. Cusano "Modal analysis and damage detection by Fiber Bragg grating sensors", Sensors and Actuators A: Physical 133(2), 12 February 2007, 415–424
21. K.K. Chau, B. Moslehi, G. Song et al. "Experimental demonstration of fiber Bragg grating strain sensors for structural vibration control", SPIE Conference Proceedings Paper, 5391, 29 July 2004
22. D.-H. Kim, J.H. Han, I. Lee, "Application of fiber optic sensor and piezoelectric actuator to flutter suppression," AIAA Journal of Aircraft, 41(2) (2004) 409–501
23. M.M. Zhang et al. "Control of vortex-induced non-resonance vibration using piezo-ceramic actuators embedded in a structure", Smart Materials and Structures 14(6) (December 2005) 1217–1226
24. L. Cheng, Y. Zhou, M.M. Zhang, "Controlled vortex-induced vibration on a fix-supported flexible cylinder in cross-flow", Journal of Sound and Vibration 292, 279–299, (2006)
25. C. Ambrosino, G. Diodati, A. Laudati, G. Breglio, M. Giordano, A.Cutolo, A. Cusano, "Active vibration control in co-located configuration using integrated piezoelectric actuators and fiber Bragg grating sensors", 2nd International Conference on Sensing Technology November 26–28, 2007 Palmerston North, New Zealand, 412–414
26. K.O. Hill, G. Meltz, "Fiber Bragg grating technology fundamentals and overview," Journal of Lightwave Technology, vol. 15, (August 1997) 1263–1276
27. A.D. Kersey, A.M. Davis, H.J. Patrick, M. LeBlanc, K.P. Koo, C.G. Askins, M.A. Putnam, E.J. Friebele, "Fiber grating sensors", Journal of Lightwave Technology, 15, (1997) 1442–1462
28. R. Kashyap, Fiber bragg gratings, Academic Press, 1999
29. A. Cusano, G. Breglio, M. Giordano, L. Nicolais, A. Cutolo, "Multifunction fiber optic sensing system for smart applications", IEEE/ASME Transactions on mechatronics, 9(1) (2004) 40–50
30. A. Cusano, A. Cutolo, M. Giordano, A. Calabrò, Sensors and Actuators A, 110(1–3) (2003) 276–281
31. Optosmart s.r.l., http://www.optosmart.com

32. R.W. Fallon, L. Zhang, L.A. Everall, J.A.R. Williams, I. Bennion, Meas. Sci. Techn., vol. 9, pp. 1969–1973, (1998)
33. A. Cusano, G. Breglio, M. Giordano, L. Nicolais, "Low-cost all-fiber Bragg grating sensing system for temperature and strain measurements", Optical Engineering 44(08), (2005) 084402
34. M. Sunar, S.S. Rao Recent advances in sensing and control of flexible structures via piezoelectric materials technology. Applied Mechanics Review, 52(1), 1–16 (1999)
35. D. Halim, S.O. Reza Moheimani "An optimization approach to optimal placement of collocated piezoelectric actuators and sensors on a thin plate", Mechatronics 13 (2003) 27–47
36. Heyelen W., Lammens S., Sas P., Modal analysis theory and testing, Kartholieke Universiteit Leuven, 1997

A Transmit Reflect Detection System for Fibre Bragg Grating Acoustic Emission and Transmission Sensors

Graham Wild and Steven Hinckley

Abstract A Transmit Reflect Detection System (TRDS) for a Fibre Bragg Grating (FBG) acoustic emissions and transmissions sensor is demonstrated. The TRDS utilises both the transmitted and reflected signals from the FBG. A narrow line width laser source is tuned to one of the 3 dB points of the FBG. The transmitted and reflected signals are detected, via two photoreceivers. The difference between the transmitted and reflected signals is then taken, doubling the total received signal. The TRDS improves the sensitivity and efficiency of the FBG acoustic emission and transmission sensor. The TRDS and FBG sensor was used to successfully receive actively generated ultrasonic signals, including acoustic communications signals, and to detect acoustic emissions. We include results for the sensitivity, frequency response, and transient response of the FBG sensor, and the acoustic emissions for low velocity impacts, and a lead pencil break test.

Keywords Acoustic emissions · fibre bragg grating · impact detection · optical fibre sensing · ultrasonics

1 Introduction

The use of Fibre Bragg Grating (FBG) sensors for the detection of Acoustic Emissions and Transmissions (AETs) has been established in the literature [6, 11]. Initially, the use of FBGs as spectral transduction elements made them immune to optical power fluctuations. As such they can be implemented in Wavelength Division Multiplexing (WDM) or Time Division Multiplexing (TDM) architectures. These applications require temporal and spectral decoding of the sensor signals, which can be costly and processor intensive. Conversely, FBGs can be used in an

Graham Wild
Physics Research Group, Edith Cowan University, Joondalup, Western Australia,
e-mail: g.wild@ecu.edu.au

Steven Hinckley
Physics Research Group, Edith Cowan University, Joondalup, Western Australia

S.C. Mukhopadhyay, G.S. Gupta (eds.), *Smart Sensors and Sensing Technology*,

intensity based edge filter detection scheme. The intensity information from the FBG can easily be correlated to a change in the measurand. Problems such as input optical power fluctuations are reintroduced into the system. However the advantages of edge filter detection, specifically the simplicity of detection, greatly outweigh the corresponding disadvantages, which can be overcome.

The use of FBGs in multiplexing systems has seen the focus of research tend towards applications with large quantities of sensors. There are, however, inherent problems with these systems, specifically the use of a single source, a single detector, and a single fibre. The failure of any of these three single components will result in the total failure of the sensing system. Optical fibre sensors offer other advantages over conventional electronics sensors, specifically in terms of sensitivity and "flexibility". FBGs, for example, are sensitive to both static and dynamic strain. This means that a single FBG can be used in applications where piezoelectric transducers and resistive strain gauges would be used together, for example, Structural Health Monitoring (SHM). The use of ultrasound, Acoustic Emissions (AE) [7] and strain monitoring [2] in SHM has been demonstrated in the literature. The ability of FBGs to measure static and dynamic strain, as well as temperature, pressure and other measurands means they are ideally suited to SHM.

Due to the high frequency, the detection of ultrasound with FBGs is typically limited to the use of intensity methods. Current reports using FBGs as intensiometric sensor for the detection of ultrasonic signals have had relatively small signal strengths, when using significant amplification on the output signal [6], and when using and amplified ultrasonic source [1]. The TRDS for FBG AET sensor previously reported [15], has significantly improved signal strength over other intensity detection systems. The use of both the transmitted and reflected signals also means the system is less sensitive to small temperature fluctuations in the FBG, or in the laser diode. These temperature fluctuations will result in a spectral shift in the FBG sensor, and the laser diode. This paper covers the theory, method, and results of work using the TRDS for a FBG AET sensor.

2 Theory

2.1 Acoustic Emissions and Transmissions

2.1.1 Acoustic Emissions

AEs are stress waves that propagate through a material, which can be generated internally by microcracks or inclusion de-cohesion under external loading [9]. A rapid local stress redistribution as a result of loading causes the material defects to release elastic energy. The energy results from crack growth, crack surface movement, or dislocations. AEs can also be generated by phase transformation or melting. These internal sources of AEs represent a passive method for the detection of damage. AE can also be generated by external sources, specifically impacts, and actively generated by actuators, such as piezoelectric transducers.

2.1.2 Acoustic Transmissions

The implementation and applications of acoustic communications has previously been reported [12, 13, 14]. Acoustic communication represents a wireless communications, with applications specifically for robotic inspection in SHM. Acoustic communication makes use of the sensing elements within a SHM system which are intended to detect and monitor acoustic signals, such as acoustic emissions or for ultrasonic non-destructive testing. Hence, additional electronics and hardware is not required to communicate wirelessly.

As with previous work [12, 13, 14], we implement Phase Shift Keying (PSK). The encoding of the PSK signal is achieved by switching the phase of the carrier wave by 180 degrees. Decoding the received signal is done with some simple mathematics, which involves;

1. multiplying the received signal by a synchronous sine and cosine (called I and Q respectively),
2. low pass filtering to remove the carrier, and then,
3. taking the arctangent of I on Q to recover the phase of the information.

The filter used was a raised cosine filter [8]. The decoding algorithm and the filtering were implemented in Matlab^{TM}.

2.2 Fibre Bragg Grating Sensor

2.2.1 Fibre Bragg Grating

A FBG is a spectrally reflective element that uses the principle of Fresnel reflection. The grating is made up of alternating regions of high and low refractive indices. The periodic grating acts as a filter, reflecting a narrow wavelength range, centred about a peak wavelength. This wavelength, known as the Bragg wavelength (λ_B), is given by,

$$\lambda_B = 2n\Lambda, \tag{1}$$

where n is the average refractive index of the grating, and Λ is the grating period.

Any measurand that has the ability to affect either the refractive index or the grating period can be measured using an FBG as a sensor. The change in the measurand will correspond to a change in the peak reflected wavelength. This is given by the Bragg grating sensitivity equation, where the change in the Bragg wavelength ($\Delta\lambda_B$) is given by [5],

$$\Delta\lambda_B = \lambda_B\left[(1 - p_\varepsilon)\varepsilon + (\alpha_\Lambda + \alpha_n)\Delta T\right], \tag{2}$$

where ε is the applied strain, p_ε is the strain optic coefficient for the optical fibre, ΔT is the change in temperature, and a_Λ and a_n are the thermal expansion and thermo-optic coefficients of the optical fibre, respectively.

2.2.2 Edge Filter Detection

Edge filter detection methods for FBG AE sensors have been outlined in a number of references [1, 3, 4]. Figure 1, shows the reflectivity as a function of wavelength of a typical FBG. Centred about λ_0, there is a linear region, $\delta\lambda$, between reflectivities of approximately 20 and 80 percent. This linear "edge" of the FBG is used as an optical filter. A narrow band laser source centred about λ_0 is then intensity modulated by the strain induced shift in the wavelength. That is, as the edge filter is shifted up and down the spectrum, the reflected optical power is also varied. The detection of the signal is then achieved using a simple, photoreceiver.

2.2.3 Transmit Reflect Detection

In edge filter detection, the reflected optical power is modulated by the measurand. This modulation also occurs to the transmitted signal. Previous FBG sensing [10] has used the transmitted signal as opposed to the reflected. When the change in the reflected optical power is positive, the change in transmitted optical power is negative, and vice versa. Since one change in measured signal is positive, and the other is negative, the two signals can be differentially amplified, giving an increased signal. Figure 2 illustrates this idea, with a FBG and a narrow band laser.

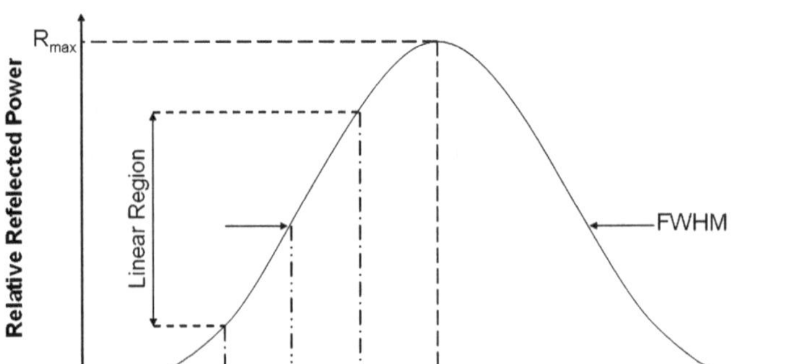

Fig. 1 Fibre Bragg Grating sensor parameters for edge filter detection

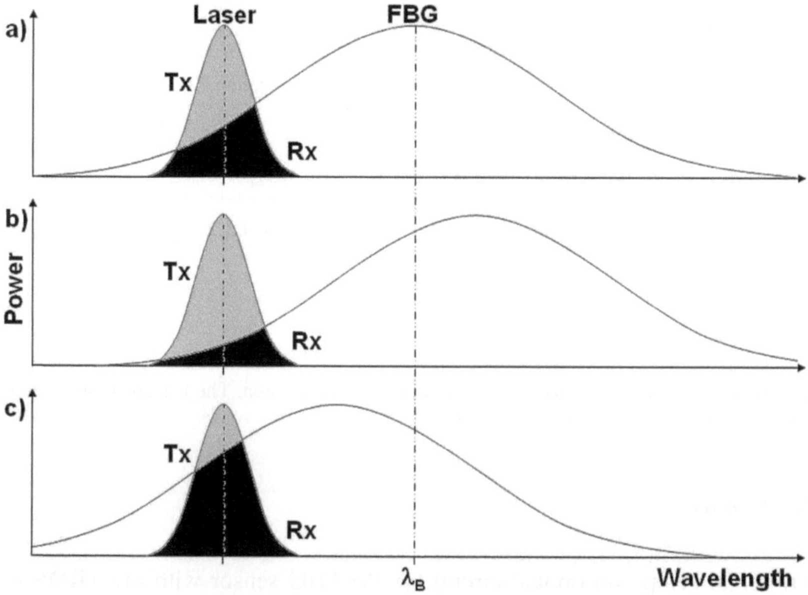

Fig. 2 Ideal spectrum for the FBG (*wide curve*) and laser (*narrow curve*) (**a**) shows the FBG with no change in measurand, (**b**) a positive change, and (**c**) a negative change, where the grey is the transmitted (Tx) optical power, and the black is the reflected (Rx) optical power

3 Experiments

3.1 FBG and Laser Characterisation

The FBG and tunable laser source required characterising to determine the operating point of the FBG sensor. The optical circuit required for the characterisation of the optical components is shown in Fig. 3. A broadband Amplified Spontaneous Emission (ASE) light source (Agilent 83438A) was used as the input to the FBG. The Optical Spectrum Analyzer (OSA) (Agilent 86140) was then used to measure the relevant wavelengths. An optical circulator (FDK YC-1100-155) was used to direct the light from the ASE to the OSA and to ensure isolation between the relevant optical paths. The output spectrum of the ASE and the tunable laser (Ando AQ 8201-13B) were also measured by connecting them directly to the OSA.

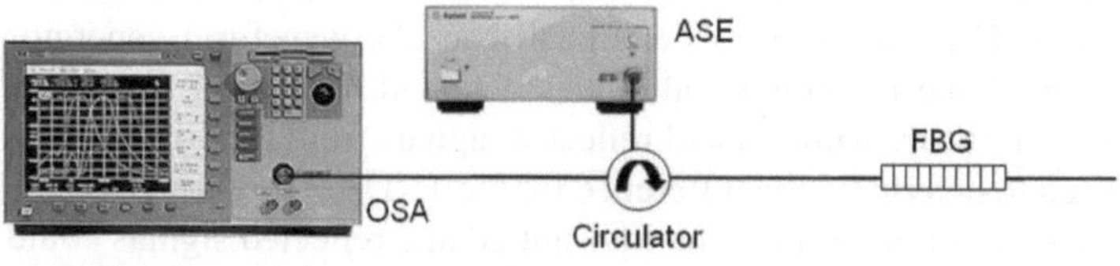

Fig. 3 Optical circuit used to characterize the FBG

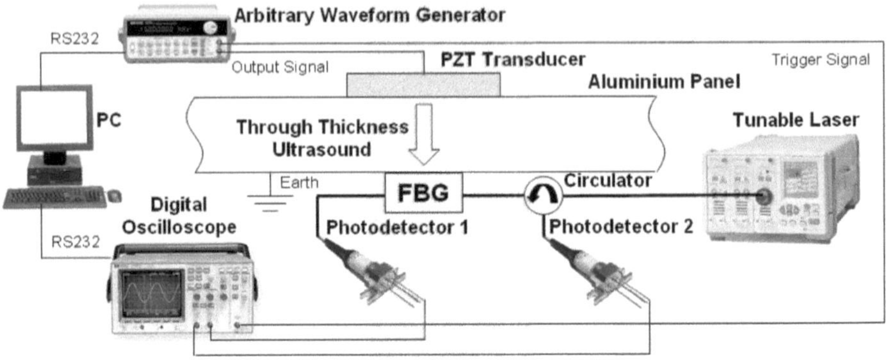

Fig. 4 Experimental setup for the detection of the acoustic transmission. The waveform generator and PZT transducer were removed for the AE tests

3.2 FBG Sensor

The experimental setup and optical circuit for the FBG sensor with the TRDS is shown in Fig. 4. The setup consists of several layers;

- a PZT (lead zirconate titanate) transducer (Stiemer Martins SMQA) as the transmitter (2.1 mm thick, 10 mm radius),
- a coupling medium (acoustic coupling gel),
- the communications medium, specifically an aluminium panel (1.5 mm thick),
- the FBG (Broptics GF-1C-1554.13-RX2), and
- the TRDS.

The optical circuit includes the FBG, tunable laser, circulator, and two photo-receivers (Fujitsu FRM3Z231KT). The FBG was coupled to the aluminium panel with acoustic coupling gel. An amplifier was not used on the input to the PZT transmitter or the output from the photodetector, nor was an optical amplifier used.

3.3 Acoustic Transmissions

Initially the FBG sensor with the TRDS was tested with an actively generated ultrasonic signal from an arbitrary waveform generator (Agilent 33120A). The output of the two photo-receivers was monitored on a digital oscilloscope (Agilent 54600A).

First a continuous sine wave, at the maximum voltage, 10 V peak, was used to interrogate the system. The frequency of the signal from the waveform generator was coarsely swept until the maximum value was obtained. The peak frequency was used to look at both the transmitted and reflected signals from the FBG. These signals were then used to determine the difference for the TRDS. The purpose of the experiment was, (1) to determine if both the transmitted and reflected signals could be detected, (2) to determine what the magnitude of the respective signals would be, and (3) to differentially amplify the signals together.

The transfer function was then measured at the resonant frequency. The amplitude of the signal from the waveform generator was varied from 500 mV to 10 V. The received voltage for both the transmitted and reflected signals, and the difference was recorded as a function of the input voltage applied to the PZT transducer.

The frequency response of the FBG sensor and TRDS was measured using the same method used to determine the peak frequency above. The frequency of the signal from the waveform generator was varied from 50 kHz to 1 MHz, recording the received amplitude at every 10 kHz, and every kilohertz about the major peaks.

The transient response of the sensor was measured using a sine wave burst from the waveform generator, set to the maximum voltage. This was done for all of the resonant frequencies. The burst count was increased from 1 to the number required to reach the maximum amplitude for a continuous sine wave, representing the onset of steady state.

The results from the frequency response were used to determine the carrier wave frequency for the acoustic transmissions. The communications signals were generated in Agilent's Waveform Editor software for the 33120A arbitrary waveform generator. The signals were then flashed to the device via the computer interface. PSK signals using a square wave carrier were generated with data rates at 1/400, 1/200, and 1/100 times the carrier wave frequency.

3.4 Acoustic Emissions

AEs were generated to determine if the FBG sensor with the TRDS could be used for their detection. Acoustic emissions were generated from low velocity impacts, and a lead pencil break test. The low velocity impacts were generated by dropping steel ball bearings of various sizes (1, 3.5, 8.3, 16.3 and 35.8 g) from a fixed position over the FBG. The lead pencil break test was located 100 mm away from, and along the optical axis of, the FBG. For the acoustic emissions tests, the PZT transmitter and the function generator shown in Fig. 4 were not required.

The five steel ball bearings were used to characterise the effect of the impactors mass on the response of the FBG sensor. The release mechanism was fixed at a height of 20 mm. The ball bearings were dropped from lightest to heaviest. The received signals were captured for each mass.

A single steel ball bearing was used to characterise the effect of the impactors energy on the response of the FBG sensor. The 1 g ball bearing was used. The release mechanism was varied from a height of approximately 7–32 mm, at increments corresponding to rotating the knob of the lab jack through one full rotation. A vernier depth gauge was used to measure the height from the aluminium panel to the lab jack. The scissor mechanism of the lab jack meant that the heights were not at regular increments. The maximum amplitude of each impact was recorded. This process was repeated several times to give the average maximum amplitude, and an error bound.

4 Results

4.1 FBG and Laser Characterisation

Figure 5 shows the spectral response of the ASE, FBG, and tunable laser set to the operating point. The location of the tunable laser peak corresponds to a value below the linear region of the FBGs spectral response, this is due to the linewidth of the tunable laser being too broad. Hence, a lower value was used to give equal transmitted and reflected optical power.

4.2 TRDS Test

The result from the FBG sensor using the TRDS is shown in Fig. 6. The figure shows the transmitted and reflected signals (the bottom two traces), and the difference between the two (the top trace). All three traces are shown on the same scale, at 1 V per division. The offset present in the two received signals is removed in the differential signal.

4.3 Sensitivity, Frequency and Transient Response

The typical transfer function of the FBG using the TRDS is shown in Fig. 7. The sensitivity (S) is given by the derivative of the transfer function, which in the case of

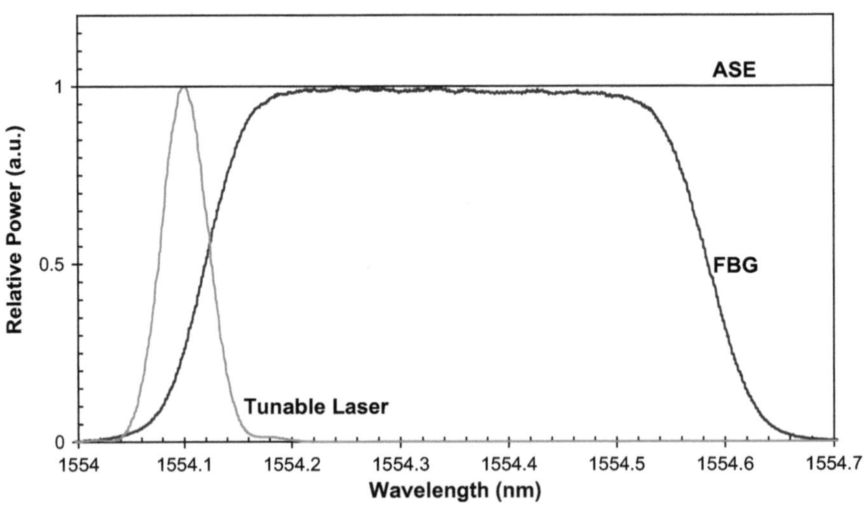

Fig. 5 Spectrum of optical components, including the ASE, FBG and Tunable laser

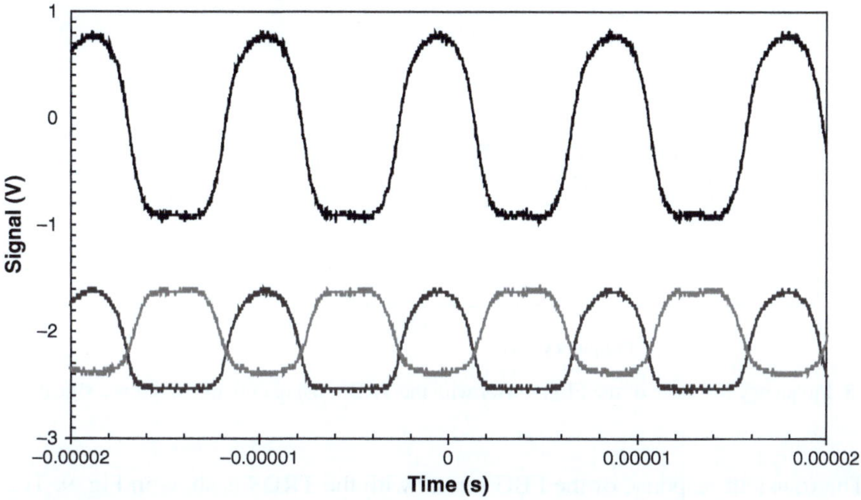

Fig. 6 Signals from two photoreceivers (*bottom traces*), and the difference (*top trace*), at 108.8 kHz

a linear response is simply the gradient. Here the sensitivity is given as 0.074 with a typical value of 0.1. Since the transfer function is from voltage to voltage, this corresponds to an efficiency of 10%.

The frequency response of the FBG using the TRDS is shown in Fig. 8. Since bit rate is related to the bandwidth, using the broadest peak would, in theory, give the highest bit rate for transmissions. Although not the broadest peak, 108.8 kHz was selected due to the fact that it had the strongest received signal. This ensures the best signal to noise ratio, which is significant for communications applications.

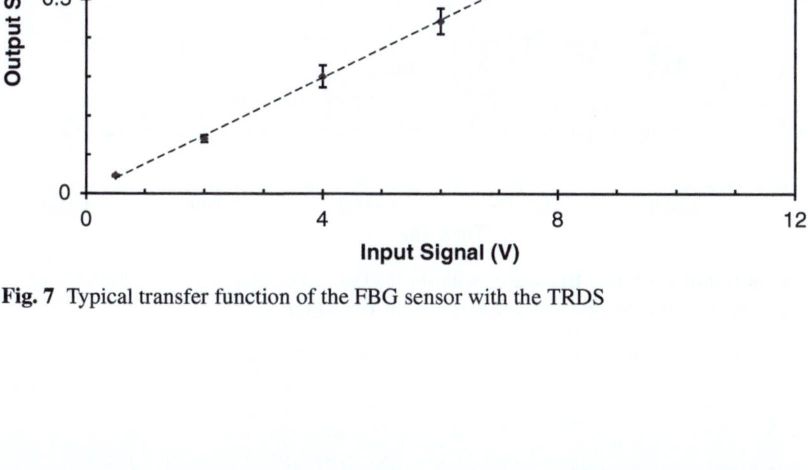

Fig. 7 Typical transfer function of the FBG sensor with the TRDS

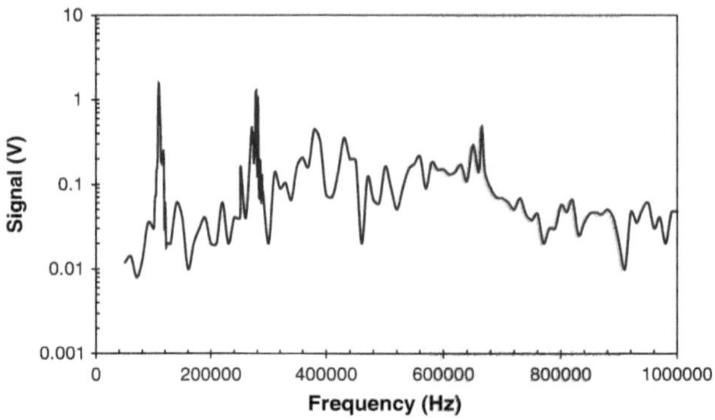

Fig. 8 Frequency response of the FBG sensor with the TRDS, using only the difference signal

The transient response of the FBG sensor with the TRDS is show in Fig. 9. The transient response was measured at 108.8 kHz, as this was chosen to be the communications carrier frequency. The transient response was 20 cycles, corresponding to a transient response time of approximately 200 μs.

4.4 Acoustic Transmissions

The received acoustic communications signal is shown in Fig. 10. The decoded acoustic communications signal is shown in Fig. 11.

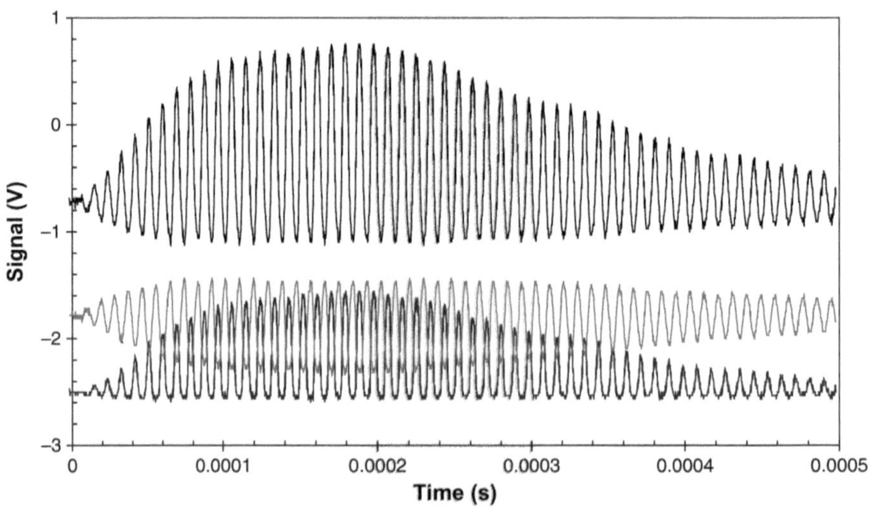

Fig. 9 Transient response of the FBG sensor with the TRDS, with the signals from both photoreceivers (*bottom traces*), and the difference (*top trace*), at 108.8 kHz

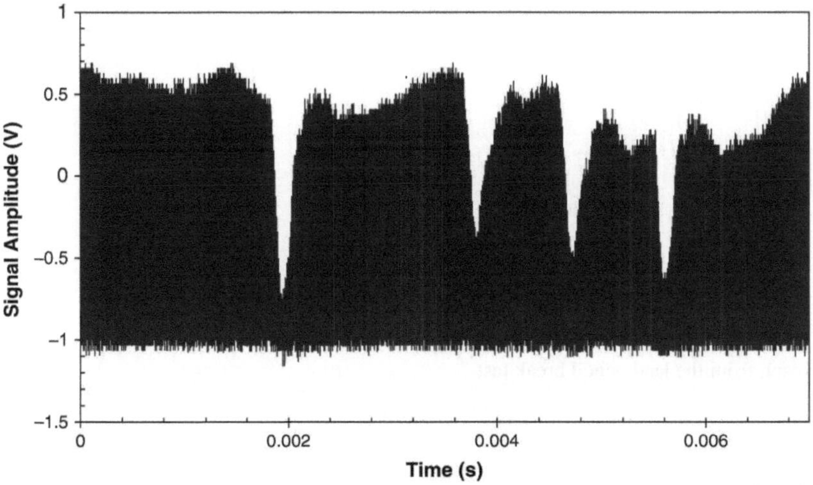

Fig. 10 Received PSK communications signal, with a carrier frequency of 108.8 kHz, and a data rate of 1.088 kbps

4.5 Acoustic Emissions

The result of the pencil break test is shown in Fig. 12. The low velocity impact results are shown in Figs. 13 and 14. Figure 13 shows the affect of varying the mass on the received signal, with a constant velocity, and Fig. 14 shows the affect of varying the velocity on the received amplitude, with a constant mass.

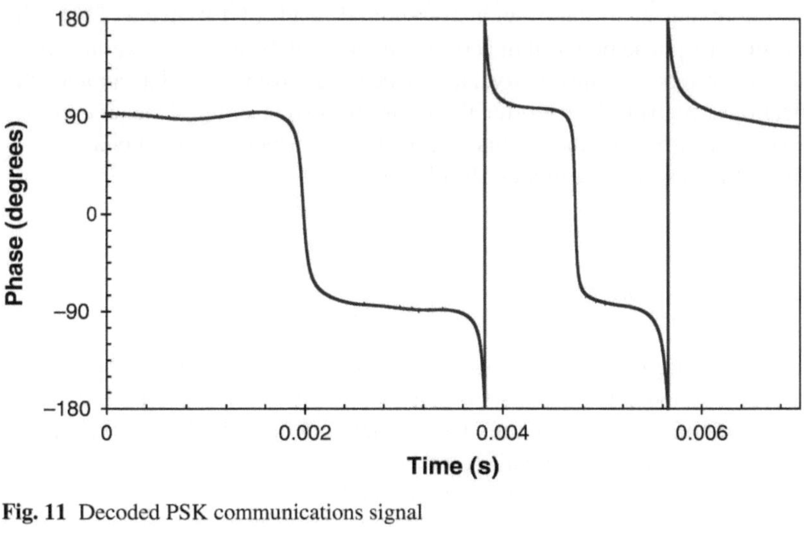

Fig. 11 Decoded PSK communications signal

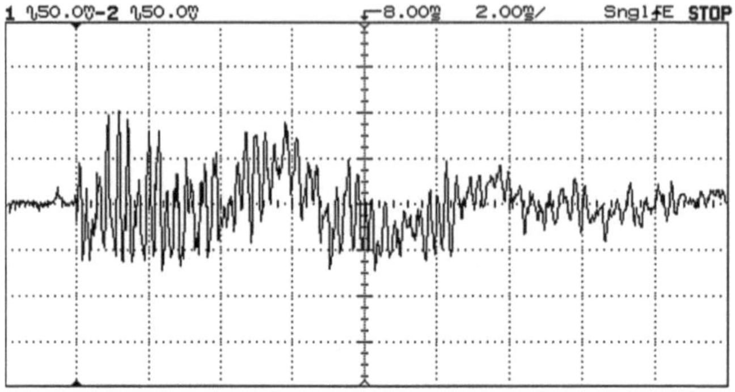

Fig. 12 Result from the lead pencil break test

5 Discussion

5.1 Frequency and Transient Response

The frequency response of the new FBG sensor with the TRDS is very different to results previously presented [14]. While there are two distinct peaks, at 108.8 and 278 kHz, the overall response of the FBG is relatively strong, with consistent values over 10 millivolts, and regular values over 100 mV. The previous FBGs used had a regular frequency response with distinct harmonics of 63 kHz. This could be due to resonant frequencies in the composite structure in which the sensor was contained.

The transient response of the new FBG sensor with the TRDS, 20 cycles, is a significant improvement to the previously reported result of 150 cycles [14]. The shorter transient response means that acoustic signals can be detected over a shorter time scale. This offers significant advantage when the sensor is used as an acoustic communications receiver. The shorter the transient response, the faster a communications signal can be switched. This applies to other encoding methods such as amplitude shift keying and frequency shift keying.

5.2 Acoustic Transmissions

The decoded acoustic communications signal, Fig. 11, shows distinct transitions and steady state periods. The results suggest that the data rate could be improved. The transient response time of 200 ms means that a value as high as 5 kbps could be achieved. A more significant result could be achieved if 665 kHz was used as the carrier frequency; 665 kHz is the final peak in Fig. 8.

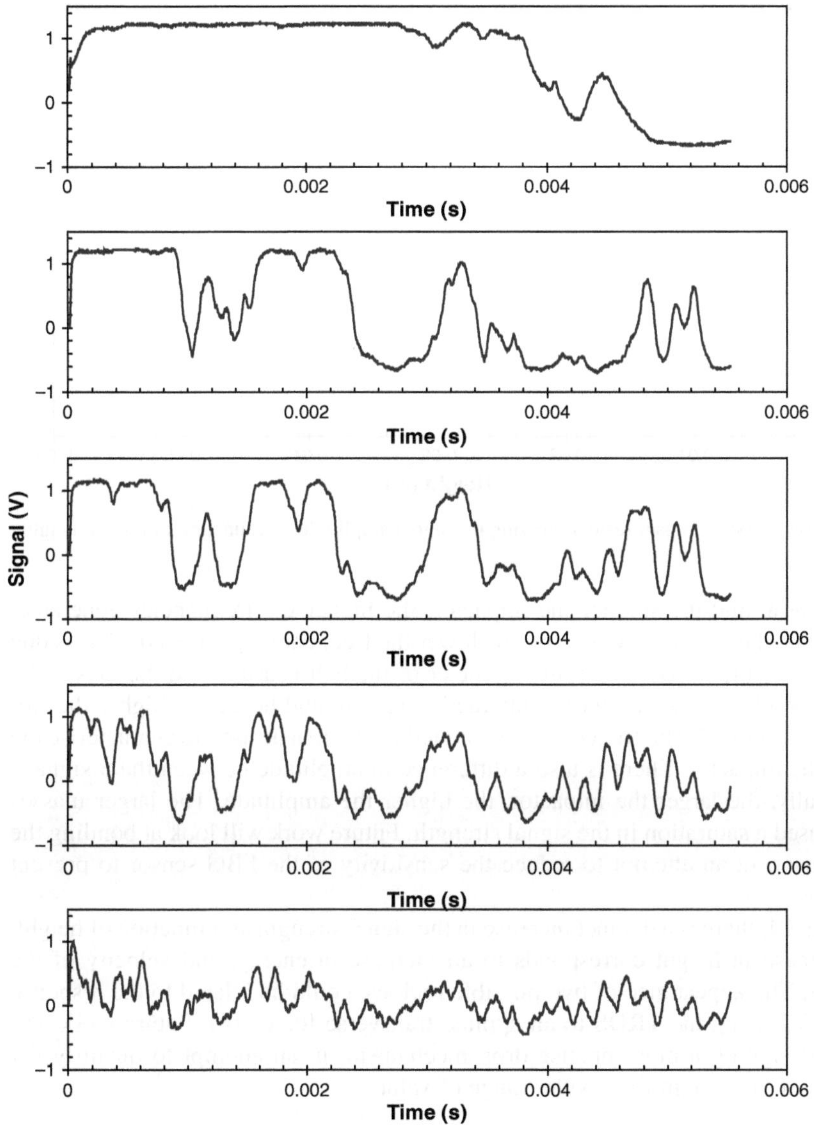

Fig. 13 Transient signals of the low velocity impact results for 5 different mass ball bearings drop from the same height. From top to bottom, 35.8, 16.3, 8.3, 3.5, and 1 g

5.3 Acoustic Emissions

The pencil break test signal gave an amplitude of 160 mV. This shows significant improvement over the previously published result of 10 mV [14]. As shown in Fig. 13, there is a distinct difference in the frequencies generated between the 5 masses. The

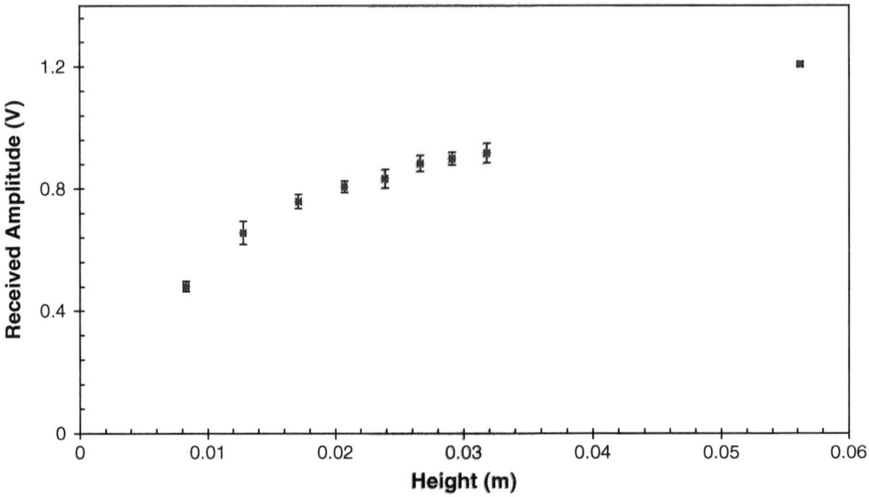

Fig. 14 Low velocity impact results showing the signal amplitude as a function of release height

signals show that the smaller the impactor, the higher the frequencies generated. Conversely, the larger the impactor, the lower the frequencies generated. This is due to the size of the impactor, i.e. the diameter of the ball bearing, and the mass. The smaller the object, the shorter the interaction region, and hence the higher the frequencies generated. The frequencies generated could then be used to characterise the size of the impactor. There is also a difference in amplitude between the 5 signals. Specifically, the larger the impactor, the higher the amplitude. The larger masses have caused a saturation in the signal strength. Future work will look at bonding the FBG sensor, in an attempt to reduce the sensitivity of the FBG sensor to prevent saturation.

In Fig. 14, there is a distinct increase in the signal strength as a function of height. The increase in height corresponds to an increase in energy, and velocity of the impactor. The appearance of two possible regions, could be related to the response of the FBG using the TRDS to an applied transverse force [16]. Future work will look at designing a more precise drop mechanism, in an attempt to quantify the response of the system for a wider range of values.

6 Conclusion

A Transmit Reflect Detection System (TRDS) has been successfully used for the detection of acoustic emissions and transmissions signals from a Fibre Bragg Grating (FBG) sensor. The FBG sensor with the TRDS has sufficient sensitivity to detect the Acoustic Emissions (AE) from a pencil break test. Also, the ability of the sensor to detect AEs from low velocity impacts was demonstrated. The AE from the impact detection suggest that the size of the impactor may be determined from the frequencies generated, and the relative energy, hence the velocity, can be determined from the amplitude. The sensor was also used for the detection of acoustic transmissions. Successful communication was achieved via the acoustic transmissions. Using Phase Shift Keying (PSK) and a carrier frequency of 108.8 kHz, a data rate of 1.088 kb per second was demonstrated, and the results suggests that up to 5 kb per second could be achievable.

References

1. Betz DC, Thursby G, Culshaw B, Staszewski WJ (2003) Acousto-Ultrasonic Sensing using Fiber Bragg Gratings. Smart Materials and Structures, 12(1):122–128
2. Ecke W, Latka I, Willsch R, Reutlinger A, Graue R, (2001) Fibre optic sensor networks for spacecraft health monitoring. Measurement Science and Technology, 12(7):974–980
3. Fomitchov PA, Krishnaswamy S (2002) Fiber Bragg Grating Ultrasound Sensor for Process Monitoring and NDE Applications. Proc. Review of Quantitative Nondestructive Evaluation, vol. 21, pp 937–944
4. Fomitchov PA, Krishnaswamy S (2003) Response of a Fiber Bragg Grating Ultrasonic Sensor. Optical Engineering, 42(4):956–963
5. Othornos A, Kalli K (1999) Fiber Bragg Grating: Fundamentals and Applications in Telecommunications and Sensing. Artech House, Norwood
6. Perez I, Cui HL, Udd E(2001)Acoustic emission detection using fiber Bragg gratings. Proc SPIE, 4328, pp 209–215
7. Price DC, Scott DA, Edwards GC, Batten A, Farmer AJ, Hedley M, Johnson ME, Lewis CJ, Poulton GT, Prokopenko M, Valencia P, Wang P (2003) An Integrated Health Monitoring System for an Ageless Aerospace Vehicle. In: Structural Health Monitoring 2003: From Diagnostics & Prognostics to Structural Health Management, DEStech Publications, Lancaster PA, pp 310–318
8. Proakis J, SalehivM (1994) Communication Systems Engineering, Prentice Hall, New Jersey
9. Staszewski W, Boller C, Tomlison G (2004) Health Monitoring of Aerospace Structures: Smart Sensor Technologies and Signal Processing, Wiley, West Sussex.
10. Takahashi N, Hirose A, Takahashi S (1997) Underwater Acoustic Sensor with Fiber Bragg Grating. Optical Review, 4(6):691–694
11. Webb DY, Surowiec J, Sweeney M, Jackson DA, Gavrilov LR, Hand JW, Zhang L, Bennion I (1996) Miniature fibre optic ultrasonic probe. Proc SPIE, vol. 2839, pp 76–80
12. Wild G (2005) Design and Evaluation of an Electro-Acoustic Communications Channel for use by Autonomous Agents in the Structural Health Monitoring of Ageless Aerospace Vehicles. Honours thesis, Edith Cowan University
13. Wild G, Hinckley S (2007) Electro-Acoustic and Acousto-Optic Communications for Robotic Agents in Smart Structures. In: Proc. of Smart Structures, Devices, and Systems III, Adelaide, Australia
14. Wild G, Hinckley S (2007) Fiber Bragg Grating Sensors for Acoustic Emission and Transmission Detection Applied to Robotic NDE in Structural Health Monitoring. In: Proc. IEEE Sensors Applications Symposium 2007, San Diego, USA
15. Wild G, Hinckley S (2007) A Fibre Bragg Grating Acoustic Emission and Transmission Sensor Utilizing a Transmit Reflect Detection System. In: Proc. 2nd International Conference of Sensing Technology, Palmerston North, New Zealand, pp 546–551
16. Wild G, Hinckley S, Jansz PV (2007) A Transmit Reflect Detection System for Fibre Bragg Grating Photonic Sensors. In: Proc. of Photonics: Design, Technology, and Packaging III, Canberra, Australia

Part V
Magnetic Sensor

A Real-Time Tracking System for an Endoscopic Capsule using Multiple Magnetic Sensors

Syed Mahfuzul Aziz, Milos Grcic and Tharshan Vaithianathan

Abstract A tracking system for visualizing the location of a magnetically marked diagnostic capsule in real time is proposed. The intended application is the gastrointestinal (GI) tract which is regarded by gastroenterologists as a black box. An endoscopic capsule capable of releasing diagnostic biomarker probes to areas of the gastrointestinal tract that are inaccessible via conventional means will enable gastroenterologists to accurately determine the changes in functionality of the GI tract due to diseases. This requires a tracking system that can show the location of the capsule in real-time as it travels down the digestive tract. This paper presents the design and implementation of such a tracking system using multiple magnetic sensors. It is free from the disadvantages most radio-frequency tracking systems suffer from. We report on the developments of the magnetic sensing hardware, tracking algorithm based on empirically developed models and experimental results. The results reveal the suitability of the proposed system for in vivo applications.

Keywords Real-time tracking · magnetic sensors · magnetic marking · capsule endoscope · gastrointestinal tract

1 Introduction

The gastrointestinal tract is approximately nine meters long. Conventional endoscopic procedures can access only about 2.5 m of the esophagus/stomach while colonoscopy procedures can access only part of the large intestine through the rectum. The small intestine, which is the longest part of the GI tract, cannot be

Syed Mahfuzul Aziz
School of Electrical and Information Engineering, University of South Australia, Mawson Lakes, SA 5095, e-mail: mahfuz.aziz@unisa.edu.au

Milos Grcic
School of Electrical and Information Engineering, University of South Australia

Tharshan Vaithianathan
Mawson Institute, University of South Australia, Mawson Lakes, SA 5095, Australia

accessed via these techniques [1]. Gastroenterologists have been looking for a reliable non-invasive means to access affected areas of the small intestine for targeted delivery of biomarker probes to determine changes in gut function due to disease. An endoscopic capsule called PillCam^{TM}, which can be swallowed by patients, is capable of capturing images of the gastrointestinal tract. It sends these images to the outside of the body using radio frequency (RF) signals and are stored on a data logger for later viewing by the physicians [1]. PillCam^{TM} is not designed for targeted drug delivery. Although it is equipped with a RF tracking system [1], it is well known that RF signals suffer significant attenuation through the human body and hence not very suitable for a real-time tracking system intended for in vivo use [2]. In contrast magnetic signals do not experience attenuation when passing through the human body. A number of research groups have used magnetic signals to develop tracking systems for in vivo use and have conducted trials producing promising results.

The tracking systems proposed in [3, 4] involve the use of the SQUID sensor system, which require the sensors and a large coil to sit above the patient's torso while the patient remains lying horizontally throughout the procedure. The tracking system presented in [5, 6] is based on a 2D array of sixteen cylindrical Hall sensors and a permanent magnet. The position and orientation of the marker (magnet) is displayed in real time for distances between the magnet and the sensors of up to 14 cm. The tracking system presented in [7] deals with the error resulting from the rotation of the magnet using large coils placed outside the patient's body. The coils would generate a magnetic field to keep the magnet (inside the capsule) oriented in the desired direction. The system presented in [8] uses a magnetically marked material inside a capsule and an array of Hall Effect sensors placed on the exterior of the body. It outputs the values of the x, y and z coordinates, as well as the angles between the X-Y and X-Z planes, but does not go as far as providing visual indication of the capsule in real-time. A similar study to localize a capsule was reported in [9].

A few other magnetic monitoring systems, used for surgical procedures [10, 11, 12], involve the use of coils attached to the surgical instruments (for example catheters), which emit magnetic signals throughout the procedure. Magnetic Resonance Imaging (MRI) is used to monitor the emitted signals and thereby the location of the instrument inside the patient. The operating table with embedded tracking technology provides the surgeon with the guidance required. In [13] magnetic markers were used on the surgical instrument instead of coils and produced real-time monitoring of surgical probes.

Other publications [14, 15] have reported on the use of SQUID sensors and magnetic marking to perform immunoassays on patients. Certain antibodies within the body are magnetically marked so that when they attach themselves to the antigens (that are to be analyzed), concentrations as well as fluid movements of these antigens can be monitored. A similar principle was used in [16] in analyzing DNA using a biochip, which is a type of magneto-resistive sensor. Certain molecules are magnetically marked by the biochip; hence the biochip can monitor the movements of these molecules, and analyze the properties of the chemical reactions when the molecules bind with particular DNAs.

Majority of the magnetic tracking systems discussed above either use complex assembly of coils, and/or require the patient to lie down during the entire procedure or are expensive. In this paper we present the design and development of a tracking system for an endoscopic capsule using only three magnetic sensors. The sensors monitor the capsule track as it passes through the gastrointestinal tract. A real-time tracking algorithm has been developed to offset errors due to orientation and rotation of the magnet. The algorithm utilizes empirical models developed for the tracking system. Laboratory tests we have conducted show very promising results.

2 The Proposed Tracking System

A block level representation of the proposed tracking system is shown in Fig. 1. A magnet is placed inside the endoscopic capsule to be swallowed by patients. Three magnetic sensors are attached to the body to detect the magnetic field emitted by the magnet. The sensors send the magnetic field values to a PC after conversion to digital form by the PC interface. Based on these field values a software program calculates the position of the capsule and displays it on the computer monitor.

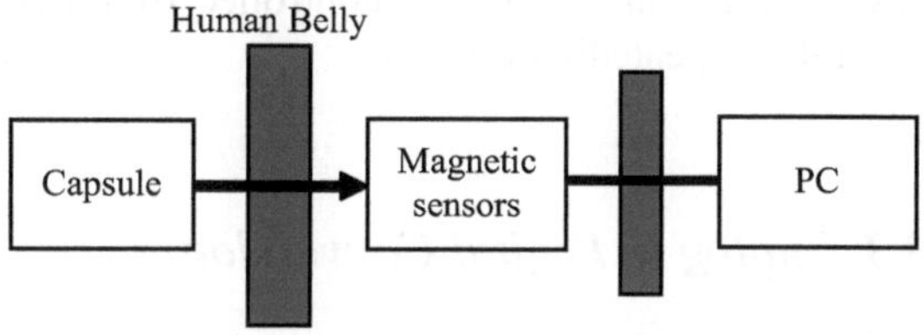

Fig. 1 Block diagram of the proposed tracking system

3 Hardware and Interface

This section of the paper focuses on the magnetic signal's path from a sensor to the PC and describes the hardware necessary to establish this link.

3.1 Magnetic Field Source

An essential element of any magnetic tracking system is a magnetic field source. A permanent magnet was selected as a suitable source. It provides a simple, effective and safe source of magnetic field. Use of a magnet allows the capsule to be free of coils, electronic components and batteries, making it friendlier to the human body. In addition, permanent magnet is a very reliable source of steady magnetic field. The accuracy and range of the tracking system depend on the remanence (strength) of the magnet, where accuracy and range increase with the remanence. The remanence increases with the size of the magnet, which is limited in our application to a size swallowable by a human. The magnet selected for our tracking system is a rare-earth Neodymium ($Nd_2Fe_{14}B$) magnet, providing the highest available remanence to size ratio [17]. The magnet used for testing the tracking system is a cylindrical one with both diameter and length of 10 mm.

3.2 Sensors

Three HMC2003 sensors from Honeywell [18] are used to measure the intensity of the magnetic field at different locations outside the human body. Each of the magnetic sensors outputs three analogue voltage signals, one for each axis (X, Y, Z). The output voltage range of the sensor is 0-5 V, where 0 V represents −200 μT of magnetic filed, 5 V corresponds to +200 μT and 2.5 V is 0 μT. The resolution of the sensor is 40 nT and it requires a power supply voltage between 6 and 15 V [18]. A set/reset circuit, as shown in Fig. 2, was designed to recover the sensor from magnetic saturation, which occurs when the HMR2003 is exposed to a field in excess of 1 mT [19]. It is controlled by a pulse from a microcontroller and is generated repeatedly every 5 s.

3.3 Analog to Digital Conversion

The nine magnetic field readings (3 from each sensor) are converted into digital values prior to being sent to the computer. We have used an 8-bit A/D converter with a maximum analogue input of 2.5 V, exactly half the maximum value of the signal

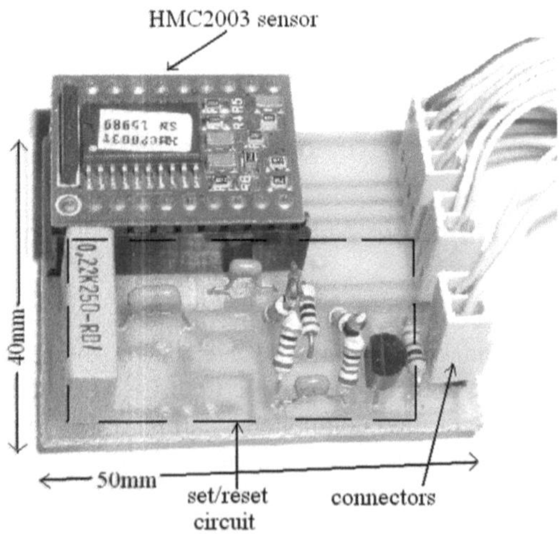

Fig. 2 HMC2003 Sensor and Set/Reset circuit

to be converted. A voltage divider was used to bring the sensor's output voltage down to the required level. The magnetic field readings of 0-5 V from the sensors are represented in 0-255 units range. The resolution of this system is 19.5 mV ($5/2^8$) which is equivalent to 1.6 μT.

3.4 Multiplexer and Microcontroller

Figure 3 shows a schematic diagram of the tracking system. The microcontroller's A/D converter cannot accept the nine magnetic field values simultaneously. A 16-to-1 analogue multiplexer was used to read the field values into the microcontroller sequentially. The multiplexer's output was connected to the A/D converter's input and the multiplexer's selection pins were addressed by the microcontroller's I/O pins. To enable successful operation of the system, a program was loaded into the microcontroller's 32 kb flash memory. This program first selects one of the sensor readings via the multiplexer, converts it to digital form, temporarily saves it in the microcontroller's memory and proceeds to read the next magnetic field value. This procedure is repeated until all nine readings are converted and stored in the microcontroller. After storing the ninth field value, the microcontroller writes all the stored values to the serial port followed by writing a carriage return. A C program running on the PC reads the field values from the serial port, calculates the position of the capsule based on these values and displays the position on the graphical user interface (GUI). Details of this program are presented in Sect. 4.

The overall tracking system hardware, as shown in Fig. 3, is composed of four main parts: three sensor boards and a custom designed enclosure accommodating the multiplexer and the microcontroller boards. Only one 18 V dc power supply is

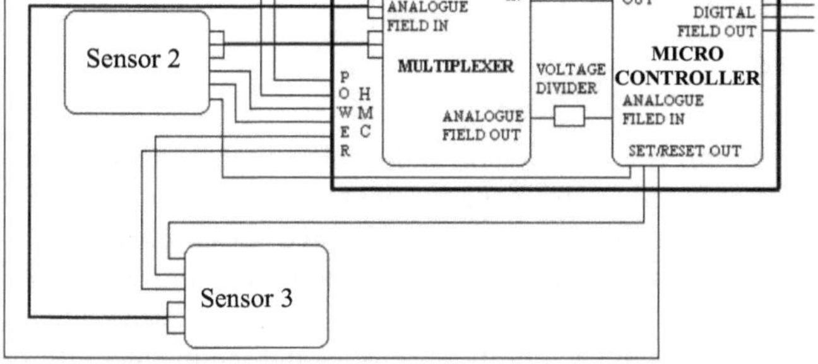

Fig. 3 Schematic of the tracking system

required in the enclosure, which will supply power to the entire system including the sensor boards. The entire tracking system draws 60 mA continuous current. It can be operated by two heavy duty 9 V batteries having a rating of 1.2 A-h for over 10 h.

4 System Software

The system software consists of the data acquisition software and the tracking software. First we present a brief description of the data acquisition software. Then we present the design of the tracking software using a single sensor along with the underlying empirical models we have developed. Finally we present the tracking software for the entire system using multiple sensors.

4.1 Data Acquisition Software

First of all, the data sent by the microcontroller has to be read by the C program. This is done in the same order in which the microcontroller has sent the values. As previously mentioned, these numbers are digital representations of voltage levels output by the sensors, where the voltage level is directly proportional to the magnetic field strength measured by that sensor. Voltage levels between 0 and 5 V represent the magnetic field in the range of −200 to 200 μT, with 0 V at −200 μT and 5 V at 200 μT. As 8-bit A/D conversion was used the values received by the computer will be in the 0–255 range, where 0 represents 0 V (−200 μT) and 255 represents 5 V (200 μT). Eleven buffers are used by the C program, nine for the digital signals and two for the carriage return character which follows the value of the ninth signal. The signals are read in the following order: |X1 Y1 Z1 X2 Y2 Z2 X3 Y3 Z3 CR|. The letters X, Y and Z represent the axes being read; the numbers 1, 2 and 3 identify the sensor. The readings are stored as variables and used to calculate the position of the magnet. The algorithm used for the calculation is explained in the next section.

4.2 Tracking Software

The fundamental idea behind this tracking system is the calculation of distance between the magnet and a sensor based on the magnetic field reading measured by the sensor. Figure 4 shows the orthogonal orientation of the three sensors used to develop the tracking system. Each sensor is responsible for determining the coordinate of the magnet's location along one axis. Positioning the three sensors orthogonally and determining the distance between each of these sensors and the magnet allows

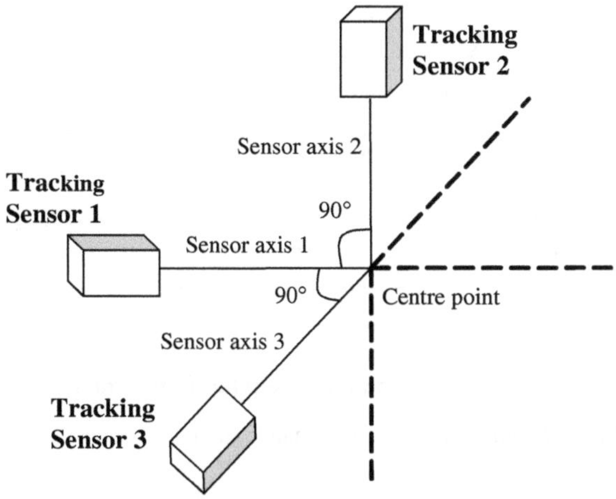

Fig. 4 Orthogonal orientation of sensors

the calculation of all the three coordinates of the magnet's location relative to the centre point. The centre point is where the three orthogonal meet in space, as shown in Fig. 4.

4.2.1 Single Sensor Tracking

Since the human body has shielding properties identical to that of the free space, it was assumed that there will be no attenuation of the magnetic field while passing through the body [20]. This means that the system operates in a homogeneous environment and is not susceptible to error due to the diverse electrical properties of the human organs, as is the case in a RF tracking system [1]. However, magnetic marking has different sources of error associated with it. Most of these issues are related to the physical properties of the magnet, such as the polarity, and the non-linear relationship between the variation in magnetic field strength with the variation in distance between the magnet and the sensor.

Non linearity: The relationship between the magnetic field and distance is not linear. The relationship, as obtained experimentally is shown in Fig. 5. The (arbitrary) magnetic field values shown on the X-axis are binary representations of the field readings (analog voltages) produced directly by the sensor along one axis. That is, the voltages have not been converted to μT. We validated this relationship by comparing it with the curve obtained by using the standard relationship given by (1) [21]:

$$Field = \frac{Br}{2}\left[\frac{L+X}{\sqrt{R^2+(L+X)^2}} - \frac{X}{\sqrt{R^2+X^2}}\right] \tag{1}$$

where, Br is the remanence of the magnet, L is the length of the magnet, R is the radius, and X is the distance between the magnet and the point where field is to be calculated [21]. Note that this equation only applies to situations where

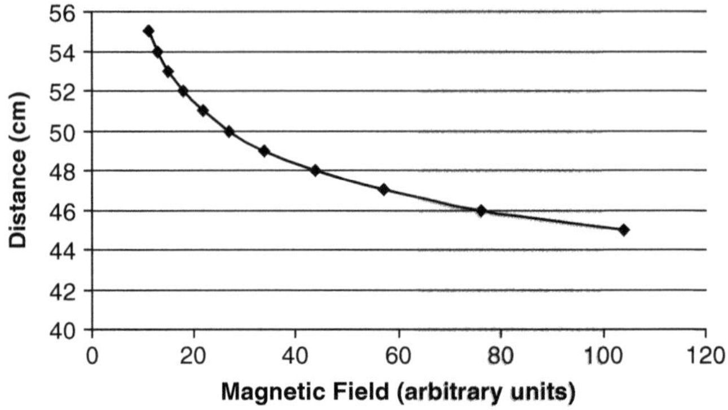

Fig. 5 Magnetic field versus distance curve

the magnet is directly facing the point at which the field is being measured. Once the curve in Fig. 5 was obtained, a mathematical model for the relationship between the magnetic field and the distance was developed. From observation, the curve is a plot of an exponential function, and can be represented by the following equation:

$$D_1 = \frac{55}{1 - a \times e^{-bx}} \tag{2}$$

where, D_1 is one of the three coordinates of the capsule's (magnet's) position with respect to the patient's feet, x is the magnetic field reading in scaled units after A/D conversion, and "a" and "b" are constants dependant on the properties of the magnet used. The numerator in the above equation is used to move the reference point of the GUI up by 55 units, from the patient's feet (original reference point) into the stomach area.

Ability to emulate the relationship between the magnetic field and distance allows the translation of the field values read by the C program into distance of the magnet from a selected reference point. For more information on this mathematical model, its accuracy and compatibility with different magnets refer to Sect. 5.1.

Rotation of magnet: A number of other issues were encountered in the early stages of development of the tracking algorithm. Overcoming the non linearity issue only enables the system to cover one axis and work effectively under very specific experimental conditions, mainly due to issues related to the polarity of the magnet. If the above relationship was obtained with the magnet's north pole facing the sensor, as shown in Fig. 6a, then it cannot be used as it is to calculate the distance of the sensor from the magnet if the latter is rotated by 180°, i.e. if the south pole is facing the sensor, as shown in Fig. 6b. In this case the sensor will register the same magnetic field value but with the opposite sign (polarity). The HMC2003 sensor is capable of reading −200 μT to +200 μT; it outputs 0 V for −200 μT and 5 V for 200 μT, with 0 T represented by a voltage in the middle of this range (2.5 V). This being the case it is necessary to subtract 2.5 V from each reading of the sensor and take

Fig. 6 Errors caused by changing magnet's orientation

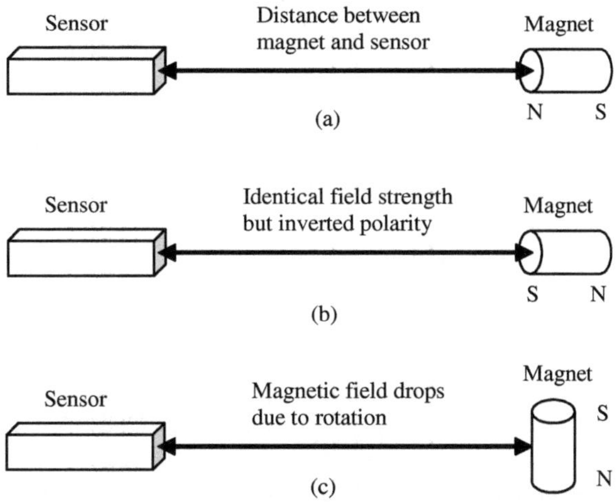

the absolute value of the difference as the magnitude of the magnetic field. This will require us to adjust any output voltage greater than 2.5 V in order to find the field magnitude only. Rotation of the magnet up to 90° is another major source of inaccuracy. Magnetic field is at its peak at 0° (a pole facing the sensor), and reaches minimum (0 μT) at 90°, as in Fig. 6c. The relationship represented in (2) was obtained with one pole of the magnet directly facing the sensor as in Fig. 6a or 6b, and it can calculate the distance accurately only for this orientation of the magnet. In a real life situation, it cannot be assumed that the magnet's pole will be facing the sensor directly at all times, i.e. maintain fixed orientation throughout the procedure.

In order to adjust the negative magnetic field values due to 180° rotation, we need to look at how the negative values are represented. In our 8-bit A/D system the magnetic field values are represented as follows: 0 represents −200 μT, 127 represents 0 T and 255 represents +200 μT. The 180° rotation issue was addressed by subtracting the value of the magnetic field reading from the mid value of the monitored axis' range (127, representing 0 T) and taking the absolute value of the result. This procedure halves the resolution of the system, leaving the system with only 128 points to represent one axis and 128^3 points to monitor the selected volume. The resolution of the tracking system can be improved by using A/D converters having higher resolution, for example a 16-bit or even a 24 bit A/D converter; however the cost and size of the system would obviously be increased.

Elimination of the above problem allows accurate tracking of the magnet for the cases when either of the poles is facing the sensor. As the magnet spins away from this orientation, the magnetic field reading for that axis falls and finally reaches zero at 90°. As the field decreases, the tracking algorithm registers the drop and as a result calculates the distance between the magnet and the sensor to be greater than it actually is. This issue was dealt with by also examining the changes in the magnetic field values of the axis towards which the magnet is rotating. Up until this stage of the algorithm development, the field value in only one axis of the sensor was used to calculate the distance of the magnet. Now, the tracking algorithm is enhanced to include the field strengths in the other two axes of the sensor in order to compensate for the error caused by the magnet's rotation. As the magnet spins away from one axis towards the 90° orientation, the magnetic field reading in that axis decreases, however the field of the axis towards which the magnet is rotating increases.

The changes in the field values of the two axes as the magnet rotates were studied and are explained in detail below. Based on the behavior observed (see below) the tracking equation was modified as follows:

$$y = \frac{55}{1 - a \times e^{-b(x + 2.3 \times y \times e^{\frac{-x}{(2.3 \times y + x)/2 + 8}} + 2.3 \times z \times e^{\frac{-x}{(2.3 \times z + x)/2 + 8)}}}} \tag{3}$$

This equation differs from (2) in that two new terms are added to the value of the reading of the monitored axis (x). The new terms are included to minimize the rotation error, which has the potential of reaching 100% in the worst case scenario. The two expressions are essentially identical, doing the same job for the two remaining axes, namely the y and z axis. The constant 2.3 represents the ratio of the maximum x reading at 0° and the maximum y reading at 90°, as obtained experimentally. This means that at 90° into y axis, the value of x is 0 and the value of 2.3*y is identical to what x would have been at the same distance at 0°.

With the incorporation of the term 2.3*y into the exponent of (2), the distance of the magnet from the sensor can be calculated accurately for the angles of 0°, 90° and 180°, but considerable amount of error still appears for the angles in between these. This error was significantly reduced by multiplying 2.3*y with an exponential function which allows the addition of 2.3*y to x in full only when the value of x is 0, meaning that the magnet is orientated at 90° to the x-axis. For angles of 0° and 180°, the value of y is zero, meaning that only the x field reading is used to calculate the distance. For all other angles only a fraction of the value (2.3*y) is added to x depending on the value of the second exponent. This exponent uses both x and y field readings rather than just the x reading, in order to more accurately estimate the distance of the magnet from the sensor. This is necessary as the variations in the x and y field readings due to rotation are also functions of the distance of the magnet from the sensor. Finally the constant 8 is used to adjust the exponential, depending on the magnet's properties. For information on the accuracy of the system refer to the experimental results presented in Sect. 5.

4.2.2 Cancellation of Environmental Fields

In the proposed real time tracking system, we envisage the use of a fourth sensor for measuring the environmental fields (including Earth's magnetic field) and cancelling the same in real time. This sensor will be positioned at a location sufficiently distant from the magnet in order to keep the field generated by the magnet negligible compared to environmental fields. Note that the magnitude of the Earth's magnetic field in South Australia is approximately 60 μT. Automatic cancellation of environmental fields will allow full mobility of the tracking system. However, we excluded the fourth sensor from initial prototype development. The cancellation of environmental fields was done by measuring the field values in the absence of the magnet (at the beginning of our experiments) and deducting it from the field value measurements by the sensors in real time.

4.2.3 System Integration and Multi-Sensor Tracking

As stated previously, the three sensors are placed in orthogonal positions. The range of the system is limited and depends on the range and resolution of the HMC2003 sensors as well as the resolution of the A/D converter. It was observed experimentally that the sensors can take meaningful field measurements starting at a minimum distance from the magnet of 5 cm to a maximum distance of 15 cm for smaller (lower remanence) magnets, and a minimum distance of 10 cm to a maximum of 50 cm for more powerful magnets. Minimum distance is determined by the range of the magnetic sensors used and the size and type of the magnet. If the distance between the magnet and the sensor is below the minimum distance then the magnetic field value at the sensor is outside the sensor's range [18], thus the position of the magnet cannot be accurately determined. Maximum distance is dependant on the resolution of the A/D converter, magnet's size and properties, and the minimum accuracy required. As the magnet moves away from the sensor, the resolution decreases, reaching the minimum allowable value at the point of maximum distance. This is a result of the exponential relationship between distance and magnetic field as depicted in Fig. 5.

The tracking software uses the three magnetic field values from each sensor and uses (3) to determine the coordinate of the magnet's location along its physical axis. It eventually determines the three coordinates from the readings of the three sensors. Thus the location of the magnet is determined within a space of 10 cm × 10 cm × 10 cm.

4.2.4 Visualization (Graphical User Interface)

After calculating the magnet's distance from each of the sensors, the three coordinates of the magnet relative to the centre point (see Fig. 4) are printed on the PC screen along with the original sensor readings. The magnet's location is graphically displayed at the same time. Graphical presentation is accomplished by displaying a red dot (marker) at the location represented by the calculated coordinates of the magnet. The marker is a plot relative to an image of a human body in 3D, which gives a visual representation of the magnet's position within the patient's body. The body is shown in the form of a mesh, where each line of the mesh is plotted by accessing the coordinates saved in a text file. The section of the code used to plot

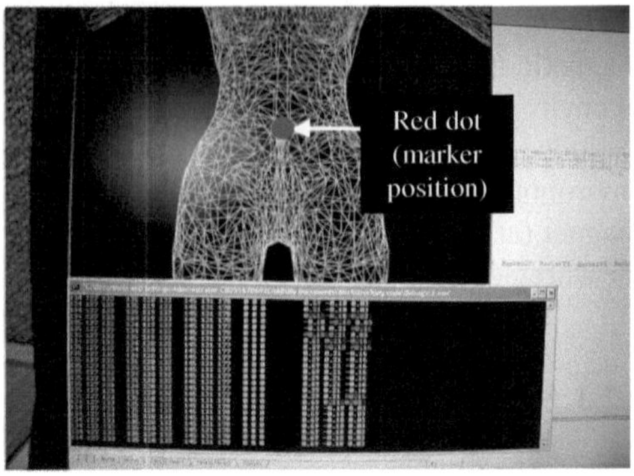

Fig. 7 Graphical User Interface: field readings, magnet coordinates and location on body mesh

the body mesh also allows us to rotate the body 360° around the X and Y axes. The Graphical User Interface is shown in Fig. 7.

5 Experimental Results

5.1 Test for Non Linearity Error

The aim of this test is to determine the percentage of error caused by the non linearity of the magnetic field after tackling this problem by applying (2). For this experiment we used the actual relationship between the magnetic field and distance as was depicted in Fig. 5. A Microsoft Excel program was then used to determine the values of the constants a and b for (2) in order to achieve a best fit with the actual relationship. Finally the amount of error was calculated by comparing the two curves. The procedure was repeated for three different types and sizes of magnets: neodymium 10 mm × 10 mm, ferrite 10 mm × 10 mm and neodymium 15 mm × 10 mm. The two curves (actual and using (2) for the 10 mm × 10 mm Neodymium magnet) are shown in Fig. 8. There is a very close match between the two curves. The remaining two magnets have shown almost identical behavior. Maximum error for each of the three cases was 0.5 cm or 5%. Average error was about half this value.

5.2 Away from Axis Test

This test aims at measuring the variation in error as the magnet moves away from the sensor's axis and/or rotates. The experimental setup is illustrated using Fig. 9a. The magnet was first placed on the sensor's axis, at a distance (D_S) of 5 cm from the centre point and the field was measured at 3 different orientations of the magnet: 0°, 45° and 90°, as shown in Fig. 9b. The magnet was then moved away from the axis

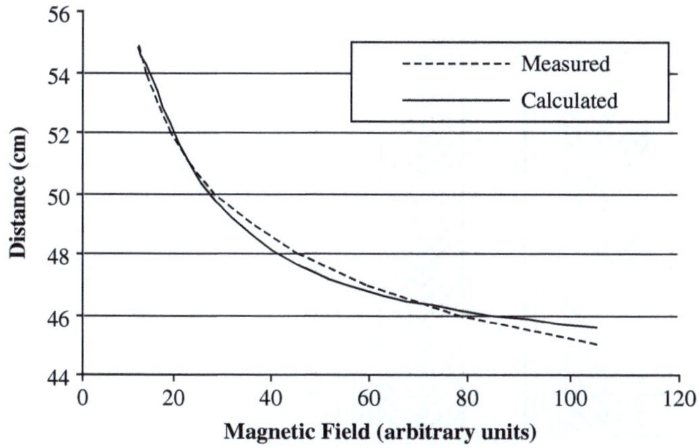

Fig. 8 Actual and modeled curves of field versus distance

to distances (D_a) of 2.5 cm and 5 cm, and the field measurements were taken for all three orientations (0°, 45° and 90°) at each of these positions. The entire procedure was repeated three times for three different distances D_s (5, 0, −5) between the magnet and the centre point.

The coordinates produced by the tracking system when the magnet is at an actual distance (D_s) of 5 cm from the reference point are shown in Fig. 10. Clearly, the accuracy of the system drops as the magnet moves away from the sensor's axis and also rotates. In the above figure, the error reaches its maximum value of around 2 cm (20%) at the maximum distance (5 cm) from the axis when the magnet is at 90° to the axis. By observing the rest of the results it was noted that out of the three worst case situations (all at 90°) at distances of −5, 0 and 5 cm,

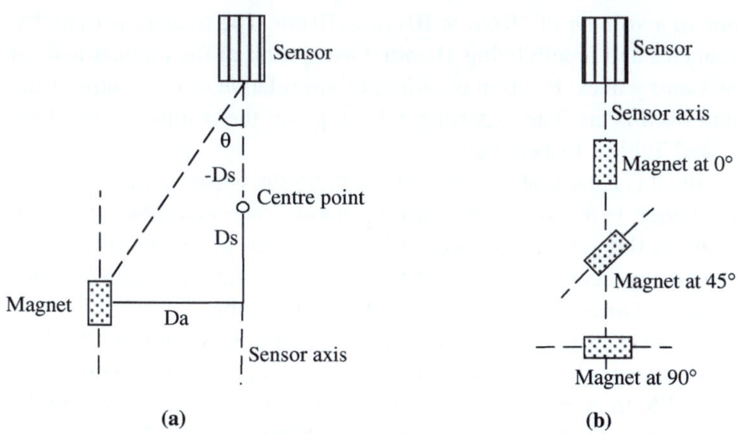

Fig. 9 Experimental setup for the away from axis test

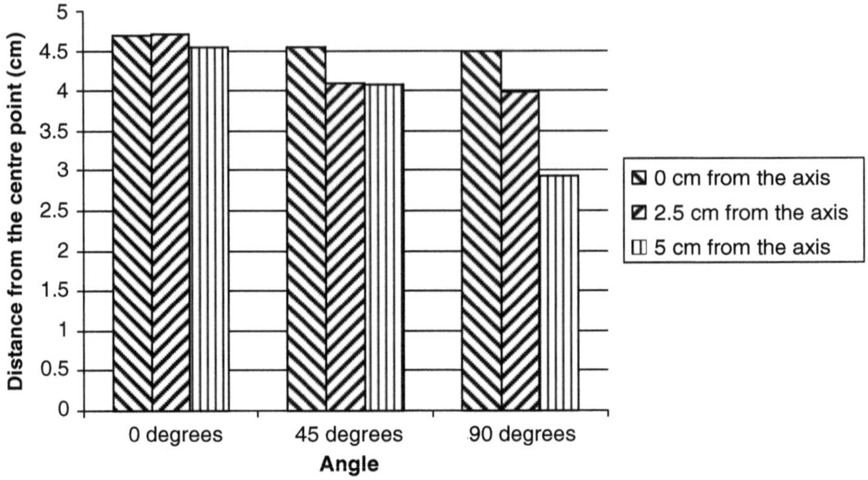

Fig. 10 Coordinates calculated by the single-sensor tracking system when the magnet is 5 cm away from the center point

the one closest to the sensor (5 cm) was the most inaccurate. This clearly suggests that the percentage of error increases with the angle (θ) the magnet forms with the sensor's axis.

5.3 Multiple-Sensor System Test

The aim of this test is to examine the accuracy of the multiple-sensor tracking system. Previous tests used only one sensor and considered only one axis. The tracking in this test is done in a volume of 10 cm × 10 cm × 10 cm. The magnet is cylindrical, with both diameter and length being 10 mm. Orientation of the magnet is 0° at z axis and 90° at x and y axes. Position coordinates are relative to the centre of the test cube. Medium used is air. The experimental setup and the results obtained are shown in Fig. 11 and Table 1 respectively.

From these results it is clear that the error of the multiple-sensor tracking system is, in most cases, under 1 cm. The worst case scenario however pushes the error to over 3 cm, meaning the system has over 30% inaccuracy under the worst case conditions. The worst case scenario occurs when the magnet is at maximum distance from all three monitored axes, i.e. in a corner of the testing cube.

In [9], test results were presented on a tracking system consisting of eight flux-gate sensors placed on a single plane. The system was tested only for one position of the magnet (x = 2.5 cm, y = 3.0 cm and z = 6.7 cm with respect to the middle sensor (5) and different orientations (angles) in the X-Y plane only. The results indicate that the error was just under 1 cm. In contrast, the results we have presented are for different physical positions of the magnet with respect to the centre point. The orientation of our magnet has always been the worst case (90°) for two axes.

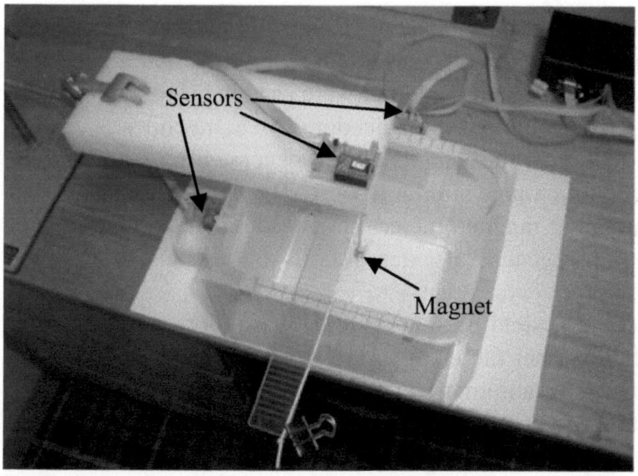

Fig. 11 Testing the multiple-sensor tracking system

Table 1 Test results of the multiple-sensor tracking system

Actual Position (cm)			Displayed Position (cm)			Error (%)		
x	*y*	*z*	*x*	*y*	*z*	*x*	*y*	*z*
0	0	0	0.29	0.21	0.03	2.9	2.1	0.3
−3	0	0	−2.89	0.21	0.3	1.1	2.1	3
−5	0	0	−4.3	−0.22	0.39	7	2.2	3.9
3	0	0	3.67	−0.22	0.8	6.7	2.2	8
5	0	0	4.83	−1.15	0.8	1.7	11.5	8
0	−3	0	0.29	−2.88	0.6	2.9	1.2	6
0	−5	0	−0.14	−4.23	0.85	1.4	7.7	8.5
0	5	0	−0.61	4.78	0.39	6.1	2.2	3.9
0	0	3	−0.54	−0.67	3.18	5.4	6.7	1.8
0	0	−5	−1.39	−1.49	−4.11	13.9	14.9	8.9
0	3	−2	−0.21	3.06	−2.01	2.1	0.6	0.1
4	0	−2	4.03	−0.75	−1.34	0.3	7.5	6.6
4	−3	−2	3.77	−2.92	−0.82	2.3	0.8	11.8
−4	−3	−2	−3.52	−3.46	−0.96	4.8	4.6	10.4
−5	−5	5	−4.82	−4.73	4.34	1.8	2.7	6.6
5	5	5	2.4	1.96	4.79	26	30.4	2.1

The volume of the magnet used in our experiment was 50% smaller than in [9]. Using magnets of the same size would allow us to accurately cover a much greater volume. Our tracking system uses only three sensors providing a much simpler and more cost effective design. In [9] the experiments were conducted in a sound isolated room and measured the ambient field at the beginning and end of the experiment for offline processing and tracking. Thus, real time tracking of the magnet was not considered.

Our tracking system also compares favorably with the one presented in [7]. The latter reported a mean resolution of around 2 mm and a maximum error of less than

1 cm. However, the complexity and cost associated with the hardware are much higher than those of our tracking system. External auxiliary coils and a spherical magnet placed inside a specially designed container were used in [7] to maintain the same orientation of the magnet throughout the procedure. An extra coil was used in front of the magnetometer to cancel the magnetic field generated by the auxiliary coils. The system was tested in the range similar to ours. The magnetic sensor used in [7] is an expensive commercial magnetometer (HMR2300). The system cannot be used for mobile applications because of external coils and associated complex hardware assembly.

The accuracy of the tracking system presented in [5] was reported to be within a few millimeters in simulations, but the error in the worst case scenario was not reported. There is no provision for automatic cancellation of environmental fields. It is necessary to recalibrate the system (sensors) before each measurement. In addition the tracking system in [5] uses 16 sensors and flux concentrators to improve sensitivity. Our system has used only three sensors, and no flux concentrator and associated circuitry. Therefore the system proposed here is much less complex and will compare favorably in terms of cost.

If we consider factors such are accuracy, range, cost, operating environment and mobility of the patient throughout the procedure, our tracking system proves to be very competitive compared to the magnetic tracking systems reported so far. Further improvements discussed later in the paper, have the potential of significantly improving the performance of our system while keeping the low level of complexity and low cost.

5.4 Attenuation Test

The aim of this test was to investigate the possibility of attenuation of magnetic field as it travels through a media emulating human tissue. The procedure used was identical to that used for the Multiple-Sensor System Test explained earlier with only one difference, namely the media through which the magnetic field was travelling. In this experiment, the magnet was tracked through 2 different media, water and NaCl + KCl solution. The two media were used in order to emulate the electrical properties of human tissue, as 80% of human body consists of water and also incorporates the above mentioned salts. The salts were used in the following concentrations: NaCl – 120mM, KCl – 4.7 mM.

The results indicated that the magnetic field did not experience any attenuation as it travelled through the emulated human tissue. Both parts of the experiment yielded results which compare closely to results of the previous test where air was used as the medium. Worst case scenario occurred at the same location (5, 5, 5) for all the three media in which tests were conducted (air, water and salt solution). Minor discrepancies are most likely caused by slight inconsistencies in experimental procedure rather than the attenuation.

6 Conclusions

The tracking system using multiple magnetic sensors proposed in this paper to determine the position of a magnetically marked capsule in real time is feasible and rational. The system's immunity to attenuation when operating in human body conditions makes it superior compared to RF systems. In contrast with many magnetic sensing systems reported to date the proposed system does not require complex coil assembly or array of a large number of magnetic sensors. Ability to calculate the position of the magnet regardless of its orientation eliminates the need for complex external coils, allowing the patient to be ambulatory during the procedure. To enhance the mobility further work is in progress for the measurement of the environmental field in real time using an additional sensor and using this to adjust the field values read by the tracking sensors in real time.

The results obtained from experiments have confirmed that the system is capable of tracking a magnet with relatively low error for most positions. Further reduction in error will be possible by improving the empirical model and the corresponding algorithm, particularly for the worst case scenario when the magnet is away from all the orthogonal axes. The accuracy and range of the tracking system will be further enhanced by increasing the resolution of the system using higher resolution A/D converters.

References

1. Shah T, Aziz SM, Vaithianathan T (2006) Development of a Tracking Algorithm for an In-Vivo RF Capsule Prototype. In: Proc 4th International Conference on Electrical & Computer Engineering, Dhaka, pp 173–176
2. Chan Y, Meng MQH, Wang X (2005) Experimental Study of Radiation Efficiency from an Ingested Source inside a Human Body Model. In: Proc 2005 IEEE Engineering in Medicine and Biology 27th Annual Conference, Shanghai, pp 1–4
3. Weitschies W, Wedemeyer J, Stehr R, Trahms L (1994) Magnetic Markers as a Noninvasive Tool to Monitor Gastrointestinal Transit. IEEE Trans Biomedical Eng, 41(2):192–195
4. Intestinal Ischemia System (Online, accessed October 2006) Model 637 Tristan Technologies, Inc. URL: http://www.tristantech.com
5. Schlageter V, Besse PA, Popovic RS, Kucera P (2001) Tracking system with five degrees of freedom using a 2D-array of Hall sensors and a permanent magnet. Sens Act A, 92:37–42
6. Stathopoulos E, Schlageter V, Meyrat B, De Ribaupierre Y, Kucera P (2005) Magnetic pill tracking: a novel non-invasive tool for investigation of human digestive motility. Neurogastroenterol Motil, 17:148–154
7. Andra W, Danan H, Walter K, Kramer H, Saupe P, Schmieg R, Bellemann ME (2000) A novel method for real-time magnetic marker monitoring in the gastrointestinal tract. Phys Med Biol, 45:3081–3093
8. Wang X, Meng MQH, Chan Y (2004) A Low-Cost Tracking Method Based on Magnetic Marker for Capsule Endoscope. In: Proc 2004 International Conference on Information Acquisition China, pp 525–526
9. Prakash N, Spelman FA (1997) Localization of a Magnetic Marker for GI Motility Studies. An In Vitro Feasibility Study. In: Proc 19th International Conference - IEEE/EMBS, USA, pp 2394–2397

10. Khan IJ, Buyer CO, Dumoulin CL, Darrow RD (2004) Magnetic Resonance System with Multiple Independent Tracking Coils. United States Patent US2004/0171934 A1
11. Peterson TH, Levine L, Anderson PT (2006) Operating Table with Embedded Tracking Technology. United States Patent US2006/0025668 A1
12. Macaulay P, Nayak A (2006) In Vivo Localization and Tracking of Tissue Penetrating Catheters using Magnetic Resonance Imaging. United States Patent US2006/0184011 A1
13. Cleary K, Banovac F, Wood B, Levy E (2003) Image Guided Liver Interventions Based on Magnetic Tracking of Internal Organ Motion. United States Patent US 2003/0220557 A1
14. Kotitz R, Matz H, Trahms L, Koch H, Weitschies W, Rheinlander T, Semmler W, Bunte T (1997) SQUID Based Remanence Measurements for Immunoassays. IEEE Transactions on Applied Superconductivity, 7(2):3678–3681
15. Enpuku K, Inoue K, Soejima K, Yoshinaga K, Kuma H, Hamasaki N (2005) Magnetic Immunoassays Utilizing Magnetic Markers and a High-T/sub c/ SQUID. IEEE Transactions on Applied Superconductivity, 15(2):660–663
16. Schotter J, Kamp PB, Becker A, Puhler A, Brinkmann D, Schepper W, Bruckl H, Reiss G (2002) A Biochip Based on Magnetoresistive Sensors. IEEE Trans on Magnetics, 38(5):3365–3367
17. Grcic M, Aziz SM, Vaithianathan T (2007) Development of an Adaptive Tracking System for an Endoscopic Capsule. In: Proc 2nd International Conference on Sensing Technology, New Zealand, pp 50–55
18. Honeywell Sensor Products (Online, accessed February 2008) HMC2003 Three-Axis Magnetic Sensor Hybrid- Data Sheet. URL: http://www.magneticsensors.com
19. Honeywell Sensor Products (Online, accessed March 2006) AN201 Set/Reset pulse circuits for magnetic sensors- Application Notes. URL: http://www.magneticsensors.com
20. Burnett GC (Online, accessed 31 June 2006) The Physiological Basis of Glottal Electromagnetic Micropower Sensors (GEMS) and Their Use in Defining an Excitation Function for the Human Vocal Tract. Department of Applied Science, University of California, 1999. URL: http://speech.llnl.gov/thesis/3_2.htm
21. Magnet Library (Online, accessed 1 May 2006) Magnet Power. 2001. URL: http://www.magnet4u.com/eng/new_lib/p1_02.htm

Part VI
Ultrasonic

High Frequency Ultrasonic Wave Propagation in Porous Aluminium Samples

A. Dawson, P. Harris and G. Gouws

Abstract The influence that microstructure has on high frequency (30–70 MHz) ultrasonic wave propagation is being investigated. In this work the ultrasound system, wave propagation simulation results and experimental results for highly structured porous aluminium samples are presented. Finite element modeling shows that a single pseudo S_0 lamb wave propagates in the porous aluminium with a velocity of 5.18 km/s. Experimentally a velocity of 5.10 km/s was measured. For angled measurements the porous aluminium samples were observed to normalise the incident waveforms allowing significant energy to be transmitted beyond angles of incidence of 40°, well past the solid aluminium P and S critical angles of 14° and 29°.

Keywords Ultrasonic · collinear microstructure · wave propagation

1 Introduction

Microstructure in materials is often only considered as a source of scattering for incident ultrasound waves. However, a fibrous or porous microstructure can have a profound influence on high frequency ultrasound propagation. This influence can also occur in materials that are considered homogeneous such as plastics [4]. In ultrasonic non-destructive testing (NDT) highly ordered microstructure may extend over large areas resulting in unsuspected guided wave modes. An example of such a material is aluminium that can oxidise at its interfaces causing a regular porous structure to develop. This can be a significant problem when welding. However, this

A. Dawson
School of Chemical & Physical Sciences, Victoria University of Wellington, P.O. Box 600, Wellington, New Zealand, e-mail: dawsonand@hotmail.com

P. Harris
Industrial Research Ltd., P.O. Box 31-310, Lower Hutt, New Zealand, e-mail: P.Harris@irl.cri.nz

G. Gouws
School of Chemical & Physical Sciences, Victoria University of Wellington, P.O. Box 600, Wellington, New Zealand, e-mail: Gideon.Gouws@vuw.ac.nz

is also the property used in anodising, historically in making humidity sensors and even in the formation of some nanostructures.

In this work broadband high frequency (30–70 MHz) longitudinal probing waves were chosen to examine collinear porous aluminium samples since these are significantly influenced by the sample microstructure [2]. In order to carry out the investigation the construction of a suitable ultrasound system for material characterisation was completed. Finite element modelling of wave propagation in immersed porous aluminium was performed and confirmed by experiment on two types of porous aluminium samples.

2 Experimental System

Frequency is a key factor determining system resolution. Increasing transducer frequency results in increased resolution, but with greater attenuation and increased transducer costs. As a trade off a focused 50 MHz immersion transducer (Panametrics V390) was chosen that possessed a beam diameter of 100 um at 50 MHz and a focal zone of 1 mm. It is a broadband device with 100% bandwidth for good axial resolution. A Precision Acoustics 75 um needle hydrophone was used as the sensor.

Figure 1 shows the experimental setup. A burst excitation waveform was computed on a PC and transmitted to the function generator via Ethernet. The excitation was amplified by the transmitter circuit driving the transducer using transformer coupling. The transducer could be positioned using micrometers, and the angle of incidence was adjustable from normal to 90°. A receiver circuit amplified and time gain compensated (when enabled) the signals from the needle hydrophone outputting the resultant waveforms to the oscilloscope. The sampled signals were finally downloaded to the controlling PC and signal processing performed.

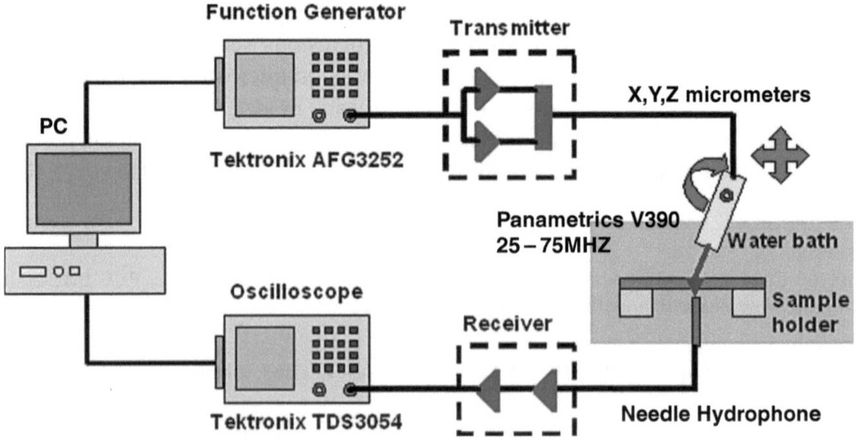

Fig. 1 Experimental setup of high frequency ultrasound system

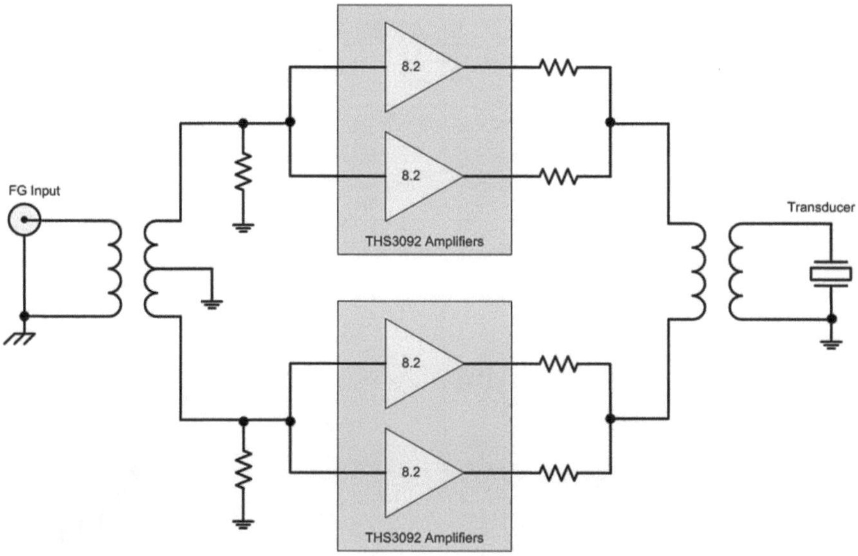

Fig. 2 Transmitter circuit schematic

The transmitter circuit [3] shown in Fig. 2 is a bridge circuit with an input matched to the excitation impedance, and a 2:1 output transformer to match to the low transducer impedance. For an excitation burst amplitude of $2.5V_p$ and op-amp gain of −8.2 an output power of 1W is delivered to the transducer. The frequency response of the circuit is shown in Fig. 3. With a source power of −45 dBm a power gain of 12 dB is observed with a 3 dB bandwidth of 100 MHz to accommodate the broad bandwidth of the transducer.

The receiver circuit [3] schematic is shown in Fig. 4. Two AD8337 variable gain amplifiers (VGA's) are used in series forming the front end of the circuit. The VGA's possess very low noise, wide bandwidth and a linear-in-dB gain i.e. a ramp with a slope equal to the attenuation (dB/mm) achieves signal compensation. An adjustable DC input to an op-amp integrator, with an analog switch for integrator reset, provides the ramp generator. The function generator sync signal is used to control the reset function. The VGA's consist of a configurable pre-amp, here set for a gain of 3, followed by a 24 dB attenuator stage (controlled via the gain pin), and a fixed 18 dB gain amplifier. The output cable driver used to drive the transmission cable to the oscilloscope is configured for a gain of 3. This results in a variable gain range for the circuit from 17 to 65 dB. The summing blocks seen in Fig. 4 control the VGA's gain. They are configured so that once the first VGA reaches its maximum gain, the second VGA gain begins to increase. In this way both amplifiers can be dedicated to DC gain if no time-gain compensation is required, all time-gain compensation, or a combination of the two [3]. The receiver circuit frequency response from minimum to maximum gain is shown in Fig. 5.

Each of the traces displays an excellent frequency response satisfying the transducer requirements up to 100 MHz with a steep roll off at higher frequencies. The

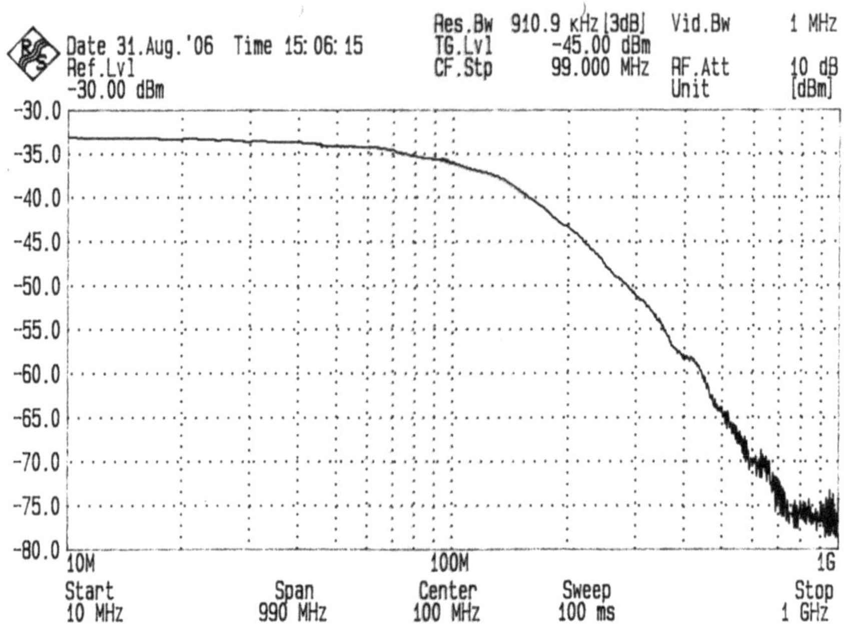

Fig. 3 Frequency response of transmitter output power delivered to the transducer

noise characteristic is analysed in Fig. 6. The lower trace shows the noise of the spectrum analyser, and the higher trace shows the noise of the receiver circuit plus the spectrum analyzer noise when configured for maximum gain. Based on the difference between the traces the noise of the receiver at maximum gain is approximately −36 dBm equating to ~2nV/√Hz i.e. the receiver noise is dominated by the VGA input noise.

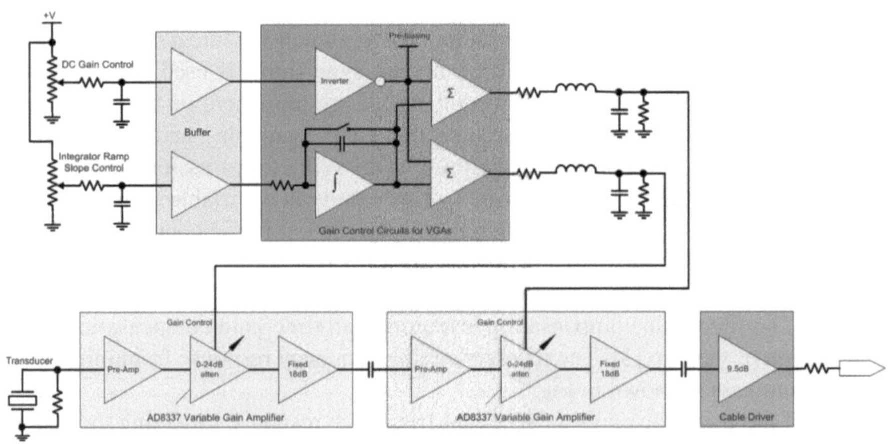

Fig. 4 Receiver circuit schematic

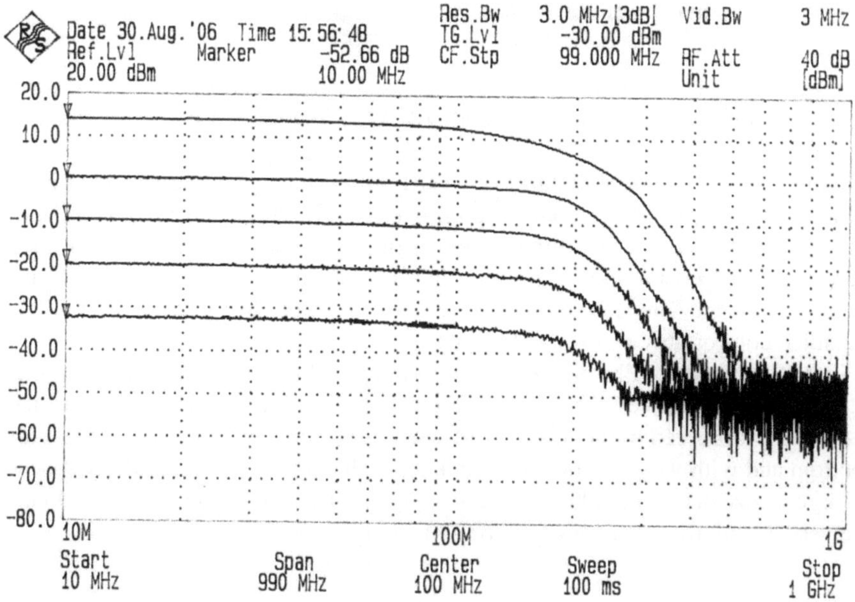

Fig. 5 Frequency response of receiver output power for five different gains

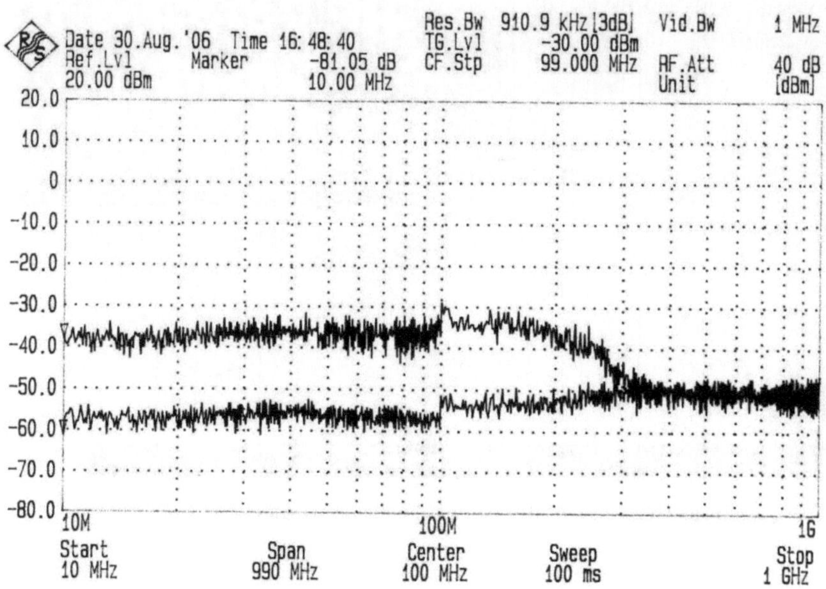

Fig. 6 Receiver circuit and spectrum analyzer noise traces

3 Porous Aluminium Samples

Two types of porous aluminium samples were used in the experiments. One type was prepared by Industrial Research Limited (IRL) [5], the second type was obtained commercially.

Figure 7 shows the SEM pictures of an IRL sample snapped to reveal the highly collinear (pores from top to bottom) porous structure. The insert picture shows a plan view of the sample highlighting the regular pore spacing. The 30 nm sized pores are spaced 60 nm apart. The thickness of the sample used was 160 um and 12 mm in diameter. The IRL samples were found to be relatively robust and able to be handled comfortably using standard tweezers.

Figure 8 shows the SEM pictures of the commercial sample. The main SEM picture shows these samples are relatively poorly collinear with many irregularities. This is easily observed in the insert SEM with many pores not connecting from top to bottom and with varying pore wall thickness. The pores are not regularly spaced and large in size at around 200 nm. These samples were only 60 um thick and found to be very brittle, fracturing easily during handling.

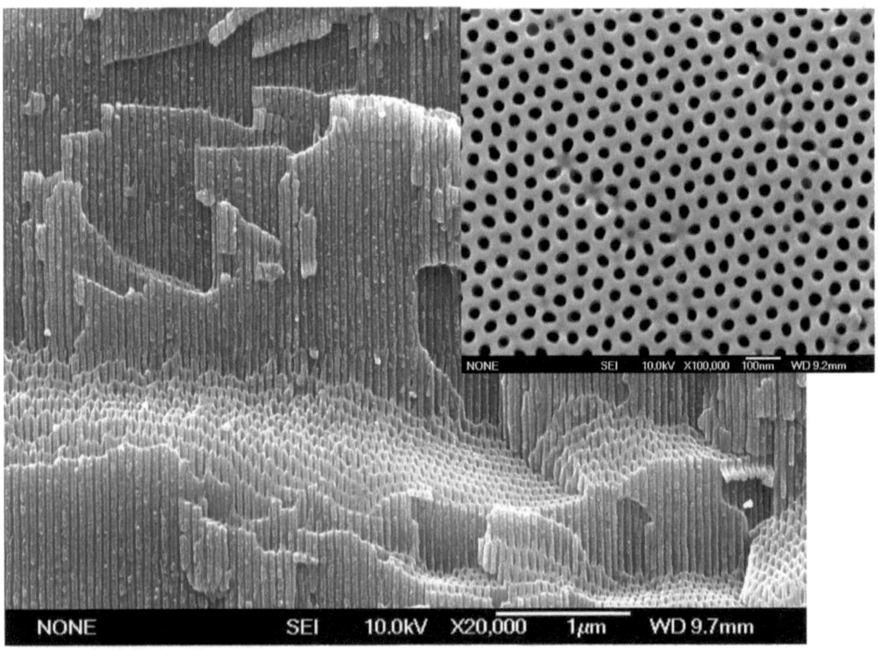

Fig. 7 Cross section and plan view of IRL porous aluminium samples

Fig. 8 Cross section and plan view of commercial porous aluminium samples

4 Finite Element Modeling

Finite element analysis (using PZFlex) was used to model wave propagation in the IRL sample microstructure. The simulation was performed in 2D with a material model comprising of 1) an axially aligned water filled porous layer containing 1000 pores representing the aluminium oxide, and 2) a homogenous block of solid aluminium. The dimensions and material properties are shown in Fig. 9. The FEA mesh size is usually calculated using the minimum velocity in the simulation i.e. the water velocity 1500 m/s, divided by the maximum possible frequency (~100 MHz) resulting in a spatial resolution of 15 um. Here, a mesh size of 15 nm ($1/1000\lambda_{H2O}$) was used in order to resolve the 30 nm pores. The temporal resolution was calculated using the spatial resolution of 15 nm divided by the maximum velocity i.e. the P wave speed in aluminium 6 km/s, resulting in a sampling interval requirement of 2.5 ps; we used 2 ps.

The excitation is a 50 MHz Blackman pulse applied over the width of a pillar at the bottom left corner of the aluminium oxide. The bottom boundary in the model is symmetrical, the left boundary free, and both the top and right boundaries absorbing.

The modeling results can be observed in Fig. 10, which shows three simulation snapshots (top) 20 ns, (center) 40 ns and (bottom) 60 ns after the excitation pulse was injected. Large concentrations of energy are represented by darker shades. During

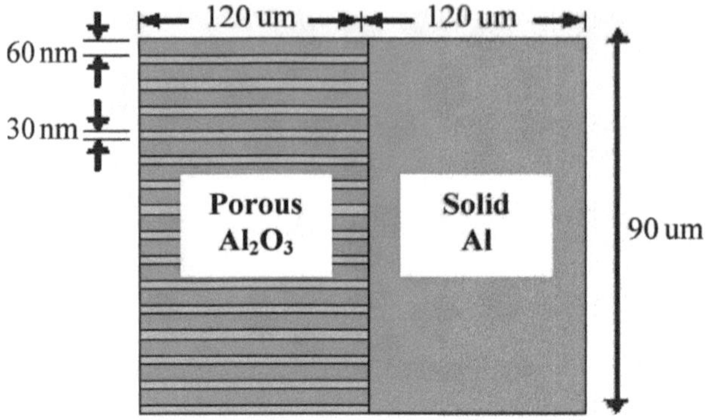

Material	Density (g/cm³)	V_P (km/s)	V_S (km/s)
Aluminium	3.0	6.0	3.0
Water	1.0	1.5	-

Fig. 9 Structure and dimensions of the simulation

the simulation it is clear from the top and center snapshots that the wave travels as a single compact wavefront in the porous aluminium being guided by the material pores. This compact wavefront, a lamb wave (or plate wave), propagates when the insonifying wavelength is comparable to the thickness of the plates or pore walls in which it propagates [1, 6]. Lamb wave velocity can vary significantly as various modes are possible. Here the wave propagates as a pseudo lamb mode with a simulated velocity of 5.18 km/s. The frequency-thickness product for the aluminium oxide suggests the wave travels as a fundamental symmetric mode (S_0) with a theoretical velocity given by

$$V_{SO} = 2V_S\left(1 - \frac{V_S^2}{V_L^2}\right)^{\frac{1}{2}} \tag{1}$$

Using (1) a theoretical V_{S0} of 5.20 km/s is calculated. An S_0 lamb mode is predominantly a longitudinal displacement and is therefore little affected by the presence of the water. Whilst in the porous aluminium the bulk of the energy is observed to be concentrated to a small region of pores as the material acts as a waveguide. However, the close proximity of adjoining pillars facilitates the transfer of some energy which forms an associated wake as can be observed in the top and center snapshots. Upon reaching the solid aluminium interface, the compact wavefront ceases to exist as, no longer guided by the material pores, the P and S components are free to travel at their respective velocities. This is shown in the bottom snapshot of Fig. 10, where the individual waves spread spherically away from the material interface.

The simulation was repeated with pore sizes of 200 and 100 nm spacing, approximating the dimensions of the commercial samples. The results observed were

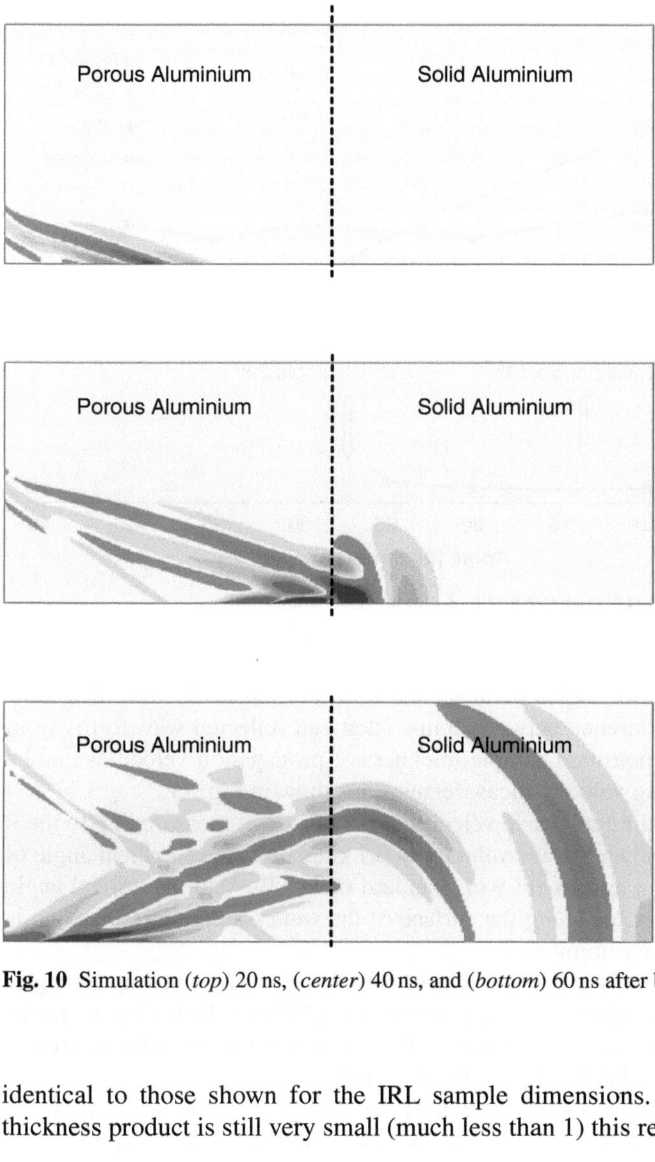

Fig. 10 Simulation (*top*) 20 ns, (*center*) 40 ns, and (*bottom*) 60 ns after beginning

identical to those shown for the IRL sample dimensions. Since the frequency-thickness product is still very small (much less than 1) this result is expected.

5 Experimental Results

Experimental work has been performed on both porous aluminium samples as well as solid aluminium samples to confirm the results obtained from the simulation. With the transducer and hydrophone vertically aligned and immersed in water, samples of known thicknesses were inserted between them (see Fig. 1). Measurements

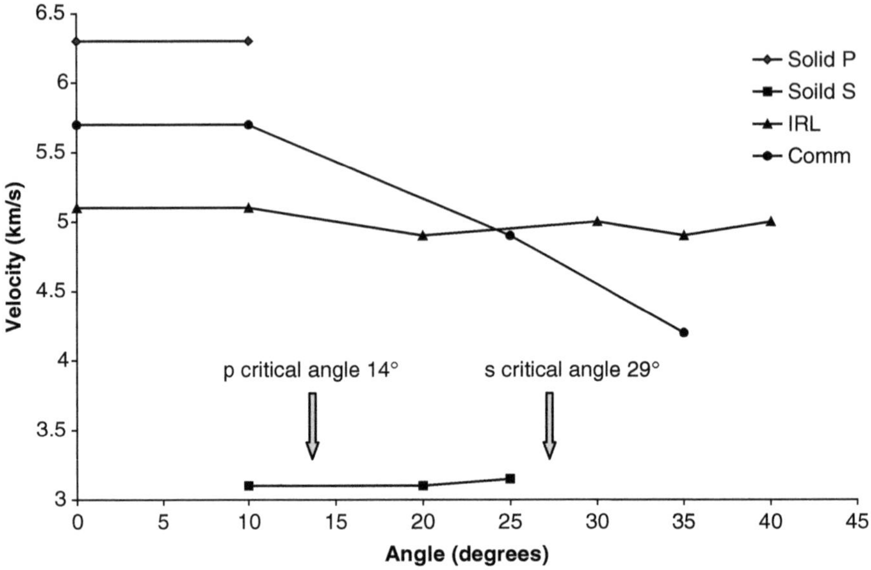

Fig. 11 Wave velocity versus transducer angle of incidence for the samples

were made for transducer angles of incidence from normal to 40° for each sample. Using the time difference between transmitted and reflected waveforms from the oscilloscope, and measured sample thicknesses, propagation velocities can be calculated. The resulting velocity measurements are show in Fig. 11.

For the solid aluminium a P wave velocity of 6.3 km/s was measured up to the P critical angle of 14°, and an S wave velocity of 3.1 km/s up to the S critical angle of 29°. These values are in agreement with standard tables. Beyond the critical angle the transmitted wave travels along the surface of the water/sample interface and is not detected in these measurements.

For the IRL sample only a single compact waveform was received up to angles greater than 40° with an almost constant velocity of 5.10 km/s. This velocity agrees well with those determined in the simulation and calculated theoretically confirming the propagation of an S_0 lamb wave in the IRL sample.

The commercial sample results are considerably different. Only a single wavefront was received, however the velocity for low angles at 5.7 km/s drops significantly with angle. Given the poor co-linearity this variation in velocity could be expected as the velocity will depend on the wave path traveled.

Figure 12 shows the influence angle has on the amplitude measured. In the solid aluminium the P wave shows a decrease in amplitude out to the P critical angle, whereas the S wave amplitude stays relatively constant out to its critical angle. The IRL sample wave amplitude initially decreases and then is constant out to 40°; the amplitude seems to follow that of the P and S amplitudes of the solid aluminium. This is not the case for the commercial sample as its amplitude is constant, seemingly tracking the shear component only.

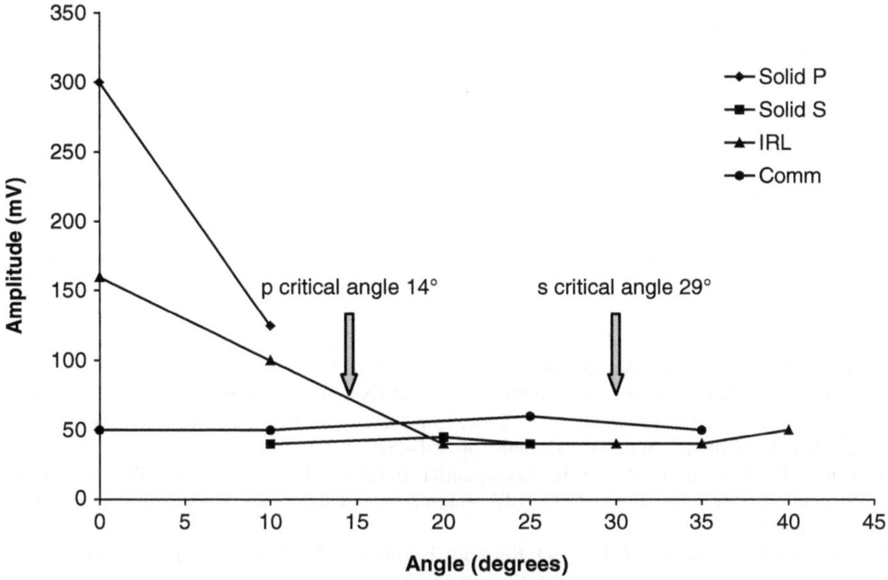

Fig. 12 Wave amplitude versus transducer angle of incidence for the samples

It should be noted that the significant amplitudes measured for the IRL and commercial samples beyond the second critical angle potentially convey important information.

6 Conclusions

Simulation confirmed by experimental work shows that high frequency ultrasonic waves in highly collinear porous aluminium samples propagate as a single spatially compact wavefront. The microstructure imposes a pseudo S_0 Lamb mode, without individual P and S modes and without the refraction (with critical angles of 14° and 29°) of the aluminium solid. The 5.10 km/s velocity measured experimentally for the IRL sample is in excellent agreement with the theoretical lamb wave velocity of 5.20 km/s and the velocity of 5.18 km/s determined in the simulation.

The IRL porous aluminium sample experimental results show significant energy is transmitted for angles from normal to 40° with nearly constant velocity. The commercial porous samples transmit energy to 40° however the velocity falls notably with increasing angle.

These results may have significant benefit in NDT contact measurements since a collinear microstructured interface layer could be used to ensure normal sample illumination and normal transducer reception.

Acknowledgements We wish to acknowledge the efforts of Dr Alexander Kirchner and Dr Ian Brown of IRL in fabricating the IRL aluminium samples, and the Foundation for Science, Research and Technology NZ for funding this work.

References

1. Cheeke D (2002) Fundamentals and Applications of Ultrasonic Waves, CRC Press, Boca Raton
2. Dawson A, Harris P, Gouws G (2007) High Frequency Ultrasound Wave Propagation in Anisotropic Materials. In: Proc. 2nd International Conference on Sensor Technology 2007 – ICST'2007, Palmerston North, New Zealand, pp 552–556
3. Dawson A, Harris P, Gouws G (2006) Design and Evaluation of Ultrasonic Transducer Circuits for Material Characterisation. In: Proc. 13th Electronics New Zealand Conference 2006 – ENZCon06, Christchurch, New Zealand, pp 261–265
4. Harris P, Dawson A, Young R, Lecarpentier F (2007) High Frequency Propagation in Structured Solids. In: Proc. 2007 IEEE International Ultrasonics Symposium, N.Y, USA, pp 690–693
5. Kirchner A, MacKenzie K, Brown I, Kemmitt T, Bowden M (2007) Structural characterization of heat-treated anodic alumina membranes prepared using a simplified fabrication process. J Membrane Sci 287:264–270
6. Rose JL (1999) Ultrasonic Waves in Solid Media, Cambridge University Press, Cambridge

Part VII
Capacitive Sensors

Capacitance-Based Sensing of Material Moisture in Bulk Solids: Applications and Restrictions

Anton Fuchs, Hubert Zangl and Gert Holler

Abstract This paper presents applications of capacitive techniques for moisture sensing of bulk solids. Based on experimental measurement data for three test materials, theoretical considerations on moisture models in bulk solids are undertaken and the reliability of these models is discussed. The sensitivity of the capacitive setup on the moisture distribution inside the particles is analyzed by means of a multilayer model and Finite Element Analysis methods. A planar sensor topology featuring spatial resolution is proposed and the cross-sensitivity on material compression as well as the influence of the inter-electrode distance are investigated. Strengths and weaknesses of capacitive moisture sensing are discussed.

Keywords Moisture measurement · capacitive sensing · bulk solids · planar setup

1 Introduction

Having good and reliable instrumentation for the determination of moisture content in gases and solid material is of importance for many applications in industry. For example, moisture content may have an effect on corrosion and decomposition of material, affects discharging behavior for solids, or may improve/worsen the feasibility of industrial transportation processes and their measurement [6].

Pneumatic conveying of granular material has become a frequently used method to transport granular material in pipelines [20]. In various industrial applications,

Anton Fuchs
Institute of Electrical Measurement and Measurement Signal Processing, Graz University of Technology, Graz, Austria, e-mail: anton.fuchs@tugraz.at

Hubert Zangl
Institute of Electrical Measurement and Measurement Signal Processing, Graz University of Technology, Graz, Austria, e-mail: hubert.zangl@tugraz.at

Gert Holler
Institute of Electrical Measurement and Measurement Signal Processing, Graz University of Technology, Graz, Austria, e-mail: gert.holler@tugraz.at

such as supply of coal powder to furnaces, loading of harvest and fertilizer, or mixing of cement powder and sand, it is essential to monitor and rate the flow of material in conveyor pipes.

A capacitive flowmeter, which has been developed by our research group, allows for the non-invasive determination of the *absolute* velocity profile and the *relative* concentration profile of a gas-solids flow [21]. For a reliable mass flow measurement, as needed for many applications, calibration strategies are required to obtain an *absolute* concentration profile as well. The main challenge for the developed capacitance-based flowmeter is its susceptibility to moisture, since the permittivity of water differs significantly from that of most dry materials [9].

Different approaches on the measurement of the moisture content in granular materials and powders have been undertaken: A standard technique is usually based on oven drying [12], which is destructive and time consuming and requires taking representative samples of the transportation medium. Microwave spectroscopy has been found to be a suitable technique to determine the moisture content in granular material and agricultural goods [5, 9], especially for frequencies above 8 GHz [13]. Electrical Time Domain Reflectometry (ETDR) is used to measure the material moisture content by means of exploiting the time needed for an electromagnetic pulse to travel forth and back through a sensing probe [8]. The moisture content can also be determined by using neutron moisture gauges, which exploit the dependency of neutron parameters on the average hydrogen concentration [15]. Infrared and laser light absorption spectroscopy are applied for the measurement of the surface moisture content in various substances [1].

Capacitive sensing is a simple and cost-effective approach for the measurement of material moisture [4]. Sensors based on capacitive principles are in use in various embodiments [10, 19]. Based on observations made during static experiments with materials of different moisture contents, this paper describes restrictions and challenges that arise for capacitive moisture metering of bulk solids. Moisture models are analyzed and compared by means of Finite Element Analysis (FEA) to investigate the impact of moisture distribution in bulk solids onto the readout of a classical, plate-capacitor, setup. Additionally, a planar moisture sensor front-end with spatial resolution is proposed and strengths and weaknesses of capacitance-based moisture sensing for bulk solids are discussed.

2 Classical Measurement Setup

Capacitive sensing has gained increased importance in the last decades and is successfully employed in various applications in industrial and automotive technologies. Aside from the non-contacting principle, capacitive technology also allows for a low-cost implementation of sensors, especially since reliable low-price devices are available for higher frequencies (e.g. due to achievements for mobile telephone systems).

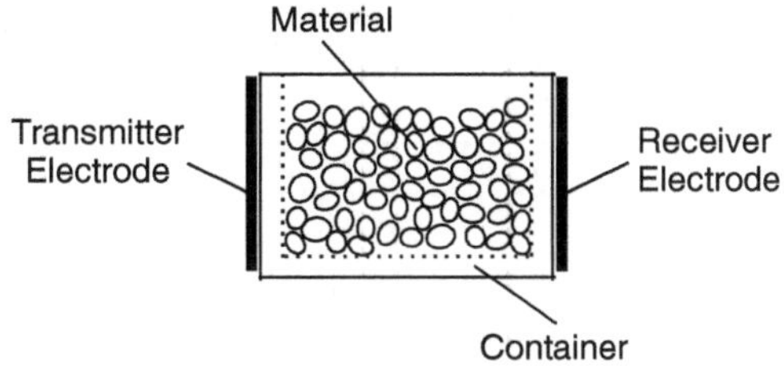

Fig. 1 Measurement setup with the filled container and the two electrodes on opposing box walls

Figure 1 shows a sketch of the measurement setup that was used for the practical experiments to determine the dependency of inter-electrode capacitance on material moisture. The dry test material is filled into a non-conductive container with a volume of 400 cm^3 and parallel side faces. Two electrodes are placed on opposing outer surfaces of the plastic box, allowing for a non-contacting measurement principle [3].

To minimize the impact of external noise, transmitter and receiver electrodes are implemented on two printed circuit boards on which the actual electrodes are surrounded by grounded surfaces and the back sides of the circuit boards are also on ground. A 40 MHz voltage signal is connected to the transmitter electrode while the receiver electrode is kept on ground, which yields the formation of a well-defined electric field inside the test material. The displacement current measured on the receiver electrode is proportional to the inter-plate capacitance and hence influenced by the material in the sensitive volume. The measurement of the displacement current is performed by using a highly sensitive circuitry with in-phase/quadrature (I/Q) channel measurement [18]. For all three test materials – fine wheat flour, sand, and gravel – the amount of added water is stepwise increased and the material is well stirred before every individual measurement to ensure approximate homogeneous moisture distribution. The material is wetted up to a moisture level where still no free water is present at the bottom layers of the material.

Figure 2 shows measurement results for the parallel-plate type capacitive moisture measurement of the three test materials. The raise of the signal amplitude of the inter-plate capacitance is quite linear as can be seen in the quality of the linear fit also shown in Fig. 2a,b,c (R-square values for the fit quality: flour 0.9852, sand 0.8882, gravel 0.9828).

Table 1 summarizes the impact of added water on the inter-plate capacitance and emphasizes that the incline of the capacitance depends on the particle size [2].

Measurement results reveal that for an estimation of the moisture content of bulk solids it might be sufficient to know the capacitance-moisture function for certain particle sizes and then measure the mean particle size. Estimation for the material moisture can then be obtained using interpolation. The problem for a capacitance-based moisture determination of bulk solids is the dependency on the moisture distribution in the particle itself, as will be shown in the following section.

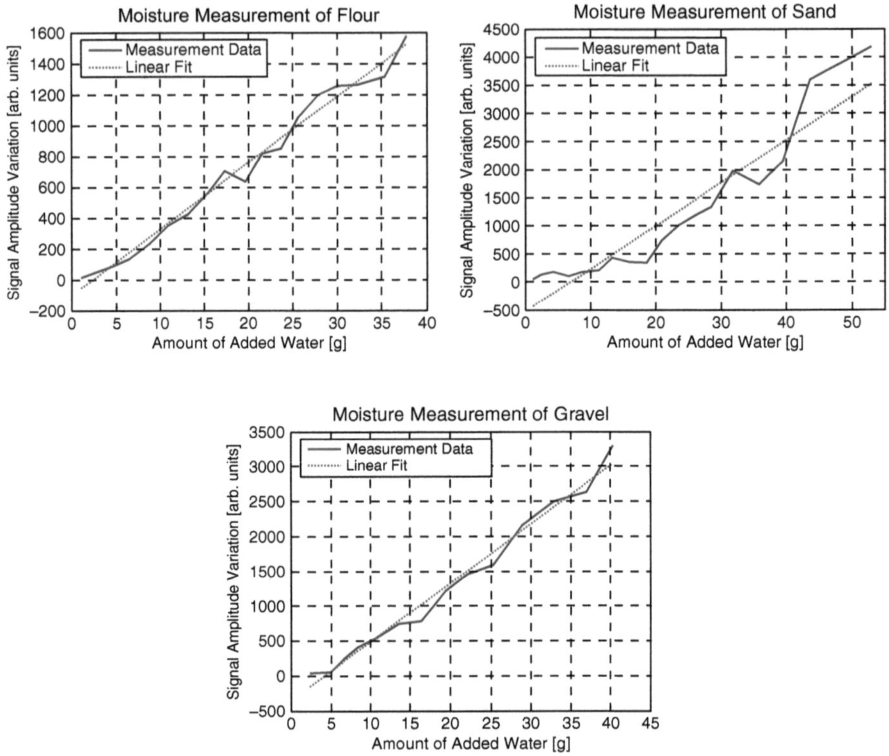

Fig. 2 Results and linear fit of the capacitive moisture measurement for (*top left*): Flour, (*top right*): Sand, and (*bottom*): Gravel; Experiments for test materials have been conducted with the same settings of the measurement unit

Table 1 Capacitance variation due to moisture

Test material	Appr. average particle size (mm)	Appr. signal variation
Flour	0.03	400 per 10 g water
Sand	2	625 per 10 g water
Gravel	15	800 per 10 g water

3 Impact of Moisture Distribution

Due to costly indoor storage space, large amounts of bulk materials have to be stored outside of buildings. The particles of such materials, especially materials with larger particle diameters, may then be dry inside and wet at surface layers (rain etc. on surfaces) or they may also be wet in the core and dry at their surface (e.g. fully wetted material that partly dried on the surface). The oven drying test [10] would yield the same moisture content although the distribution of the moisture is totally different in the particles.

The bulk material permittivity is affected by the volume fractions of bound and free water, but also by the geometries of the soil components and the shape of water inclusions. Water near solid surfaces is restricted in its rotational mobility, and therefore, the dielectric constant of bound water is lower than that of free water [16].

Four different moisture models are investigated by means of 2D-FEA to examine the use of such models in capacitive moisture sensing. For all models it is assumed that the permittivity of the dry material, ε_{mat}, is 4 and that a constant volume fraction, f, of 6.56% of each particle is wet. The permittivity of the wet media, ε_{wet}, is varied from 4 to 80 to simulate the moisture impact from bound water to free water.

Particles with a diameter of 3 mm are placed in-between transmitter and receiver electrodes. The electrodes are chamfered to avoid numerical inaccuracy due to sharp edge effects.

3.1 Maxwell-Garnett Model

The Maxwell-Garnett mixing formula is used to calculate the effective material permittivity ε_{mat} with randomly positioned inclusions ε_{wet} [17]

$$\varepsilon_{eff} = \varepsilon_{mat} + 3 \cdot f \cdot \varepsilon_{mat} \cdot \frac{\varepsilon_{wet} - \varepsilon_{mat}}{\varepsilon_{wet} + 2 \cdot \varepsilon_{mat} - f \cdot (\varepsilon_{wet} - \varepsilon_{mat})} \qquad (1)$$

The volume fraction f of the moisture inclusions is 6.56%. For an increase of the wet material permittivity from 4 to 80, increase of the effective permittivity ε_{eff} for the entire material from 4 to 4.72 can be derived from the Maxwell-Garnett model in (1).

3.2 Polder-van Santen Model

The Polder-van Santen formula, also named as the de Loor formula, is widely used in geophysics for the calculation of the real permittivity of a mixture from its components [11]

$$f \cdot \frac{\varepsilon_{wet} - \varepsilon_{eff}}{\varepsilon_{wet} + 2 \cdot \varepsilon_{eff}} + (1 - f) \cdot \frac{\varepsilon_{mat} - \varepsilon_{eff}}{\varepsilon_{mat} + 2 \cdot \varepsilon_{eff}} = 0 \qquad (2)$$

Also for this model, the volume fraction f is 6.56% and the wet material permittivity ε_{wet} is also varied from 4 to 80. According to (2) the effective permittivity for the Polder-van Santen moisture model ranges from $\varepsilon_{eff} = 4$ to 4.80.

3.3 Ring Moisture Model

If we assume that the moisture is not randomly distributed over the particle volume but concentrated close to the particle surface, a ring moisture model can be set up. For all particles in the computational domain the permittivity of the dry "core" material is set to 4 while the permittivity of the moisture ring is stepwise varied in the simulations. The 3 mm particles exhibit a 0.05 mm wet layer in the simulations for the ring moisture model (i.e. the volume fraction of the moisture inclusion is again 6.56%).

Figure 3 shows simulation results of the normalized electric field strength for particles with a dry surface layer (left) and with a wet surface layer (free water layer, right). The electric field strength maximum between particles for the wet case is more than twice the inter-particle field strength maximum in the dry case.

3.4 Equivalent Material Model

For the equivalent material model, the ring geometry of Sect. 3.3 is utilized but the moisture is distributed over the entire particle volume using additive superposition of the dry and wet regions. With r_1 as the inner radius of the wet layer and r_2 as the particle radius, the following equation for the effective equivalent material permittivity can be derived:

$$\varepsilon_{eff} = \varepsilon_{mat} \cdot \frac{r_1^2}{r_2^2} + \varepsilon_{wet} \cdot \left(1 - \frac{r_1^2}{r_2^2}\right) \tag{3}$$

For an increase of the wet material permittivity from 4 to 80, an effective permittivity ε_{eff} from 4 to 8.9822 can be calculated for the entire particle permittivity according to (3).

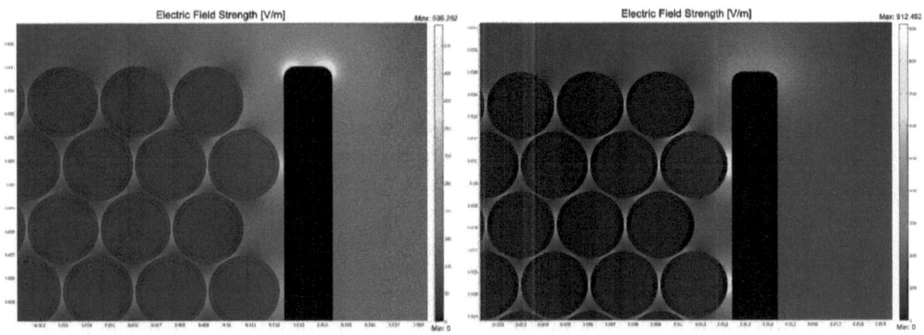

Fig. 3 FEA simulation results for the electric field strength of particles with (*left*): A dry surface layer and (*right*): A wet surface layer

3.5 Comparison of Moisture Models

Figure 4 summarizes simulation results for the inter-plate capacitance using the four moisture models. The permittivity ε_{wet} of a volume fraction of 6.56% is stepwise increased from 4 to 80 to simulate moisture impact from bound to free water. As can be seen, the models differ significantly for higher moisture permittivity and the distribution of the moisture strongly affects the inter-plate capacitance in such a setup (compare ring moisture model and equivalent material model).

3.6 Multi-Layer Moisture Model

Due to deviating trends of investigated models, a more reliable, detailed multilayer moisture model is investigated. For this model it is assumed that increasing moisture content for porous media means that more and more concentric layers are affected and that the moisture is penetrating the particle from outside. As shown in Fig. 5, the area of the particle is subdivided into ten annuli and a core circle. All annuli have the same area (i.e. wetting of particles in equidistant steps).

To analyze the sensitivity of different moisture layers, for all particles in the computational domain the ten layers are subsequently set to a permittivity value of 80 (wet) while the other layers and the particle core feature a permittivity of 4 (dry). In the simulations, wet layer number 1 represents the outermost layer and wet layer 10 the innermost.

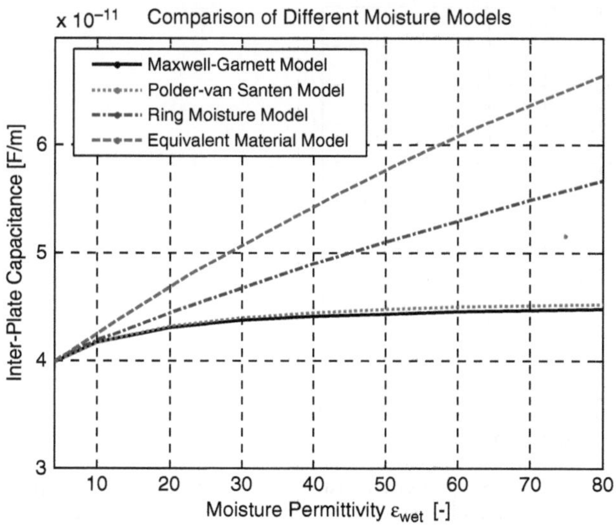

Fig. 4 Comparison of different moisture models used for inter-plate capacitance calculation

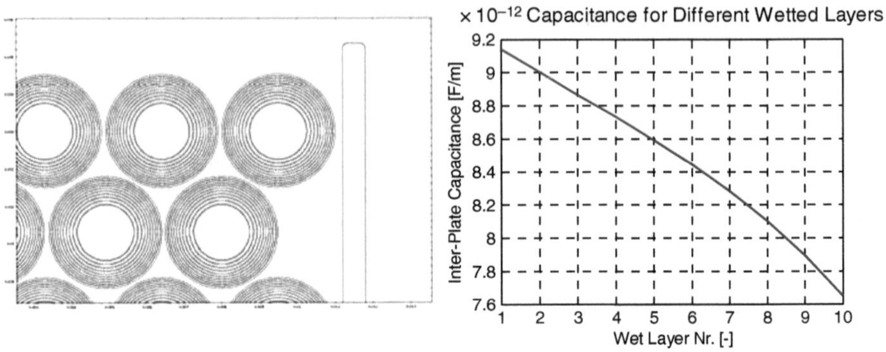

Fig. 5 Multilayer moisture model with ten annuli per particle; (*left*): Geometry and (*right*): Simulation results.

The strong dependency of the inter-plate capacitance on the wet layer depth in the particles can be observed as pointed out in Fig. 5. Although the same amount of water is used, the capacitance decreases by 16.3% in relation to the outermost layers when all innermost layers are wetted.

Beginning from layer 1 towards the particle center, the annuli permittivities of one layer by the other is set to "wet" and the inter-plate capacitance is calculated (i.e. inter-capacitance for accumulated wetted layers). Figure 6 shows simulation results for a permittivity of $\varepsilon_{wet} = 80$ and $\varepsilon_{wet} = 8$. It can be seen that the characteristic in the high permittivity case is nonlinear while it is nearly linear for the low permittivity

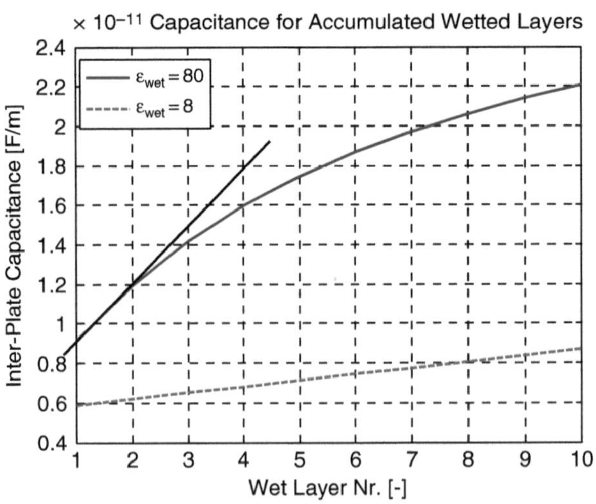

Fig. 6 Simulation results for high and low permittivity moisture ($\varepsilon_{wet} = 80$ and $\varepsilon_{wet} = 8$ respectively) for accumulated wetted layers. A straight line is plotted in the diagram to show the linear characteristic for layers 1 and 2

case. An estimation of the water mass per particle in the practical experiments on the basis of sphere packing density and the knowledge of material and added water mass makes clear that only the two outermost layers can be affected by moisture (moisture penetration depth 0.094 mm for a 3 mm particle). In this range, also the high permittivity simulation shows a rather linear behavior that could be observed in the practical experiments in Fig. 2.

4 Planar Measurement Setup

For a variety of applications in industry, a measurement setup with a one-sided, planar structure is favoured over a setup sensing the moisture through the cross-section of a pipe or a channel. Reasons therefore are limited access to opposed sides of a conveyor piping system, weak sensitivity and long wiring between the two electrodes (especially for pipes with larger diameters), and costly fabrication and Teflon coating for two devices.

A planar capacitive sensor topology can be understood as a bent open parallel plate capacitor, where transmitter and receiver electrode are in the same plane [14], even though the calculation and the model for the inter-plate capacitance are more complex since fringe effects can not be neglected. The electric field develops between active transmitter and receiver and the permittivity and hence the moisture content of the dielectric material affects the measurement signal.

For our practical experiments we produced an electrode topology comprising 16 transmitter segments (height 11.8 mm, width 5.17 mm, gap between neighbouring segments 0.4 mm) and one continuous receiver electrode with a width of 7.41 mm on a printed circuit board. The gap between transmitter and receiver electrodes was 11.8 mm for the first set of experiments. The printed circuit board was mounted on a Perspex container. Multiple electrodes are used to exploit spatial averaging and also to investigate possible moisture gradients along the columnar Perspex container. The transmitter electrodes are subsequently excited with an AC signal (time division multiple access, TDMA) while the remaining transmitter electrodes are kept on ground potential. The inter-electrode capacitance between active transmitter segment and the receiver is determined using suitable circuitry with I/Q demodulation [7], operated at a measurement frequency of about 1 MHz. Since the topmost and the bottommost transmitter segments only have one neighbouring segment each and hence feature a deviant capacitance to the receiver, these capacitances are not recorded. After applying temporal averaging, 14 measurement values are recorded for one individual measurement. Various bulk solids like corn, flaxseed, millet, and different types of sand have been tested with the sensor. Figure 7 shows the setup with planar electrode topology used for the experiments. The geometry and the topology in this setup differ significantly from the parallel-plate approach presented in Sect. 2. Signal amplitudes and –variations are hence different as well.

In capacitive moisture sensing basically two effects have to be considered: On one hand the increased material permittivity due to moisture causes improved

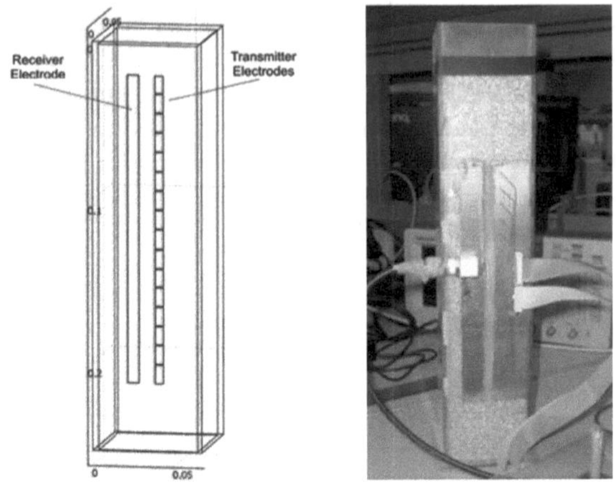

(a) (b)

Fig. 7 (a): Sketch of the measurement setup with planar topology and **(b):** Photo of the setup filled with bulk solids

coupling between a transmitter and a receiver electrode, resulting in increasing inter-electrode capacitance. On the other hand higher material moisture also improves the coupling to surrounding ground potential and hence a "suction" of electric field lines from transmitter to receiver, resulting in decreasing inter-electrode capacitance. For improper electrode topologies both effects may interfere. Particular attention should therefore be paid on reliable grounding and careful design of the front-end to allow for robust sensing with monotonous signal trends.

4.1 Sensitivity on Material Compression Ratio

In a first set of experiments for moisture measurement of bulk solids with planar topology, the cross-sensitivity of the sensor on different compression ratios of the material is investigated. Due to irregular shaped particles and air gaps between them, bulk solids like sand show different levels of compression, from loose to a highly compressed state with compression ratios of more than 15% compared to loose.

To determine the sensitivity of proposed sensor on material's compression ratio and for comparison with the obtained sensitivity on moisture, material is loosely inserted into the Perspex container and all 14 inter-electrode capacitances are recorded. The compression ratio is increased up to a level of maximum achievable compression. Figure 8a shows measurement results for the test material sand with a polydispers particle distribution from a few μm to about 2 mm. The thick solid line represents the mean capacitance value for the 14 electrodes.

It can be seen that for a compression ratio of more than 10% (i.e. the fill level of the bulk solid material decreases by more than 10%), the mean capacitance variation is only about 25 digits even though this variation of compression ratios is very high for an industrial processes. Note that due to local material permittivity distribution

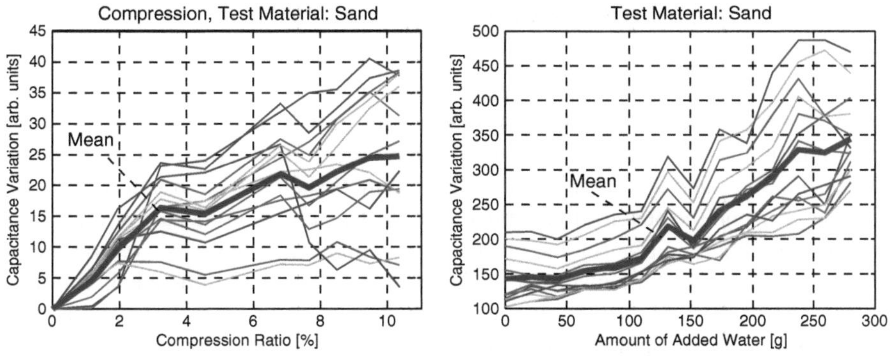

(a) (b)

Fig. 8 **(a)** Measurement results for different compression ratios of sand. Uncompressed sand yields a capacitance value of 143 digits. **(b)** Measurement results for increasing moisture content of sand. In both diagrams the thick lines represent the mean value of all 14 measurement values

in the container caused by irregular compression but also due to minor imperfection in the geometry of the transmitter electrodes and the feed wiring (can be calibrated) the signal trends for individual segments differ.

For the determination of the moisture sensitivity, the dry test material is loosely inserted into the container to allow for a measurement of the 14 inter-electrode capacitance values. The material is then poured in another container where distilled water is admixed. For each increased moisture step the capacitance values are determined in the setup. 2049.4 g of sand are mixed with a total amount of 280.8 g distilled water, resulting in a moisture content [4] of 12.05%. Figure 8b shows measurement results for the inter-electrode capacitances between transmitter segments and receiver for increasing moisture content of the test material sand. The mean value is again shown as a thick solid line. The mean capacitance variation turned out to be more than 200 digits, indicating that the setup is much more sensitive on moisture and shows less undesired cross-sensitivity on material compression.

The absolute change of the dielectric constant of the material due to moisture has been found by means of a parallel-plate capacitor evaluated with a commercial LCR bridge: the permittivity of the dry sand could be determined to be $\varepsilon_{mat} = 3.4$ while the material with a moisture content of 12.05% showed a permittivity of $\varepsilon_{eff} = 12.7$. Similar results could be obtained for other bulk solids with particle sizes below 1–2 mm. When we convert this mass moisture content into a volume fraction and use the Maxwell-Garnett and the Polder-van-Santen model, we obtain effective permittivity values of slightly more than $\varepsilon_{eff} = 5$, indicating that both models underestimate the effective permittivity of the material caused by increased moisture and are hence not suitable for capacitive moisture sensing.

The increase of inter-electrode capacitance due to moisture is less significant for the chosen planar topology compared to a parallel-plate structure. The influence of fringe effects is a major reason therefore.

4.2 Matching Electrode Distance with Particle Size

When we test particles with particle sizes in the range of the distance between transmitter segments and the receiver, we observe an unsteady trend for the moisture determination. The corn particles for example used in the experiments are about 12 mm in length and about 6 mm in width. Figure 9 sketches the relation between electrode gap and the penetration depth of the electrical field: For a narrow electrode topology, as shown in Fig. 9a, the electrical field and hence the sensitive volume is close to the vicinity of the electrodes. The inter-electrode capacitance is highly dependent on the position of the few particles in sensing range and on their moisture content. For a wider topology, as shown in Fig. 9b, the penetration depth is deeper into the material and the sensitive volume is enlarged, resulting in better averaging over a larger set of particles.

Figure 10a shows measurement results for the determination of corn with the setup presented in Fig. 7 (i.e. distance between transmitter segments and receiver is 11.8 mm). Figure 10b shows measurement results for moisture sensing with the same test material but with a distance between transmitter segments and receiver of 29.6 mm.

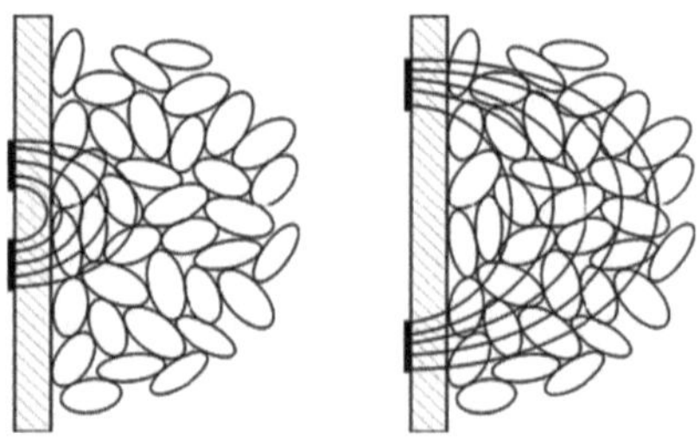

(a) (b)

Fig. 9 (a): Narrow electrode topology resulting in low penetration depth of the electrical field and hence a reduced measurement (and averaging) volume and **(b):** Broader topology with increased measurement volume

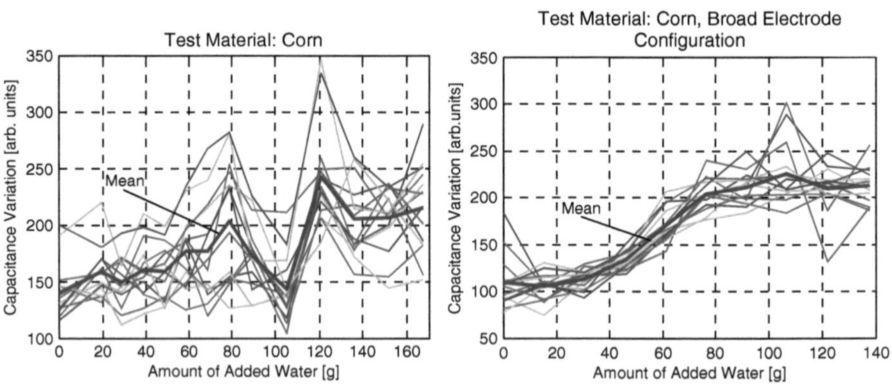

(a) (b)

Fig. 10 Measurement results for moisture sensing of corn for **(a):** 11.8 mm and **(b):** 29.6 mm distance between transmitter segments and receiver

The unsteady trend for increasing moisture content in Fig. 10a is the result of a small sensitive volume in the close vicinity of the electrodes, while the enlarged measurement volume in Fig. 10b results in a smoother trend of the results.

5 Conclusion

In this paper two different approaches for capacitance-based sensing of the moisture content in bulk solids are discussed. We show that standard models for the determination of the effective permittivity are not reliably applicable for bulk solids. Four models with different moisture distribution in the particles are analyzed by means of FEA. The models show that the moisture distribution strongly affects the inter-plate capacitances of the setup. We demonstrate that a planar capacitive setup is capable of sensing material moisture even though the required modelling is more complex.

References

1. Edwards C, Barwood G, Bell SA, Gill P, Stevens M (2001), A tunable diode laser absorption spectrometer for moisture measurements in the low parts in 109 range, Meas. Sci. Technol. 12: 1214–1218
2. Fuchs A, Holler G, Zangl H (2007), Moisture calibration strategies for a capacitive mass flow measurement system in gas-solid flows, Proc. of the 7th International Conference on Electromagnetic Wave Interaction with Water and Moist Substances (ISEMA), Hamamatsu, Japan, pp. 81–88
3. Fuchs A, Zangl H, Holler G (2007), Approaches and Restrictions for Capacitance-Based Moisture Sensing of Bulk Solid Materials, Proc. of the 2nd International Conference on Sensing Technology (ICST 07), Palmerston North, New Zealand, pp. 562–567
4. Fuchs A, Zangl H, Holler G, Brasseur G (2008), Design and Analysis of a Capacitive Moisture Sensor for Municipal Solid Waste, Meas. Sci. Technol. 19: 025201
5. Gradinarsky L, Brage H, Lagerholm B, Björn I, Folestad S (2006), In situ monitoring and control of moisture content in pharmaceutical powder processes using an open-ended coaxial probe, Meas. Sci. Technol. 17: 1847–1853
6. Gundelach V, Litz L (1999), Moderne Prozessmesstechnik – Ein Kompendium, Springer-Verlag, Berlin, Heidelberg, New York
7. Hrach D, Zangl H, Fuchs A, Bretterklieber T (2007), A Rapid Prototyping Platform for Capacitive Measurement Systems, Proc. of the 2nd International Conference on Sensing Technology (ICST 07), Palmerston North, New Zealand, pp. 441–446
8. Khoshbakht M, Lin M (2006), Development of an electrical time domain reflectometry (ETDR) distributed moisture measurement technique for porous media, Meas. Sci. Technol. 17: 2989–2996
9. Kraszewski A, Trabelsi S, Nelson S (1997), Moisture content determination in grain by measuring microwave parameters, Meas. Sci. Technol. 8: 857–863
10. Li X, Zyuzin A, Mamishev A (2003), Measuring moisture content in cookies using dielectric spectroscopy, Proc. of the Conference on Electrical Insulation and Dielectric Phenomena, pp. 459–462
11. Liu N (2007), Soil and Site Characterization using Electromagnetic Waves, Dissertation, Virginia Polytechnic Institute and State University, USA

12. McFarlane I (2006), Control of final moisture content of food products baked in continuous tunnel ovens, Meas. Sci. Technol. 17: 241–248
13. Meyer W, Schilz W (1980), A microwave method for density independent determination of the moisture content of solids, J. Phys. D.: Appl. Phys., 13: 1823–1830
14. Mukhopadhyay S, Gooneratne C (2007), A Novel Planar-Type Biosensor for Noninvasive Meat Inspection, IEEE Sens. J. 7 (9): 1340–1346
15. Nagy A, Vertes P (1968), Correction for dry bulk density in measurements with neutron moisture gauges, J. Sci. Instrum. 2 (1): 1097–1100
16. Regalado C (2006), A geometrical model of bound water permittivity based on weighted averages: the allophone analogue, J. Hydrol. 316: 98–107
17. Sihvola A (1999), Electromagnetic mixing formulas and applications, The Institution of Electrical Engineers, London, UK
18. Wegleiter H (2006), Low-Z Carrier Frequency Front-End for Electrical Capacitance Tomography Applications, Dissertation, Graz University of Technology, Austria.
19. Wobschall D, Lakshmanan D (2005), Wireless soil moisture sensor based on fringing capacitance, Proc. of IEEE Sensors 2005, pp. 8–11
20. Wypych P (2001), Dilute-phase pneumatic conveying problems and solutions, Handbook of Conveying and Handling of Particulate Solids, A. Levy and H. Kalman (eds), Elsevier Science, Amsterdam, The Netherlands, pp. 303–318
21. Zangl H, Fuchs A, Bretterklieber T, Brandstätter B (2005), A Novel Approach for Spatially Resolving Capacitive Sensors, Proc. of the 4th World Congress on Industrial Process Tomography, Aizu, Japan, pp. 36–41

Part VIII
Sensors Based on CNT

The Design and Performance of Tactile/Proximity Sensors Made of Carbon Microcoils

X. Chen, S. Yang, N. Sawada and S. Motojima

Abstract By mimicking Meissner's corpuscles, carbon microcoils (CMCs) were embedded into an elastic resin to produce biomimetic proximity/tactile sensors. The CMC sensors were found to have a high elasticity, high sensitivity, high discrimination ability, and a high performance, as well as being easily made in a micron size. These sensors have potential applications in robotic surgery, medical treatment, and diagnosis, etc.

Keywords Carbon microcoil · helical carbon · proximity sensor · tactile sensor · robotic surgery

1 Introduction

The carbon microcoils (CMCs [1, 2, 3], Fig. 1a), which were prepared by the catalytic pyrolysis of acetylene, have an interesting 3D-helical/spiral structure with a coil diameter of 1–30 μm and a coil length of 50–5000 μm, depending on the chemical vapour deposition conditions. The CMCs can be elastically extended to more than 1.5 time (usually 5–10 times) of their original coil length by applying a small load and then contract to the original length after releasing the load (Fig. 1b, Fig. 2). The CMCs are very interesting as highly sensitive tactile sensors, electromagnetic wave absorbers, etc. On the other hand, Meissner's corpuscle, which is formed by coiling the terminal of nerves with a coil diameter of 40–80 μm and a

X. Chen
Department of Pure and Applied Chemistry, Faculty of Science and Technology, Tokyo University of Science, Yamazaki 2641, Noda, Chiba, e-mail: xqchen39@hotmail.com

S. Yang
Gifu University, Gifu 501-1193, Japan, e-mail: smyang@apchem.gifu-u.ac.jp

N. Sawada
Gifu University, Gifu 501-1193, Japan, e-mail: motojima@apchem.gifu-u.ac.jp

S. Motojima
Gifu University, Gifu 501-1193, Japan, e-mail: motojima@apchem.gifu-u.ac.jp

S.C. Mukhopadhyay, G.S. Gupta (eds.), *Smart Sensors and Sensing Technology*,

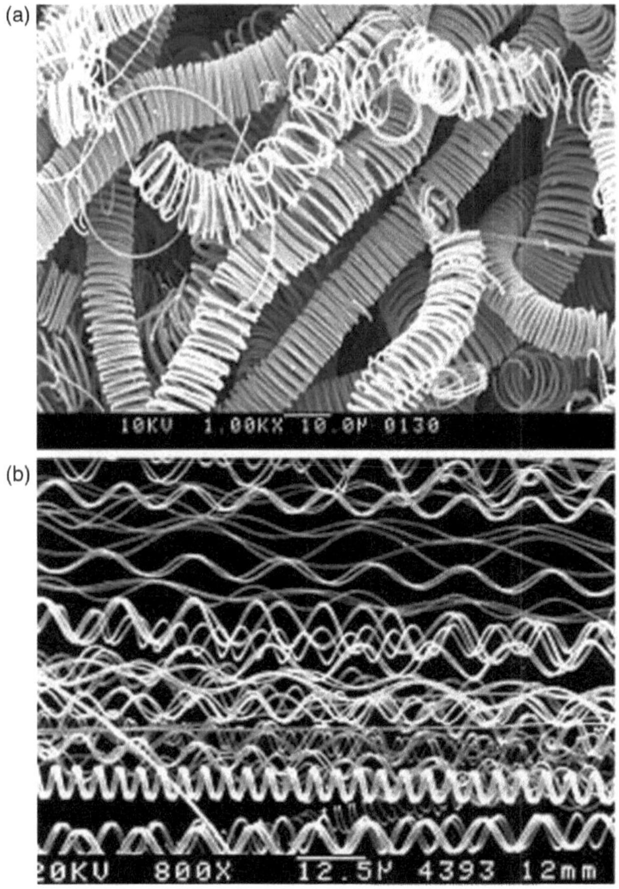

Fig. 1 Source carbon microcoils (CMCs) and their extension state

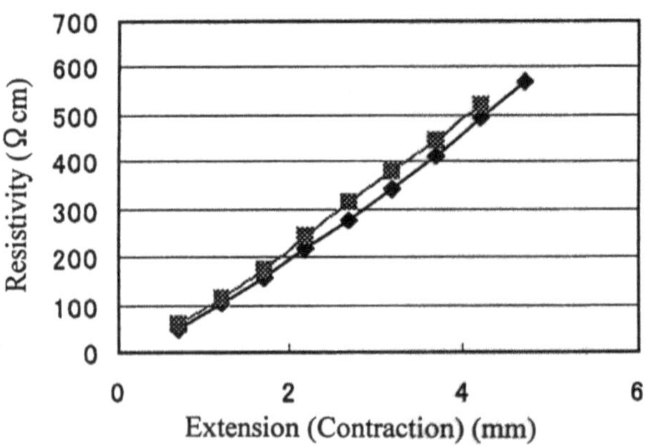

Fig. 2 The relationship between the extension/contraction of bulk CMCs and electrical resistivity

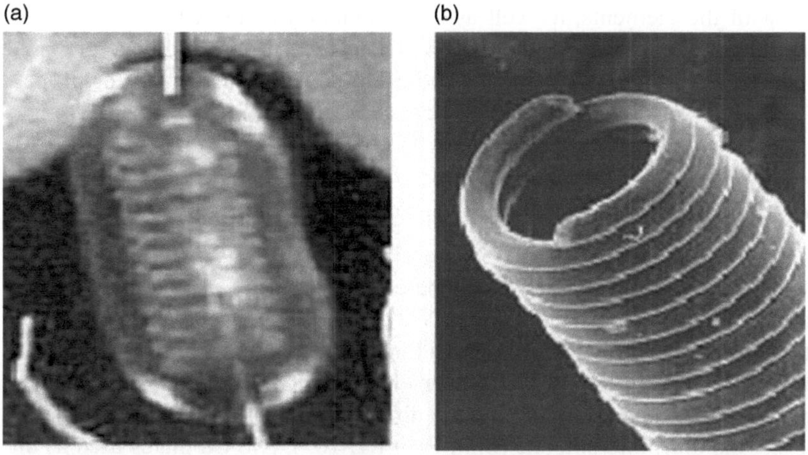

Fig. 3 Comparison between the morphology of the Meissner's corpuscles (**a**) and carbon micorcoils (**b**)

coil length of 20–150 μm, is one of the most important tactile sensing receptors under the human skin. Meissner's corpuscle and the CMCs have many similarities (Fig. 3, Table 1). Accordingly, to mimic Meissner's corpuscles, the CMCs were embedded into an elastic resin to produce the CMC biomimetic tactile sensors, and they were found to have a high elasticity, high sensitivity and high discrimination ability [4, 5, 6, 7, 8, 9, 10, 11, 12], they also have proximity sensing properties at the same time.

On the other hand, in the past several dozens years, various tactile sensors have been developed [13]. These sensors utilize strain gauges, capacitive sensors, piezoelectric sensors, resonation sensors, etc., while there have been very few reports on novel sensing materials. However, during actual use of these artificial tactile sensors, dynamic stimuli as well as static ones have to be detected with a high sensitivity and discrimination ability, which may be affected by the surface conditions; (i.e., morphology, roughness, hardness, elasticity, etc.,) of the sensor elements as well as those of the objects contacted by the sensor elements. Furthermore, the softness and high elasticity of the sensor elements are required properties.

CMC sensors perform in novel sensing mechanism,which is different with traditional tactile sensors. The deformation of the CMCs is strongly affected by the morphology and roughness of the surface of the CMC sensor elements and objects

Table 1 Comparison between Meissner's corpuscles and CMCs

Items	Meissner's corpuscles	CMCs
Material	3D helical nerve fibers	3D helical carbon fiber
Spiral shape	Single-helix	Double-helix or single-helix
Helix dia.	Helix dia. 40–70 μm	Coil diameter 1–20 μm
Helix length	Helix length 20–150 μm	Coil Length: 50–500 μm

in contact with the elements, as well as the orientation of the CMCs in the bulk matrix and also the elasticity of the matrix. Therefore, in this study, we prepared novel biomimetic CMC sensor elements and examined their tactile sensing properties, the proximity sensing property is also concerned.

2 Experimental Details

In this study, CMCs which have a coil diameter of ca. 15 μm were synthesized by a conventional chemical vapor deposition (CVD) process using acetylene as a carbon source; the representative figure is shown in Fig 1. The detail preparation conditions are reported elsewhere [2, 3]. CMC sensor elements (CMC/silicone rubber composite sheets) were prepared by mixing the different amount of CMCs (0–10 wt %, length of 0.3–0.5 mm) into silicone (Shin-Etsu KE-103), two Cu plates used as the electrodes were inserted into the composite sheets, and 0.5V AC from 40 Hz to 30 MHz. The electrical parameters were measured by an impedance analyzer (Agilent 4294A) when some loads were vertically applied on the whole CMC sensor elements. Usually, outputs at 100 KHz were plotted vs. the loads applied on the sensors.

Preparation of randomly dispersed CMC double-layer sensor element for broadening dynamic range: the CMC samples were uniformly embedded into a polysilicone matrix (top layer using GE Toshiba Silicone YE 5623 with a higher softness and bottom layer using Shin-Etsu KE-103) to prepare double sensor elements with a surface area of $10\times10\,mm^2$ and different thickness ranging ratios as shown in Table 1. The CMCs are relatively uniformly dispersed in the matrix. The CMC element was set on parallel electrodes (Cu plate, $3\times10\times0.1\,mm^3$, the separation of two plates: 2 mm), which was placed on an insulator stage (The frequency was from 1 KHz to 100 MHz). Some stresses were then vertically applied over the whole CMC sensor elements, which were placed between two electrodes, and the output change in the electrical parameters, i.e., impedance (ΔZ) was measured using the same analyzer.

3 Results and Discussion

3.1 Basic Characteristics of CMC Sensors

The Young‘s modulus of the polysilicone resin (matrix) is $9.36\,kgf/cm^2$ (Durometer-24). The tensile modulus of the CMCs is about 10 GPa. The elasticity of CMCs is far more than that of the polysilicone matrix. It was found that the lowest content of CMCs for producing apparent output was 0.1 wt%. With the increase in the addition amount up to 7 wt%, the sensitivity increased. However, when the addition amount was more than 10 wt%, percolation phenomena occurred, the circuit between the CMCs, the matrix and the electrodes changes from capacitance type to resistance

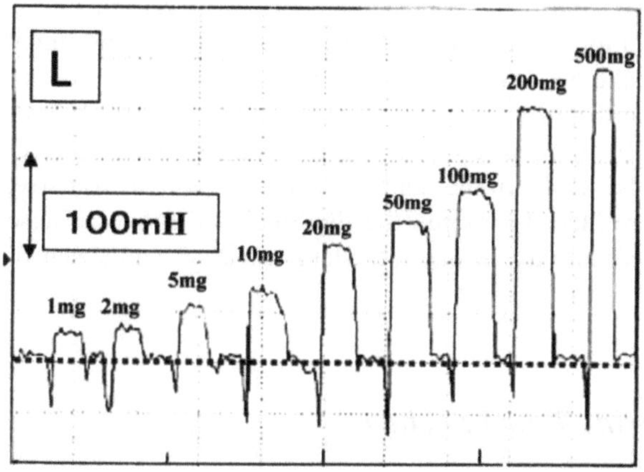

Fig. 4 The change in L (inductance) parameter of CMC/polysicilone tactile sensor elements under the application of loads.Addition amount of CMCs in polysilicone: 1 wt%, thickness of elements: 0.1 mm, separation of electrode: 2.5 mm

type. At this addition amount, the contents of the numbers of CMCs was about $10^7/\text{cm}^3$, which is much higher than the content ($1.5\times10^3/\text{cm}^3$) of the Meissnor's corpuscles under human skin. It was found that the suitable sensor thickness was about 1–2 mm, and with a decrease in the thickness, the sensitivity decreased. However, by aligning CMCs among the matrix, the element can be made to be very thin (without dramatic loss in the sensitivity).

The CMCs/polysilicone YE-103 sensor element can detect applied load by the change of electrical parameters with short response times of 0.3–0.5 s. The electrical resistivity increased logarithmically with increasing the applied load and decreased with increasing the hardness of matrix. In details, for closely understanding the influence of as small load as milligram orders on the change of L, C, and R parameters, respectively, various weights less than 200 mgf (and the smallest was 1 mg) were vertically applied on the CMC sensor element of $10\times10\times0.2\,\text{mm}^3$. The electrical parameters L (inductance) changed when a load as small as 1 mgf was applied, and with the increasing load, the output increased (Fig. 4). The C (Capacitance), and R (Resistivity) parameter also experienced a similar change tendency. The sensitivity of the SE-CMC sensor element was about 1mgf, while the sensibility of conventional capacitive sensors are 13-32 pF/100 gf or 200–250 V/100 gf [11], accordingly, the SE-CMC sensor elements have three to four orders magnitude higher sensitivity than that of conventional capacitive sensors.

3.2 Effect of the Miniaturization and CMC-Alignment of the CMC Sensor Element

The CMC tactile sensors could be easily miniaturized by successively cutting the large sheets to obtain micron sheets or blocks using an ultrasonic cutter. It was observed that the $0.1 \times 0.1 \times 0.1\,mm^3$ miniaturized elements showed the same signal levels. In order to miniaturize the SE-CMC sensor elements further, the effect of arraying SE-CMCs on the sensing ability was then investigated by preparing micro-sensor elements of $80 \times 80 \times 80\,\mu m^3$ with the arraying of about 50 pieces (corresponds to about 1 wt%) of the SE-CMC in the matrix. Because of the micron size, the micro-sensor elements were placed between the two micro-electrodes. These micro-elements have the same sensitivity as that of the large $2 \times 2 \times 0.1\,mm^3$ sensor elements.

3.3 Effect of the Fingerprint-Typed Surface

A higher sensitivity was also obtained by using the sensor element with a rough surface. The sensor elements with a fingerprint-type surface were prepared using moulds with fingerprints (the moulds were obtained by printing fingerprints on glass plates with a two times enlargement and then carved using a micro-grinder). The dynamic load was applied by a soft brush which was rotated by a motor, while the static load was vertically applied to the surface of the sensor elements using a manipulator. It was found that finger print-type sensor elements showed a higher sensitivity to the applied dynamic loads than that of static loads. It was also found that the sensitivity for the fingerprint-type sensor element showed an anisotropic sensing property in the direction of the applied loads.

3.4 Effects of Multiple Layers Sensors

It was considered that the hardness and the thickness of the sensor element are the key parameters to affect the dynamic ranges. Thus, we tried to use a soft matrix YE5623 to investigate the dynamic range by changing the sensor element thickness. Changing the applied load, the impedance, phase angle and resistivity of CMC/polysilicone tactile sensors from 0 gf to 3 kgf were measured in frequency range of 40 Hz-30 MHz at room temperature. Figure 5 shows the dependence of impedance of CMC tactile sensor with different composite thickness (3, 5 and 7 mm) on the applied loads as a function of sensor element thickness. When the applied load was increased, the impedance was also increased dramatically, and reached some saturated value for the sensors with a thickness of 3 and 5 mm. on the other hand, the impedance change in the thick composite (thickness 7 mm) wasn't saturated with increasing the applied load slowly. From these results show that for CMC/YE5623 sensor with single-layer, the optimum composite thickness of the CMC tactile sensor is 3–5 mm for obtaining a wide dynamic range, these CMC tactile sensor can detect the applied load of 0–1500 gf.

We further design double-layer sensors as Table 2. Changing the applied load from 0 gf to 3 kgf, impedance of CMC/ polysilicone YE5623 (a soft matrix) tactile

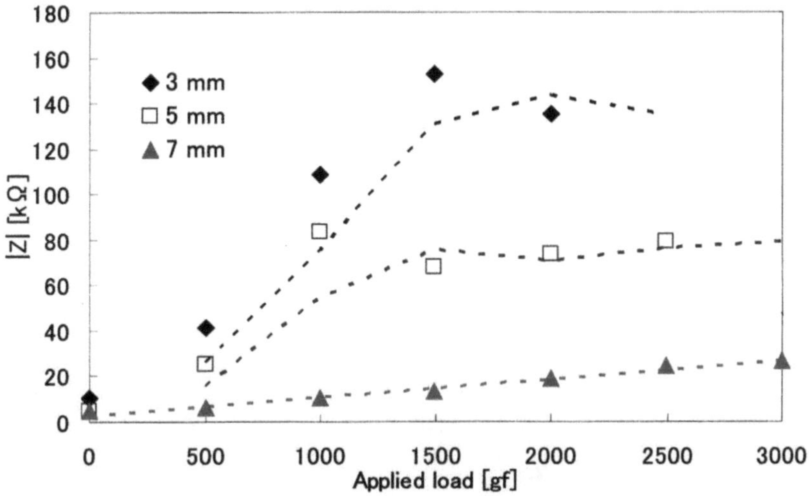

Fig. 5 Dependence of |Z| for CMC tactile sensor with different thickness as a function of applied load at 1 kHz, using a soft matrix YE5623

sensors were measured by employing the same impedance analyzer. For double-layer sensors with different ratio of the upper layer (soft) to the under layer (hard), the dependence of the impedance change |Z| on frequency changed in different ways. Sensor A and Sensor B (top and middle in Fig. 6, respectively), in which the ratio is 1:3 and 2:2, responded the output changes unstably and irregularly with loads changes. However, for Sensor C, (bottom in Fig. 6,), in which the ratio is 3:1 responded the output changes regularly with loads changes, indicating that Sensor C can detect the objects in the force range 0-3000 gf.

3.5 CMC proximity-tactile sensor

Furthermore, it was found that the CMC sensors also had the proximity sensing property, as shown in Fig. 7, the nearer is the hand get close, the stronger is the signal. The continuous proximity and tactile sensing properties of the CMCs sensor

Table 2 Preparation conditions of double-layer CMC tactile sensor elements

Matrix		CMC length	CMC content [wt%]	Composite thickness [mm] (A)	(B)	(C)	Applied load [gf]
Upper layer	YE 5623	0.3~0.5 mm	7	1	2	3	0~3000
Under layer	KE 103	0.3~0.5 mm	7	3	2	1	0~3000

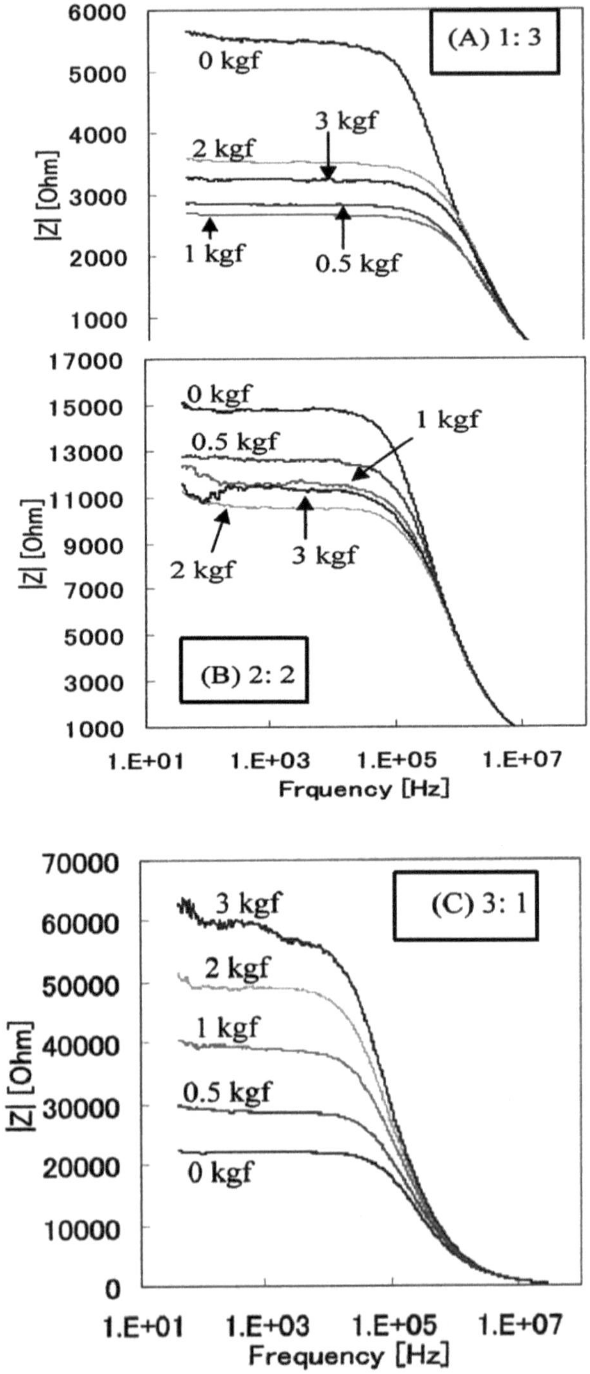

Fig. 6 Dependence of impedance Z on frequency for the double-layer sensor elements with varying thickness ratios, which were prepared according to the conditions shown in Table 2

Fig. 7 Real part of impedance Z for an approaching and releasing palm to the sensor element at intervals of 10 s

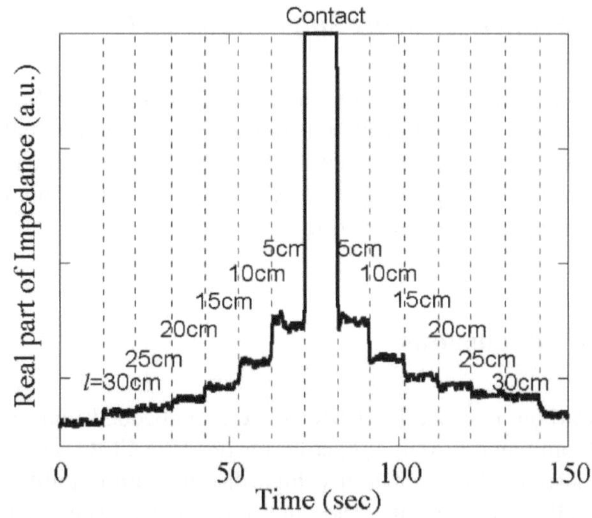

elements (e.g. the change of Z when Cu plates, hand palms approaching to the sensor element, etc.) were examined. The optimum CMC content as proximity sensor element was 3 wt%, and the tactile sensor properties were increased in intensity with CMC content increased to over 7 wt%. The CMC sensor element with surface size of 300 × 300 mm and with a CMC content of 3 wt% is able to detect the existence of a human palm which is about 30 cm away from CMC sensor element. Furthermore, the dramatic change of impedance Z was observed when the object contacted with the sensor (Fig. 7). In another word, we succeeded in the preparation of CMC sensor elements with tactile and nearness sensing properties.

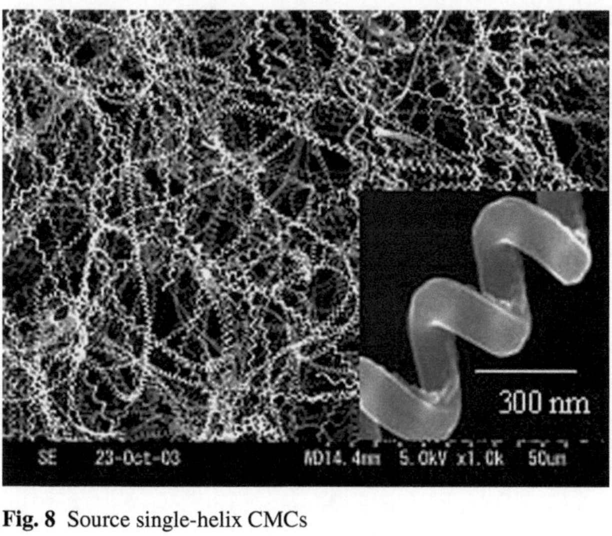

Fig. 8 Source single-helix CMCs

3.6 Effect of using single-helix CMCs

Protein-like single-helix carbon microcoils (SH-CMCs, Fig. 8) were used as the sensing material, with the replace of the double-helix carbon microcoils (DH-CMCs). It was found that the SH-CMC sensor elements are more stable and sensitive than that of the DH-CMC sensor elements. This may be caused by the larger coiling gaps, better wetting.

4 Conclusions

Carbon microcoils (CMCs) were embedded into an elastic resin to produce biomimetic proximity/tactile sensors. The CMC sensors were found to have a high elasticity, high sensitivity, high discrimination ability, and a high performance, as well as being easily made in a micron size. These sensors have potential applications in robotic surgery, medical treatment, diagnosis, etc.

Acknowledgements This research is supported by the Knowledge Cluster Initiative Project of Japan:Gifu Ogaki Robotics Advanced Medical Cluster; also partly by REFEC Research Foundation for the Electrotechnology of Chubu of Japan, and Japan Society for the Promotion of Science (P04418).

References

1. X. Chen., T. Saito, M. Kusunoki and S. Motojima, Three-dimensional vapor growth mechanism of carbon microcoils, J. Mater. Res., 14, 4329–4336(1999).
2. S. Motojima, X. Chen, Nanohelical/sprial materials, by H. S. Nalwa, Editor, Encyclopedia of Nanosci. and Nanotech. American Science Publisher, California 6: 775–794(2004).
3. X. Chen, S. Motojima, The growth patterns and morphologies of carbon micro-coils produced by chemical vapor deposition. Carbon, 37(11):1817–1823 (1999).
4. S. Yang, X. Chen, and S. Motojima, Tactile sensing properties of protein-like single-helix carbon microcoils, Carbon 44(15), 3352–3355(2006).
5. X. Chen, S. Yang, M. Hasegawa, K. Takeuchi, S. Motojima, Novel tactile sensors manufactured by carbon microcoils. Proc. Int. Conf. on MEMS, NANO, and Smart Systems, Banff, IEEE, 2004:486–490.
6. X. Chen, S. Yang, M. Hasegawa, K. Kawabe, S.Motojima, Tactile microsensor elements prepared from arrayed superelastic carbon microcoils. Appl. Phys. Lett., 87(5): 054101-1~3 (2005).
7. X. Chen, S. Motojima, J. Sakai and S. Yang, Biomimetic tactile sensors with knot-type or fingerprint-type surface made of carbon microcoils/polysilicone, Jpn. J. Appl. Phys., 45, L1019–L1021(2006).
8. S. Yang, N. Matushita, A. Shimizu, X. Chen and S. Motojima, Proc. of the 2005 IEEE, Int. Conf. on Robotics and Biomimetics, (June 29–July 3, Hong Kong and Macau), pp. 41–44(2005).

9. H. Natsuhara, T. Katsuno, X. Chen, S. Yang, S. Motojima, The 5th IEEE conference on sensors, Daegu Exhibition & Convention Center (EXCO), Daegu, Korea, Dates: Oct. 22–25, (2006).
10. S. Yang, X. Chen, H. Aoki and S. Motojima, Smart Mater and Structures, 15, 687–694 (2006).
11. Application of CMC sensors in medical robotics autonomous system X. Chen, S. Yang, H. Natuhara, K. Kawabe, T. Takemitu and S. Motojima, In: Proc. Fourth International Conference on Computational Intelligence, Robotics and Autonomous Systems November 28–30, Palmerston North, New Zealand CIRAS'200, Palmerston North, New Zealand, pp. 132–136 (2007).
12. X. Chen, S.Yang, H. Natuhara, T. Sekine, and S. Motojima, Novel tactile/proximity sensors made of vapor grown carbon microcoils (CMCs). In:Proc.2nd International Conference on Sensing Technology ICST'2007, November 26–28, Palmerston North, New Zealand pp. 446–449 (2007).
13. Eltaib MEH, Hewit JR. Tactile sensing technology for minimal access surgery—-a review, Mechatronics, 13(10): 1163–1177 (2003).

Part IX
Novel Sensors Techniques

Post-Earthquake Pipeline Leak Detection Technologies

Winncy Y. Du and Scott W. Yelich

Abstract Pipelines carry water, fuel, waste, communication and power cables, all of which are important to the operation of residential, hospital, military, and industrial complexes. Often, pipelines are severely damaged after catastrophes such as earthquakes. The immediate inspection of pipelines is critical to prevent fires, explosions, and contamination from broken gas, water, or sewage lines. This paper reviews the conventional and emerging technologies used in pipeline leak detection. These technologies are discussed in terms of the phenomena employed, principles, system applicability, and positive and negative virtues of each when applied to various pipeline defect detection.

Keywords Pipeline inspection · leak detection · post-earthquake pipeline damage evaluation

1 Introduction

Lifeline systems are basic public facilities which include water supply, natural gas, sewer systems, electricity, and telecommunication. The integrity of these pipelines is critical to the operation of residential, hospital, military, and industrial complexes. During earthquakes, utility pipelines can rupture and break due to the passage of seismic waves through the soil, permanent ground displacement (such as faulting, landsliding or liquefaction), or collapsing building structures (see Fig. 1). This damage can cause significant post-earthquake disasters such as fires, explosions, personal property loss, water damage, and environmental pollution. Some examples include: the natural gas fires resulting from broken pipelines during the 1994 Northridge earthquake, and the greatly hampered firefighting ability due to water line breakage following the 1906 San Francisco earthquake. Power outages can

---

Winncy Y. Du
San Jose State University, San Jose, USA, e-mail: winncy.du@sjsu.edu

Scott W. Yelich
San Jose State University, San Jose, USA, e-mail: swyelich@sbcglobal.net

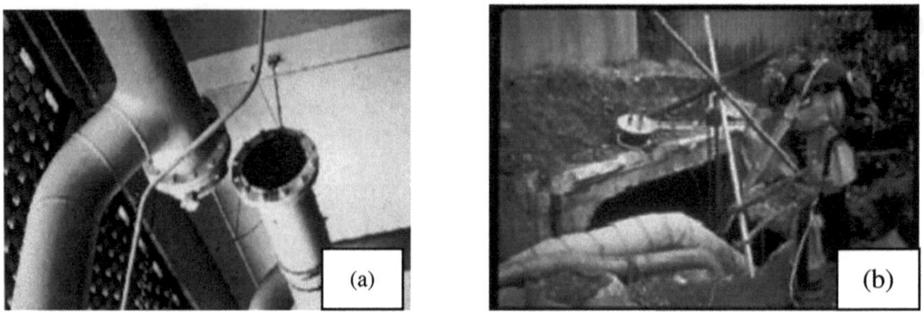

Fig. 1 (**a**) Hospital waste water pipe broken at the joint during the 1971 San Fernando Earthquake (Photo by J. Marx Ayres); (**b**) Main sewage treatment conduit rupture from the 1995 Kobe Earthquake (Source: Earthquake Engineering Research Center, UC Berkeley)

affect the operation of life support equipment in hospitals and cause patient deaths. Power surges caused by earthquakes can also damage circuits and result in irretrievable loss of important data. The economic losses caused by these secondary disasters are often higher than the initial building structural damage directly caused by the earthquake itself. Thus, the immediate assessment of pipeline integrity is critical to prevent further post-earthquake damage. Pipeline leak evaluation technologies have also become increasingly important due to the requirements for environmental protection and greater emphasis on energy and water conservation. In the U.S., significant resources are expended annually on the inspection of pipeline systems.

Statistics have shown that the most vulnerable pipelines are typically those carrying sewage because they are made of the most brittle materials and do not have sealed joints. The next most vulnerable are water pipelines followed by natural gas pipelines. Extensive research efforts have been focused on assessing the condition of pipelines after earthquakes [1, 2] and on leak detection technologies [3, 4, 5, 6, 7, 8]. These methods can be divided into two categories, externally based (direct) or internally based (inferential). Externally based methods detect leaking product outside the pipeline through visual inspection, or sensing via fiber optic or dielectric cables. Internally based methods use instruments to monitor internal pipeline parameters (i.e., pressure, flow, temperature, etc.). This paper overviews pipeline leak detection methodologies, sensor development, and inspection technologies. Section 1 is an introduction, Sect. 2 presents the existing leak detection technologies, Sect. 3 outlines the technology selection criteria and Sect. 4 gives a summary.

2 Technology Overview

There are many established technologies in use for the detection of various leaks in pipelines; each associated with physical phenomena which occur either at a leak site or in the system due to the leak.

2.1 Leak Detection Cables

Leak detection cable is laid in probable leak sites and monitored for changes in impedance, conductivity or refractive index [3].

- Impedance – the outer layer of a cable is coated with a permeable material. Since the cable has a known length and dry impedance, when a leak changes the dielectric properties at a specific location within the cable, it increases the impedance which can be detected and located along the length of the cable.
- Conductivity – two conductors (typically stainless steel) are woven into a specially treated cable or ribbon. When a conductive liquid forms a continuity path between the two wires, a leak condition is identified.
- Refractive Index – fiber optic cable is clad with a material that interacts with a fluid, this fluid changes the refractive index and thus the light transmission inside the core, providing a means of detecting and locating a leak.

These cables are usually attached to control panels which determine if a leak occurs and pinpoint the exact location of the leak along the cable route. This technology can be used to detect aqueous leaks or (with special polymers) leaks of fuel and gas. For area applications, the cable is laid in a serpentine pattern. In multi-branch pipe systems, individual cables can be run from branching junctions. Figure 2 shows a SeaHawk water leak detection cable and LD5100 Distance Read – a control panel with a LCD (liquid crystal display) that displays the location of the leak. The cable is constructed of non-conductive polymers (for leak detection) and an abrasion resistant polymer core (for cable strength). The flexible cable allows for small bend radii, and the mating connectors make installation and replacement of damaged cable quick and easy.

2.2 Acoustic Leak Detectors

When a pressurized pipeline leaks, fluid inside the pipe sprays through the rupture and into the surrounding area at a high speed. As a result, (1) the pipeline vibrates

Fig. 2 SeaHawk water leak detection cable and its LD5100 Distance Read (Source: http://www.rletech.com/products/cable.html)

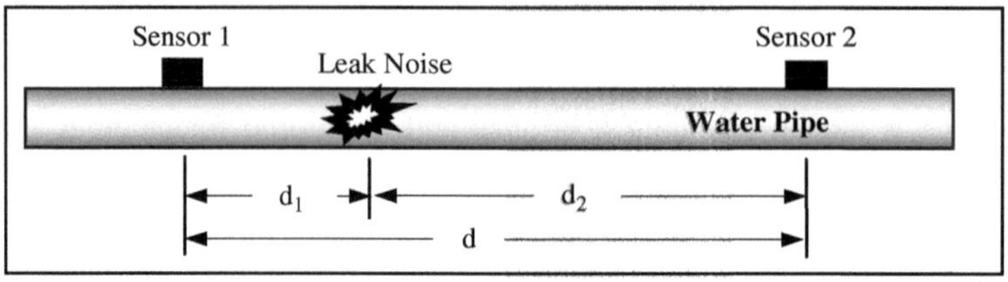

Fig. 3 A typical setup for locating a leak using the acoustic method

and generates noise at the leak location. This noise is transmitted by the pipe and can be picked up even at remote points (valves, hydrants, meters, etc.) by sensitive microphones; (2) the fluid jets and pipes induce vibration in the soil which transmits to the surface, where it can be picked up as ground-borne noise. Typically, two sensors (either *piezoelectric transducers* or *hydrophones*) are mounted on the pipe to locate the leak (see Fig. 3). A set of these acoustic sensors is placed on either side of the suspected leak area. The position of the leak, as referenced from sensor 1, can be found by [9]:

$$d_1 = \frac{d - c\Delta t}{2} \tag{1}$$

where d is the distance between the two sensors; c is the wave speed at which the leak noise propagates through the pipe; Δt is the difference in arrival times of the noise at the two sensors.

The piezoelectric transducers or hydrophones translate waves generated by leaking fluid into an electrical signal. A circuit is often required to process the detected signal and make a "leak" or "no leak" determination based on a pre-set threshold in the memory. Acoustic leak detection techniques are effective and are commonly used by water companies. They are applicable to any system that carries fluids under pressure. In leak detection surveys using acoustic methods, the most widely used approach involves the cross-correlation of the measured acoustic signals. This has proven to be reasonably effective in detecting and locating leaks within metal pipes, but has been problematic when used on plastic pipes [10]. Recent work at the University of Southampton has focused on improving the acoustic leak detection technique for plastic pipes [11]. This technology can be used on pipes that are buried or in open air. Since wave speed is highly affected by the pipe material, thickness, diameter, and temperature, it is difficult to estimate the wave speed, c, accurately. The detecting system may give false results due to flow noise around valves, geometry changes, and noise interference from the surroundings.

2.3 Magnetic Flux Leakage (MFL)

When a powerful magnet is placed next to a steel pipe, it magnetizes the steel, and most of the flux lines pass through the pipe wall. However, in a damaged region,

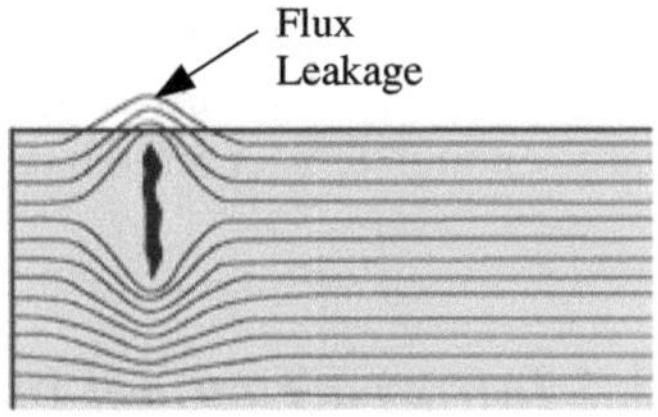

Fig. 4 Flux field around a defect in a steel or cast iron pipe

the flux carried by the broken or cracked section is less than that carried in the uncompromised wall, thus the magnetic flux "leaks" out of the pipe wall, as shown in Fig. 4. A Gauss meter can detect this flux leakage at the magnetic "poles" created by the flaw. The presence of an anomaly is indicated when the meter senses a higher flux density or magnetic field. The strength of the leakage field depends on the radial depth, axial length, circumferential width, and shape of the anomaly, as well as the magnetic properties of the pipe material.

Pipeline inspection gauges (or *pigs* as they are known in pipeline industry) are measuring tools that are sent down a pipeline, propelled by the pressure of the fluid in the pipeline. Most pigs use the MFL method (although some pigs use the ultrasound method, see "*2.6 Ultrasonic Leak Sensors: Elastic Wave Vehicles*") to perform their inspections. Figure 5 shows an MFL pig. A strong magnetic field is established in the pipe wall using either magnets or by inducing electrical current into the steel. Areas of the pipe containing flaws will have magnetic flux leaks out of the pipe wall. An array of sensors (e.g., Hall sensors or eddy current sensors) around the circumference of the pig detects the magnetic flux leakage and locates the damage area.

MFL is the most commonly used method for inspecting in-service pressurized pipelines. It is often used to inspect extensive pipe networks. The technique provides robust, reliable, accurate, and repeatable measurement of defects in pipelines, thus minimizing the need for expensive excavations (i.e. digging up the pipe to find the leak). Its low-noise and low-temperature sensitivity make it suitable for applications requiring tenths of pico Tesla resolution and nano Tesla accuracy. The main

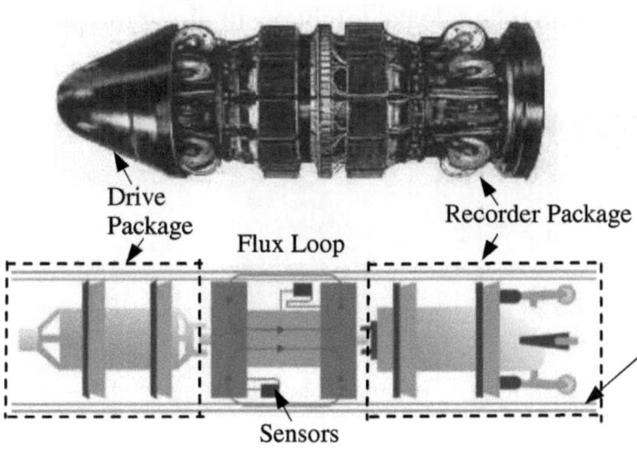

Fig. 5 A photo of a pig and its schematic illustration inside a pipe (source: NDT Resource Center, www.ndt-ed.org/.../PipelineInspection.htm)

disadvantage of the MFL method is that it can only be used to inspect pipe systems that are made of steel or cast iron. Also, pipeline diameters and bend radii must be large enough to accommodate the pig.

2.4 Infrared Thermal Imaging (IR Thermography)

IR Thermography is a technique that converts invisible heat energy emitted, reflected, or transmitted from a pipe and its surroundings into a visible image to reveal defects. By utilizing an infrared imaging system equipped with a camera, thermal patterns along the inspected pipeline can be recorded and shown in a color spectrum. More sophisticated cameras can actually measure the surface temperature of any object in the image and indicate leaks with a defined color pattern. Figure 6 shows the thermograms of service pipes acquired using IR cameras.

Computerized IR Thermographic pipeline inspection is now a refined process having been thoroughly proven to be an accurate, cost effective, and efficient technology. The IR technique has been used in both aerial and ground based applications to test water supply systems, natural gas pipelines and sewer systems. It is a non-destructive, non-contact process that can be quickly performed without interruption to service. The primary limitation is that it cannot be used on pipe systems that do not exhibit a temperature differential. Performing inspections of underground pipelines also presents challenges due to variation in surface makeup, burial depth of the pipe, and flow rate and temperature of the fluid.

2.5 Fiber Optic Leak Detectors

Fiber optic sensors are attractive options for detecting and locating the presence of fluids. An optic fiber can serve the role of both sensing element and signal delivery channel. Distributed fiber optic sensors, in particular, are well-suited for monitoring pipelines. Figure 7 illustrates two configurations. In (a), the fiber optic cable is laid in a location where leaks might occur along the pipe. If a leak occurs, the cladding of

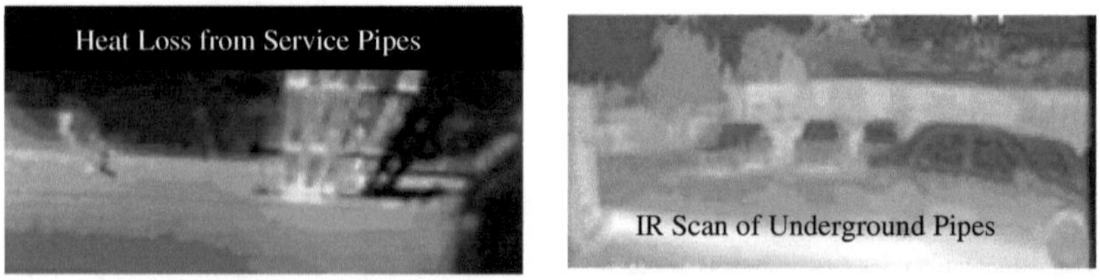

Fig. 6 Thermograms of pipes acquired using IR camera (source: Sierra Pacific Corp., www.x20.org)

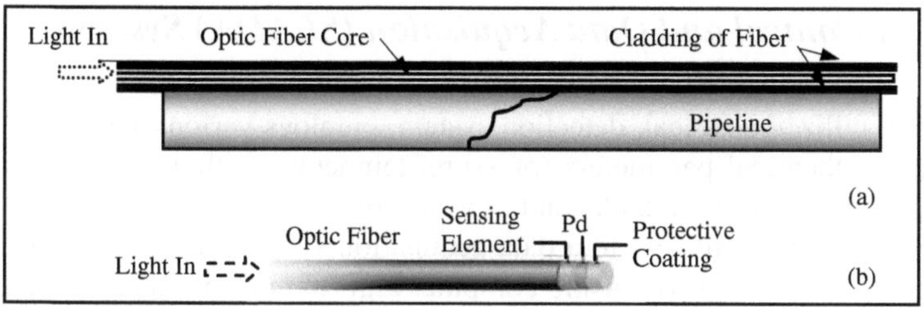

Fig. 7 Two fiber optic sensor configurations for leak detection

the fiber changes its refractive index. Leakage is identified by monitoring the light transmission characteristics of the fiber (e.g., increasing light losses in the core). This refractive index variation can be used for gas or liquid leak detection, as long as the cladding is sensitive to gas or liquid. In (b) [12], the sensor consists of a thin film of sensing material, deposited on the end of a fiber cable and used to detect gas or liquid. In the presence of leaks, the sensing material undergoes optical changes – either a color or transmittance change through the film contacting the gas or liquid. The intensity of a beam of light, either transmitted or reflected down the cable, is measured to indicate a leak.

Fiber optic sensors have several advantages, they: can withstand shock and high temperatures (> 1000°C); are immune to electromagnetic interference and are environmentally rugged; allow data to be transmitted over long distances; can be multiplexed with many fiber optic sensors to obtain multifunctional capabilities. The sensors may need to be isolated from unwanted agents, and light sources and suitable instrumentation are required during the measurement.

2.6 Ultrasonic Leak Sensors: Elastic Wave Vehicles (EWVs)

EWVs were first developed in 1993 to detect and measure defects in gas pipelines. EMV design embodies a unique adaptation of ultrasound technology that makes defects visible by viewing them from the side in high profile. Ultrasonic waves are induced into the pipewall circumferentially around the pipe. When they encounter an axial crack or other defect within the pipewall, the waves are reflected and the echo is detected by transducers. Because ultrasonic waves do not propagate in gas, each transducer is enclosed in a fluid-filled wheel mounted on the outer surface of a pig (pipeline inspection gauge). These special wheels run in contact with the inner surface of the pipewall. The material used for the wheels is permeable to ultrasonic waves, and is sufficiently robust to withstand impact with weld beads, bore changes and other features of the pipe surface. The EWV performs equally well in liquid-filled or dry pipelines.

2.7 Supervisory Control and Data Acquisition (SCADA) Systems

A SCADA system, utilized as a leak detection system, employs various sensors to measure physical or chemical parameters (pressure, temperature, flow rate, level, and concentration) in pipelines or tanks and converts these parameters into electronic signals. These signals are sent to a data acquisition system, a signal conditioning unit, and are finally analyzed using computer algorithm to produce a leakage determination. SCADA systems are used to control large networks of pipelines such as water distribution and heating ventilation and air conditioning systems. A central computer monitors the system for changes in pressure, flow rate, and volume of liquid. More than one leak detection algorithm can be integrated into a SCADA system to monitor the entire pipeline network. This allows for accurate leak detection, including leak location and flow rate identification. Figure 8 shows a WaveAlert® VII SCADA system. It monitors the pipeline for a line pressure drop due to a sudden leak. The pressure wave travels at the speed of sound in the pipeline. By accurately determining time of arrival at two pressure sensor locations, the leak is located.

The multiplicity of sensors requires an effective SCADA system to handle the data fusion, data interpretation and decision making algorithms. Leak detection algorithms encompass the following areas: pressure point analysis, mass-balancing, level monitoring, and real-time transient analysis. The cost of sensors and data communication equipment necessary for implementing a SCADA system on a large scale pipeline system is still relatively high.

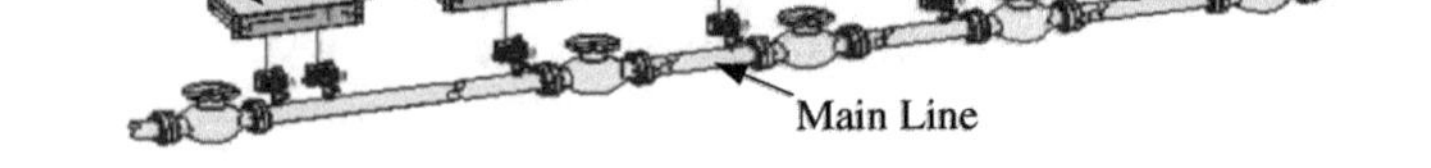

Fig. 8 WaveAlert® VII leak detection system (Source: SAS Tech Solution CO., LTD, www.sastechsolution.com/index.htm)

2.8 Pressure Point Analysis (PPA)

Pressure Point Analysis (PPA) operates on the principle that when a leak occurs in a pipeline, the pressure in the line drops. Using various pressure sensors, a decrease in the mean value of a pressure is detected. If the decrease is larger than a predefined level, a leak alarm is triggered. The relationship between leak rate and pressure drop is described by [13]:

$$\text{Leak rate (cm}^3\text{/s)} = \frac{\text{Total Test Volume (cm}^3\text{)} \times \text{Pressure Decay (mbar/s)}}{\text{Atmospheric Pressure (mbar)}} \qquad (2)$$

PPA can also be used to find the location of leaks in pipelines [14]. The PPA method is usually applied to pressurized pipelines ranging from 3 to 42 inches in diameter, without limitations on pipe material or liquid properties. This method has a high degree of sensitivity and a high response rate. Pressure transmitters with minimal damping and high-turndown flow meters are ideal. Since a PPA system measures internal pressure, external factors such as ground water, contamination, or location of the pipeline (above or below ground) have no effect on performance. The pressure sensors are plumbed directly into the pipeline and can be spaced miles apart, unless special conditions exist (like steep hills, where sensors must be placed on the hilltop). The size of the leak that can be detected is dependent upon pipeline volume. For instance, a 0.05 gph leak is detectable in a 5,000 gallon system, and a 0.1 gph leak is detectable in a 17,000 gallon system [3]. PPA has its limitations; normal operations (such a valve movement) can mimic leaks and flow noise can mask a leak indication.

2.9 Mass Balance, Volumetric Testing, and Real-time Transient Analysis

The *Mass-balance method* is based on conservation of mass in each pipeline segment. A potential leak is indicated when the mass leaving a segment is less than that entering it. The mass-balance method is the most widespread method in use and it is more effective for liquid carrying pipelines than for gas carrying pipelines. Multiple averaging periods are used to enable the quick detection of large leaks while still maintaining sensitivity for small leaks.

The *Volumetric testing* technique measures the volume of liquid per hour that must be added to a pipeline system to maintain a constant pressure, thus providing the leakage flow rate in gallons per hour. Since volume or pressure may change with temperature, a temperature compensation unit is incorporated with the system to prevent false alarms.

Real-time transient analysis compares actual and predicted flow rates and pressures, and identifies significant differences as leaks. Leak thresholds are continually adjusted for different flow rates so that they are above instrumentation noise levels. This method, therefore, is adaptive to transient conditions and minimizes false alarms. However, costly instrumentation of sufficient accuracy is required and tuning of steady-state and transient models may also require full-time SCADA support.

The following sensors are often used for leak detection in the above thre methods.

(1) *Calorimetric Flow Meters*
Calorimetric flow meters utilize the calorimetric principle – a means of measuring flow rate through thermal energy dispersion. Typically the measurement involves two thermistors (see Fig. 9) embedded in the probe tip or sensor housing. One of the thermistors (the reference thermistor) monitors the temperature of the fluid medium being measured, while the other is heated a fixed amount above the reference thermistor. This establishes a differential temperature between the two. If the fluid in a pipe is not moving, the differential temperature will remain stable. However, when the fluid begins to flow, it cools the heated thermistor, thus changing the differential temperature between the two thermistors proportional to the flow speed. The differential temperature can be used to derive the flow velocity, $\bar{V}$, with *King's Law* [15]:

$$\overline{V} = kl\left(1 + \sqrt{\frac{2\pi\rho c d v}{k}}\right)(T_s - T_0) \tag{3}$$

where k, c – thermal conductivity and specific heat of fluid at a given pressure; ρ – fluid density; l, d – length and diameter of the sensor respectively; T_s – temperature measured by the heated thermistor; T_0 –temperature at the reference thermistor. Calorimetric flow meters can be used to measure any medium (liquid or gas) that is thermally conductive. They provide highly accurate flow detection.

(2) *Magnetic Flow Meters*
Magnetic induction flow meters, shown in Fig. 10, operate based on the conductive properties of liquids. When a conductive fluid flows through a magnetic field, a voltage is generated (*Faraday's Law*) proportional to the average flow velocity. Knowing the conductivity of the liquid, the strength of the magnetic field, and the cross-sectional flow area, the flow velocity can be determined by measuring the voltage potential. These types of sensors are mounted at two monitoring locations (typically the inlet and outlet points of a pipeline) to accurately

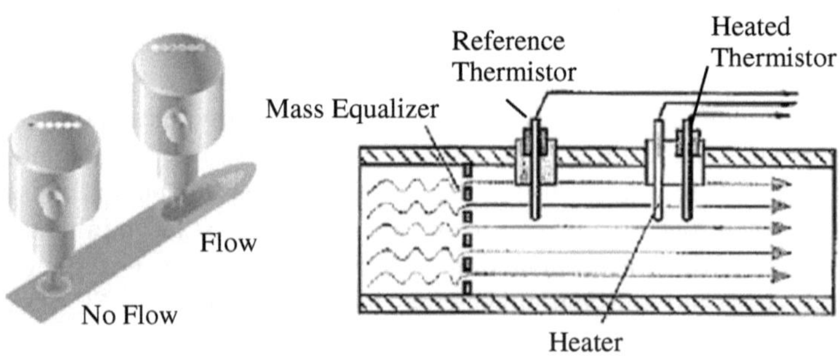

Fig. 9 Calorimetric principle for flow rate measurement [16]

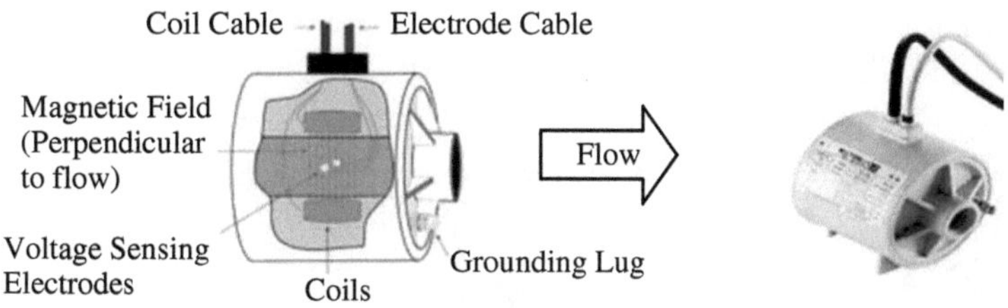

Fig. 10 A magnetic flow meter (Flomotion Systems Inc.)

detect fluid loss. There are velocity restrictions which depend on the liquid being pumped.

(3) *Ultrasonic Flow Meters*
There are two types of ultrasonic flow meters: *transit-time* (*time-of-flight*) and *Doppler flow meters*. The former is based on the principle that sound waves traveling in the direction of fluid flow require less time than those traveling in the opposite direction. The difference in transit times of the ultrasonic signals is an indication of the flow rate of the fluid. Transit-time flow meters utilize two transducers (e.g., *piezoelectric* transducers), which function as both ultrasonic transmitters and receivers. Figure 11 shows this principle in *reflection mode* (*V mode*). In Fig. 11(a), the transducer *A* emits a sound wave in the direction of flow. This wave travels inside the pipe and is reflected by the pipe wall and finally is received by transducer *B*. The time that the sound wave travels is recorded as t_1. In Fig. 11(b), the transducer *B* emits a sound wave opposite to the direction of flow. This wave travels inside the pipe and is reflected by the pipe wall and then is received by transducer *A*. The time that the sound wave travels is recorded as t_2. The difference, $\Delta t = t_2 - t_1$, is proportional to the flow velocity. The transducers can also be arranged in a *diagonal mode* (*Z mode*), where they are mounted on opposite sides of the pipe (e.g., top and bottom).

The operating principle of *Doppler flow meters* employs the frequency shift (*Doppler Effect*) of an ultrasonic signal when it is reflected by discontinuities (suspended particles or gas bubbles) in a flowing liquid. Ultrasonic sound is emitted

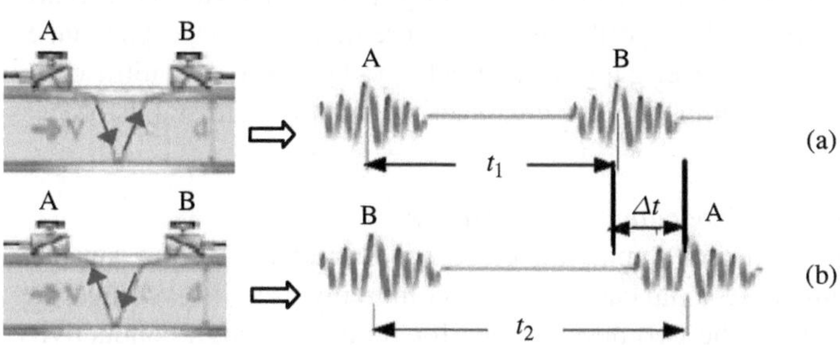

Fig. 11 Transit-time principle for flow rate measurement (Source: JLC International)

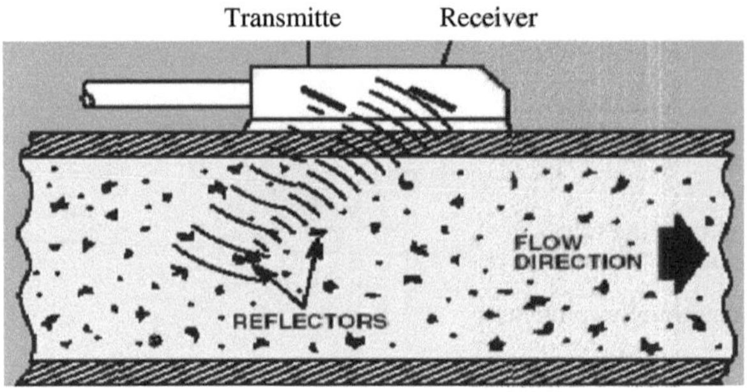

Fig. 12 Doppler principle for flow rate measurement (OMEGA Engineering Technical Reference, www.omega.com)

by a transmitter into a pipe, and the reflected wave is received with a different frequency. The frequency difference is directly proportional to the flow rate of the liquid (Fig. 12). Current technology requires that the liquid contain at least 100 parts per million of 100 micron or larger suspended particles or bubbles.

Both types of ultrasonic flow meters feature clamp-on designs with transducer assemblies that can detect flow rates from the outside the pipe. Installation can be made without breaking open the line or interrupting flow. Ultrasonic sensors can be used for a wide range of fluid from water to natural gas, to sewage. The systems provide high measuring accuracy, create no pressure losses or flow obstructions, and have low power consumption.

2.10 Level Measurement Methods

Level monitoring methods use high precision liquid level measurements to determine the volume of liquid contained in tanks. Holding tanks are found in a wide variety of plumbing systems. In a closed system, when there is no leak, the tank volume should remain constant for a given pressure and temperature. Any variation in the level indicates possible leaks. There are a variety of technologies available for level measurement (see Fig. 13), such as sight glasses, float switches, hydrostatic devices (displacers, bubblers, differential pressure transmitters), capacitance-, conductance-, and magnetic-level gauges, load cells, radar, laser, and ultrasonics. The method selected is determined by the characteristics of the fluid and its process conditions.

(1) *Radio Frequency (RF) Capacitance Level Meters*
RF capacitance level meters employ the electrical characteristics of a capacitor for level measurement. Coaxial capacitors are commonly used in level measurement. Figure 14 shows the two designs: one for fluids with high conductivity, the other for fluids with low conductivity. In Design (a), the metal probe (coated with a thin isolating layer to prevent an electric short through the liquid) acts

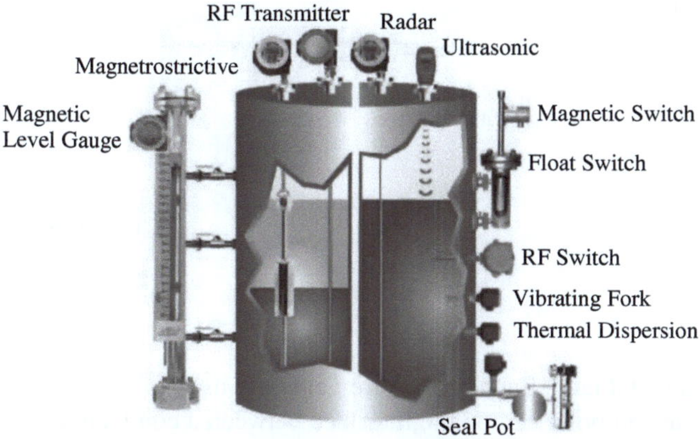

Fig. 13 Variety of technologies for level measurement [17]

as one plate of the capacitor, while the grounded conductive liquid acts as the other. The dielectric medium in this case is the probe's sheath. In Design (b), a bare probe and the metallic vessel wall form the electrodes of a capacitor; the dielectric medium is the liquid. The vessel wall (or reference probe) is grounded. When the liquid level changes, capacitance changes proportionally, and can be measured.

RF capacitance level measurements are suited for detecting the level of liquids, slurries, and granular products and can be designed for level measurement at a specific point, at multiple points, or continuously over the entire vessel height. Radio frequencies for all types range from 30 kHz to 1 MHz [18].

(2) *Conductance-Based Level Meters*

The conductance method is based on measuring the electrical conductance of a liquid using a low-voltage power source (normally <20 V) [18]. Conductance is a relatively low-cost, simple method to detect level in a vessel. One common way to set up an electrical circuit is to use a dual-tip probe that eliminates the

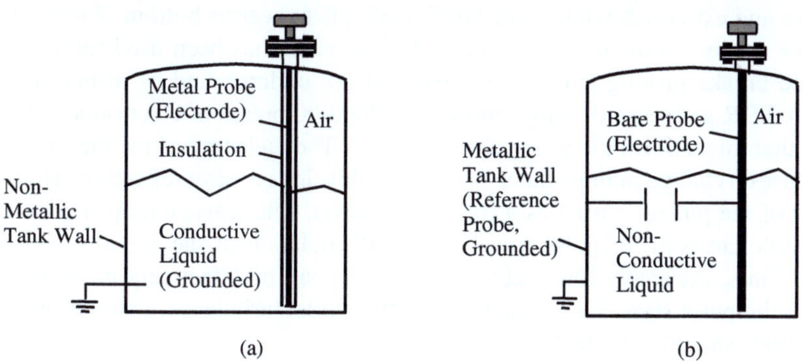

Fig. 14 RF capacitive liquid level sensors

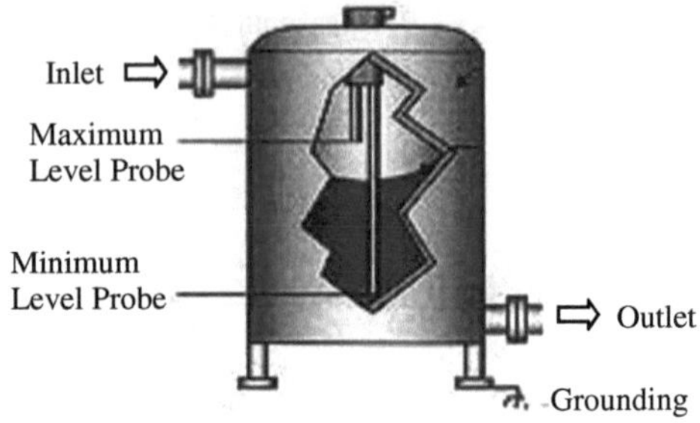

Fig. 15 A conductance-based liquid level sensor [18]

need for grounding a metal tank. Such probes are generally used for point level detection, and the detected point can be the interface between a conductive and nonconductive liquid. Figure 15 shows an arrangement with two dual-tip probes that detect maximum and minimum levels. A low tank condition is sensed when the liquid level falls below the bottom probe.

(3) *Radar or Microwave Level Meters*

Radar methods use electromagnetic waves to detect leaks (typically in the 10 GHz microwave band range, thus referred to as *microwave* types) [18]. There are two primary noninvasive types and one invasive type of radar level meters. One noninvasive method uses *frequency-modulated continuous wave* (FMCW) technology. From the top of the tank, an oscillator sends down a linear frequency sweep at a fixed bandwidth and sweep time. The reflected radar signal, with a frequency different from that of the transmitted signal, is delayed in proportion to the distance traveled to the fluid. These two signals are used to accurately determine the liquid level. *Pulsed radar* (*pulsed time-of-flight*) is another noninvasive type. A radar pulse is sent to the liquid's surface and the transit time of the pulse's return is used to indicate level. Because pulsed radar is lower power than FMCW, its performance can be affected by obstructions in the tank as well as foam and low-dielectric materials. The choice of one or the other depends on application factors such as tank obstructions that may serve as reflectors, the presence of foam, and fluid turbulence.

Guided-wave radar (GWR) is an invasive method. It uses a rod or cable to guide the wave and extends down into the tank's contents to near its bottom. The basis for GWR is time-domain reflectometry (TDR), which has been used for years to locate breaks in long lengths of cable that are underground or in building walls. A TDR generator develops more than 200,000 pulses of electromagnetic energy that travel down the waveguide and back. The dielectric of the measured fluid causes a change in impedance that in turn develops a wave reflection. Transit time of the pulses is used as a measure of level. The waveguide provides a highly efficient path for pulse travel so that degradation of the signal is minimized. Thus, extremely low dielectric materials can be effectively measured. Because the pulse signals are channeled by the guide, turbulence, foam, or tank obstructions should not affect the measurement.

(4) *Sonic and Ultrasonic Level Meters*
Both sonic (≤ 10 kHz) and ultrasonic (20~200 kHz) level instruments use sound waves to determine fluid level. A piezoelectric transducer located at the top of the tank emits sound waves downward onto the measured surface. Echoes of these waves return to the transducer and the travel time is used to determine the fluid level. Some problems that may be encountered when using this method are [18]: (a) the speed of sound through the medium (usually air) varies with the medium's temperature – a temperature sensor to compensate for changes may be needed for accurate measurement. (b) turbulence of the liquid can cause fluctuating readings – use of a damping adjustment in the instrument may help overcome this problem. (c) heavy foam on the surface of the fluid can act as a sound absorbent – a waveguide may be needed to guide the wave to enhance performance.

(5) *Magnetic Level Gauges and Magnetostrictive Level Transmitter*
Magnetic level gauges (MLG) are similar to conventional float devices, except the MLG's float carries a set of strong permanent magnets that ride in an auxiliary column (float chamber) connected to the vessel. As the float rides up and down with the fluid level, a magnetized follower moves with it, showing the exact position of the float and thereby the fluid level.
Magnetostrictive level transmitters (MLT) are based on the magnetostrictive principle. Figure 16 shows a KSR Magnetostrictive level transmitter [19]. A current impulse generates a magnetic field, 3, along the length of a wire, 1. The wire is made of magnetostrictive material and it is held under tension inside a guide tube. A float, riding on the liquid surface, is fitted with permanent magnets, 4. When a pulse reaches the float, the two magnetic fields interact and a torsional force results, which induces a torsional stress wave, 5, in the wire. A piezoceramic converter at the end of the wire in the transmitter housing, 2, converts this stress into an electrical signal. By measuring the elapsed transit time, it is possible to determine the start point of the torsional stress wave

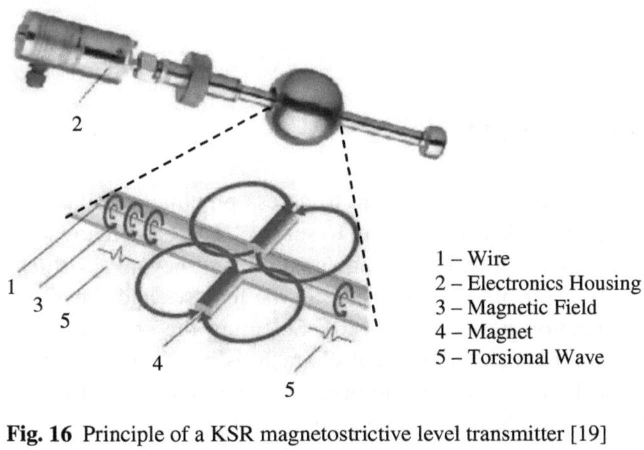

Fig. 16 Principle of a KSR magnetostrictive level transmitter [19]

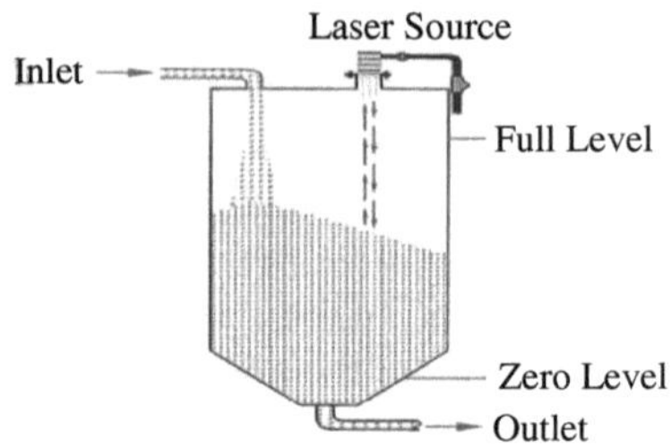

Fig. 17 A laser transmitter uses a reflected light beam to measure level [17]

and therefore the float position with a high degree of accuracy. This system can work only if the auxiliary column and chamber walls are made of nonmagnetic material.

(6) *Laser Level Transmitters*
Lasers operate on a principle similar to that of ultrasonic level sensors. Instead of using sound waves to find the level, a light beam is used. Figure 17 demonstrates this principle. A laser transmitter at the top of a vessel emits a short pulse of light down to the liquid surface, and the light is reflected back to the detector. The elapsed time is recorded to determine the distance. The laser method is very accurate and gives no false echoes.

2.11 Chemical-Based Leak Sensors

Chemical sensors are used to detect changes in chemical concentration. In this application, electrochemical sensors are calibrated to background concentration levels of a particular compound of interest. In some situations this compound may be added to a liquid or gas to serve as a tracer and make detection feasible. The sensors are then placed in potential leak locations, where readings can be taken and compared with threshold values to determine either a "leak" or "no leak" condition.

3 Technology Selection Criteria

Immediately following earthquakes, it is important to assess pipelines for damage in order to facilitate repairs and quickly return them to service. There are many leak detection methods available and each technique has positive and negative virtues. No one method is universally applicable or possesses all the features and functions required for perfect leak detection performance [20]. Selecting the optimal technology depends on many factors such as physical characteristics of the pipeline, location and accessibility of the pipeline, and properties of the fluid carried in the pipeline. Some key attributes of a robust leak detection system are:

(1) **Sensitivity:** the ability to detect small leaks.
(2) **Locating Capability:** the ability to identify the location of leaks.
(3) **Rate Estimation:** the ability to measure leak rates. Some leak detection techniques, such as temperature compensated pressure testing, can provide a volumetric measure of the leak rate, while other techniques, such as liquid sensitive cables, can identify leak location, but not leak rate.
(4) **Accuracy/Reliability:** the system must have enough sensitivity to accurately detect leaks but at the same time it must not generate an excess number of false alarms during leak-free operations.
(5) **Robustness:** the ability to perform accurate measurements in severe environments or rough operating conditions. Without this, some tests may require temporary suspension of operations. For example, temperature compensated pressure tests must be conducted when a pipeline is "quiet" and acoustic techniques can be disturbed by vibrations generated from heavy traffic in the surrounding area.
(6) **Efficiency:** the ability to quickly and accurately perform detection tasks. This includes sensor installation time, detection cycles (the number of times an operation must be performed in order to achieve satisfactory results), and effectiveness of the algorithms in making leak determinations.
(7) **Calibration/Configuration Simplicity:** hardware and software that has minimum configuration and tuning requirements in order to achieve precise measurement results. In addition, once adjusted, the equipment should maintain calibration for extended periods.
(8) **Power Independence:** the ability to operate without external communication or power sources. During and following earthquakes these services may not be available.
(9) **Operational Simplicity:** detection tools that can be installed and operated with minimum training and skill requirements.
(10) **Adaptability:** the ability to perform accurate inspections under various conditions. For example, acoustic techniques may be affected by the type of soil around the pipeline; tracer gas migrates more quickly in dry, porous soil than in wet soil; and older, small diameter pipelines containing sharp bends may be unsuitable for pigging.

In some cases, more than one detection technique may be required in order to conduct a complete and accurate inspection. For example, a leak might first be identified with one technique (e.g., pressure testing), and then located with another technique (e.g., tracers), and finally a third technique might be used to measure the leak rate (e.g., flow meters).

4 Summary

The main objectives of post-earthquake pipeline inspection are to assess pipeline damage, prioritize repairs, and return the pipeline to operation in the shortest time possible. Investigation of pipeline failures also provides valuable data for designing safer, more reliable pipeline systems that have greater resistance to earthquake damage. This paper reviews existing pipeline leak detection technology. These technologies apply a broad range of principles from the fields of physics, material science, electrochemistry, electronics, and computer engineering. Each technique provides a unique means for detecting leaks. Selecting the best technology depends on many factors such as the pipelines' physical characteristics, properties of the fluid, and environmental conditions.

References

1. Hwang H, Lin H, Shinozuka M (1998) Seismic Performance Assessment of Water Delivery Systems. In: Journal of Infrastructure Systems, Vol. 4, pp. 118–125.
2. Kitaura M, Miyajima M, and Namatame N (1998) Damage to Water Supply Pipelines during the 1995 Hyogeken Nambu Earthquake and Its Seismic Response Analysis. Proc., 3rd China-Japan-US Trilateral Symposium on Lifeline Earthquake Engineering, pp. 81–88.
3. U.S. Army Corps. of Engineers (2001) Leak Detection, Public Works Technical Bulletin 420-49-36.
4. Park G, Cudney H, Inman D (2001) Feasibility of Using Impedance-based Damage Assessment for Pipeline Systems. In: Earthquake Engineering & Structural Dynamics Journal, Vol. 30, No. 10, pp. 1463–1474.
5. Albert T (2004) Technology Assessment for Delivery Reliability for Natural Gas Inspection Technologies: RFEC. Gas Technology Institute, Des Plaines, IL.
6. Olga D, Kaspar A, Lakmal S (2002) State of the Art in Sensor Technologies for Sewer Inspection. In: IEEE Sensors Journal, Vol. 2, No. 2, April, pp. 73–81.
7. Lee D, Ban S, Lee M, Lee D (2006) Micro Gas Sensor Array with Neural Network for Recognizing Combustible Leakage Gases. In: IEEE Sensors Journal. Vol. 5, No. 3, June, pp. 530–536.
8. Fenner R, Zdankiewicz E (2001) Micromachined Water Vapor Sensors: A Review of Sensing Technologies. In: IEEE Sensors Journal, Vol. 1, No. 4, December, pp. 309–317.
9. Gao Y, Brennan M, Joseph P, Muggleton J, Hunaidi O (2004) A Model of the Correlation Function of Leak Noise in Buried Plastic Pipes, Journal of Sound and Vibration, Vol. 277, pp. 133–148.
10. Hunaidi O, Chu W (1999) Acoustical Characteristics of Leak Signals in Plastic Water Distribution Pipes, Applied Acoustics, Vol. 58, pp. 235–254.
11. Brennan M, Joseph P, Muggleton J, Gao Y (2008) Some Recent Research Results on the Use of Acoustic Methods to Detect Water Leaks in Buried Plastic Water Pipes. Retrieved on Feb. 8, 2008 at: www.isvr.soton.ac.uk/DG/Water%20Leak%20Detection%20in%20Plastic%20Pipes.pdf.
12. Roland P (2007) Thin Film Hydrogen Sensors. National Renewable Energy Laboratory, retrieve on Feb. 8, 2008 at: www.nrel.gov/hydrogen/adv_materials.html.
13. www.uson.com/content/usefulinfo/leakrate, retrieved on Jan. 14, 2008.
14. Silva A, Buiatti M, Cruz L, Pereira R (1996) Pressure Wave Behavior and Leak Detection in Pipelines. In: Computers and Chemical Engineering, Vol. 20, pp. 491–496.
15. King L (1914) On the Convention of Heat from Small Cylinders in a Stream of Fluid. Phil. Trans. Roy. Soc., A214, 373, 1914.
16. Fraden J (2003) Handbook of Modern Sensors: Physics, Designs, and Applications. Springer-Verlag, New York 2003.
17. Kevin H (2004) A Dozen Ways to Measure Fluid Level and How They Work. In: Sensors, December.

18. Gabor V (2000) The Principles of Level Measurement. In: Sensors, October.
19. KSR Magnetostrictive Level Transmitters Catalog, www.ksr-usa.co/sensors.htm.
20. Mo M, Jason S (2004) Pipeline Operation & Maintenance: A Practical Approach. Thomas Van Hardeveld, pp. 554–555.

Multiple-Rangefinders Calibration Based on Light-Section Method Using Spheres

Yusuke Yachide, Makoto Ikeda and Kunihiro Asada

Abstract This paper shows a camera and projector calibration scheme based on the light section method using spheres for multiple rangefinders. The proposed calibration fits the scanning data based on the light-section method to the shape of each sphere. The procedure of the proposed calibration consists of two steps in camera and projector calibration. First, the locations of spheres are estimated by the projected spheres' images in an image sensor. Second, the camera and projector parameters are obtained using the locations of spheres. We propose an objective function with both of the locations and the shapes of spheres. The estimated location of spheres only using a camera has an error. Therefore we calibrate all parameters changing the ratio of the locations and shapes in the function. And also, we extend the calibration method to the multiple-rangefinder system. The range accuracy is similar to the other non-self calibration method. We calibrate the camera and projector parameter only using spheres without known position, which contributes low cost calibration.

Keywords Calibration · light-section method · sphere · multiple viewpoints

1 Introduction

Calibration is an essential factor for the accurate 3-D data reconstruction. Calibration methods have been studied for camera, projector parameter acquisition. The calibration parameters have an internal and external parameter. For example, in a

Yusuke Yachide
Department of Electronic Engineering, The University of Tokyo,
e-mail: yachide@silicon.t.u-tokyo.ac.jp

Makoto Ikeda
Department of Electronic Engineering, VLSI Design and Education Center (VDEC), The University of Tokyo, e-mail: ikeda@silicon.t.u-tokyo.ac.jp

Kunihiro Asada
Department of Electronic Engineering, VLSI Design and Education Center (VDEC), The University of Tokyo, e-mail: asada@silicon.t.u-tokyo.ac.jp

camera, the internal parameters consist of focal length, the position of the optical axis on an image sensor, a pixel aspect ratio and pixel scale. The external parameter includes a rotation matrix and a parallel translation vector. High-accuracy calibration parameter acquisition is a key for the high-accuracy 3-D data reconstruction, since the accuracy of the 3-D reconstruction depends on the accuracy of the calibration parameters.

The camera calibration method can be categorized into two classes: self-calibration and non-self-calibration. The self calibration [1, 2, 3] does not use a calibration target. Many images of a static scene are captured by a moving camera. The internal parameters do not change during the camera moves in the scene. The camera parameters are obtained by corresponding points in captured images. However, the accuracy strongly depends on the image noise. And also the solution in the calibration is often unstable. It is feasible for flexible range-finding systems. However, it is not feasible for the high-accuracy range-finding. On the other hand, the non-self calibration method [4, 5, 6, 7] uses a calibration target. It extracts calibration parameters using corresponding point between a point in target's coordinate and a point in a camera's/projector's coordinate. The camera parameters can be obtained stably and accurately compared with the self calibration. For example, the paper [8] reported that about 1~3mm range error inside the measurement distance of around 850~2450 mm can be achieved using the non-self calibration methods. However, many corresponding points are required to obtain high-accuracy camera parameters using a specific calibration target, which causes the high calibration cost. There are trade-off between the accuracy and the calibration cost which mean the detection of the corresponding points. Moreover the cost increases in case of a multiple-vision system. The scalability of the calibration is important for the multiple-viewpoint systems [9, 10].

The calibration method using spheres has been reported [11, 12] for low cost calibration. They calibrated the camera parameters utilizing the ellipticity of projected sphere's images. Their method estimates the imaged sphere center and the camera parameters. They only captured the sphere's images at various locations in target space for the calibration. And the correspondences are not needed. Therefore, the calibration using spheres is feasible for the calibration in multiple-viewpoint system. However, the accuracy of the sphere's locations and the camera parameters depends on the noise of the sphere image. For example, there is large focal length error of 1–20% in 0-2 pixel image noise. This error causes an increase in range error. Moreover, it only calibrates camera parameters (not projector calibration). Our primary motivation is high-accuracy range-finding in a multiple-viewpoint system with low calibration cost. Therefore, we employ the light-section method which has a capability of high-accuracy range finding. The projector also has to be calibrated in the light-section method. For the purpose, we propose the calibration method based on light section method using spheres. Our method uses the many captured 3-D positions without explicitly correspondences, which contributes the high-accuracy calibration with low calibration costs.

In this paper, we present the calibration method using spheres for multiple rangefinders each of which consists of cameras and projectors. Our method is feasible for multiple-viewpoint range-finding system. Firstly we show the calibration method for a rangefinder. In the calibration, we estimated location of each sphere utilizing the ellipticity of imaged spheres. After the estimation, we calibrate the camera and projector parameters fitting the shapes of spheres assisted by the estimated sphere's location. Secondly, we describe the extension of the multiple-viewpoint calibration utilizing sphere locations which are captured in multiple viewpoints.

2 Light-Section Method

The light-section method is based on the triangulation method. Figure 1 shows the principle of the light-section method. The 3-D position is detected by a range finder which consists of a camera and a scanning sheet beam projector. The sheet beam is projected on a target, and the camera detects the incident beam positions. The distance between the projector and the camera is defined as *d*. The scanning mirror in the projector provides the direction $\alpha 1$ of projected light. θ'corresponds to the angle between the X-Z plane and a point on the target while $\alpha 2$ is derived from the projector position on the image sensor. When a target object is placed at *m(Xi, Yi, Zi)*, a position sensor detects a reflected beam at *e(ui, vi)* on the sensor plane. The target position *m* is calculated by following equations,

$$
\begin{aligned}
X_i &= \frac{d(\tan\alpha_1 - \tan\alpha_2)}{2(\tan\alpha_1 + \tan\alpha_2)} \\
Y_i &= \frac{d\tan\alpha_1 \tan\alpha_2}{\tan\alpha_1 + \tan\alpha_2}\sin\theta' \\
Z_i &= -\frac{d\tan\alpha_1 \tan\alpha_2}{\tan\alpha_1 + \tan\alpha_2}\cos\theta'
\end{aligned}
\tag{1}
$$

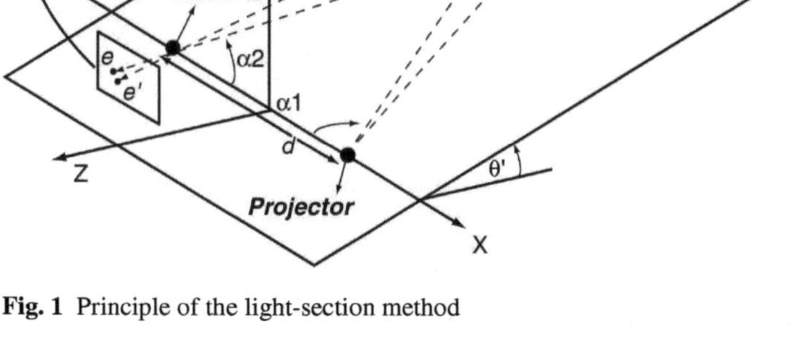

Fig. 1 Principle of the light-section method

3 Calibration Method

We describe the proposed calibration method based on the light-section method using spheres. First, we describe the calibration model. Secondly we show the calibration method. The proposed technique has two steps in the calibration process. In the first step, the locations of spheres are estimated by the projected sphere images in an image sensor. In the second step, the camera and projector parameters are obtained using the estimated location of spheres as the reference. We propose an objective function with both of the locations and the shapes of spheres because the estimated location of spheres only using a camera has an error. Therefore, we calibrate all parameters changing the weighted ratio of the factors of the location and sphericity in the function.

3.1 Calibration Model

Figure 2 shows the calibration model in the 3-D coordinate. We ignore the lens distortion. In our calibration model, the origin of the 3-D coordinate is defined as the center of the lens and X-, Y-axis are defined as the direction of the image sensor's row, column address, respectably. And Z-axis is also defined as the optical axis. We list the definitions as below. And also the calibration method is based on the assumptions described as below.

- Definitions
  1. Origin: center of the lens
  2. Defined X-, Y-axis in the direction of row and column of the image plane
  3. Defined Z-axis as optical axis

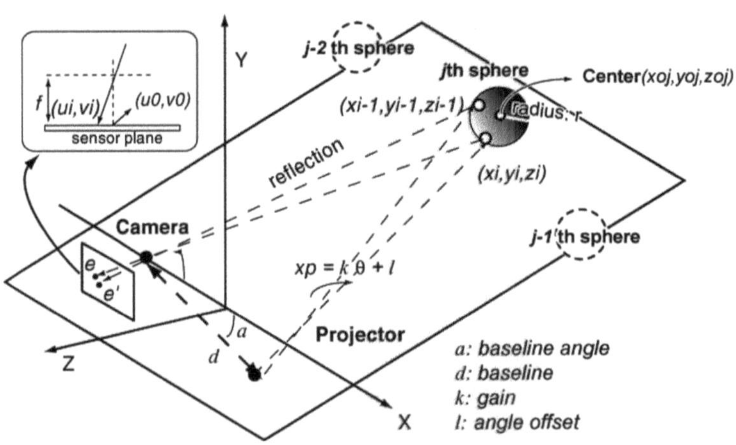

Fig. 2 Calibration model based on light-section method

- Assumptions

1. Lens is parallel to the image plane
2. Mirror rotation axis is same as the projected position on the mirror

The first assumption means that the Z-axis is vertical to the image plane since the X-, Y-, Z-axis are at right angles to each other. The second assumption means that the sheet beam is projected to the rotation axis of the mirror. Note that the pixel scale (pitch) is known. And we listed the calibration parameters as below.

- Camera parameters

1. $(u0, v0)$: Optical axis position address on image sensor
2. f: Focal length

- Projector parameters

3. k: Mirror angle gain of DA converter
4. l: Mirror angle offset
5. d: Baseline length
6. a: Baseline angle

Here, the projection angle is calculated with a first-order approximation using k and l,

$$\theta = k \cdot xp_i + l \tag{2}$$

where, xp_i is a operational signal of the projection angle for DAC. The DAC controls the projection mirror. The ith 3-D position *(xi, yi, zi)* is an intersection position of the sheet beam and the incident light. The sheet beam is defined as a plane equation. The incident light is defined as a line passing through the origin and a projected position $(ui, vi, -f)$ on the image sensor. The intersection position, which is the 3-D position based on light-section position, is calculated by following equation,

$$\begin{aligned} x_i &= \frac{u_i(d\cos\alpha\sin\theta - d\sin\alpha\cos\theta)}{u_i\sin\theta + f\cos\theta} \\ y_i &= \frac{v_i(d\cos\alpha\sin\theta - d\sin\alpha\cos\theta)}{u_i\sin\theta + f\cos\theta} \\ z_i &= \frac{f(d\cos\alpha\sin\theta - d\sin\alpha\cos\theta)}{u_i\sin\theta + f\cos\theta} \end{aligned} \tag{3}$$

3.2 Sphere Center Location Estimation (1st step)

We describe the first step of the sphere's location estimation. The sphere is projected as an ellipse on the image sensor in general. First, the image sensor captures a sphere image and the binary image of the sphere is generated. After the generation of the binary image, we estimate the location of the sphere. Figure 3 shows the projection figure of the sphere on the image plane and 3-D space. The number of the pixels

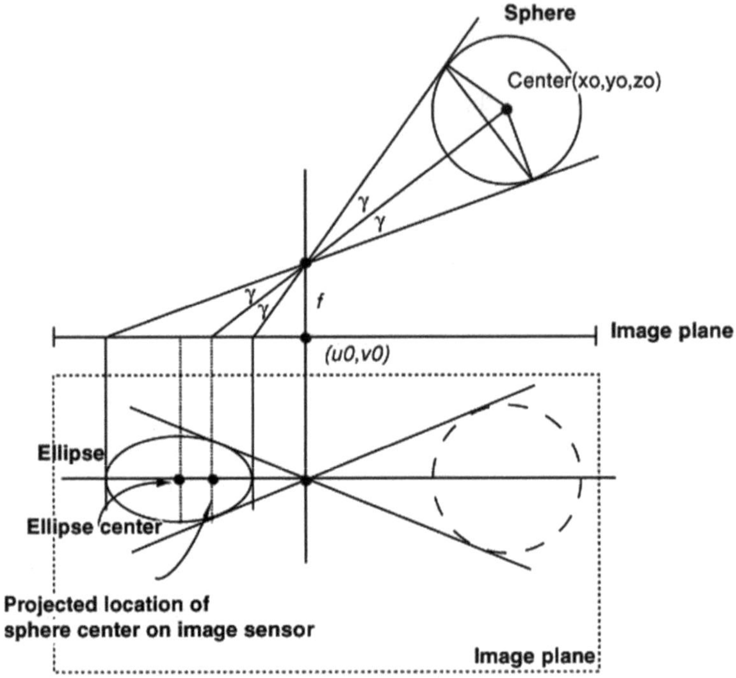

Fig. 3 Projected sphere image on an image sensor

is counted by searching all angles around the center $(u0, v0)$. We decide the major axis of the ellipse images according to the largest counted number of the pixels. The edge pixels through the major axis of the ellipse are detected. The farthest pixel and the nearest pixel are defined as (uf, vf) and (un, vn), respectively. We can obtain the angle using the focal length f and the farthest and the nearest pixels of the ellipse by following equation,

$$\gamma = \frac{1}{2}\left(\arctan\left(\frac{\sqrt{\left(u_f - u_0\right)^2 + \left(u_f - u_0\right)^2}}{f}\right) - \arctan\left(\frac{\sqrt{(u_n - u_0)^2 + (u_n - u_0)^2}}{f}\right)\right) \quad (4)$$

Note that, in general, the center of the sphere does not correspond to the center of the ellipse as shown in Fig. 3. The location of the sphere center is estimated by the angle γ and sphere radius r. The above step is processed in all pixels and focal lengths. Note that we search around the focal length of the lens. Finally we decide the location of the jth sphere center (xoj, yoj, zoj) minimizing the below function,

$$F'(x_i, y_i, z_i; k, l, d, a) = \left(\frac{1}{M}\right)\sum_{j=1}^{M}\left(\frac{1}{N_j}\right)\sum_{i=1}^{N_j}\left(\sqrt{\left(x_{oj} - x_i\right)^2 + \left(y_{oj} - y_i\right)^2 + \left(z_{oj} - z_i\right)^2} - r\right)^2 \quad (5)$$

where, N_j is the number of the scanning points in jth sphere, M is the number of the spheres, r is the radius of the sphere.

3.3 Camera, Projector Calibration (2nd step)

Here, we show the camera and projector calibration using the estimated sphere center location in 1st step. We assume that the estimated sphere center location contains an error. Therefore, we calibrate the camera and projector parameters with the sphericity around the estimated position. The estimated location of the sphere center assists the calibration. We define the functions F, G, H(xi, yi, zi; $u0$, $v0$, f, k, l, d, a, r, $xooj$, $yooj$, $zooj$) (they are represented as F, G, H, respectively) for the calibration

$$\begin{aligned} F = &\alpha\left(\frac{1}{M}\right)\sum_{j=1}^{M}\left(\frac{1}{N_j}\right)\sum_{i=1}^{N_j}\left(\left(x_{ooj}-x_{oj}\right)^2+\left(y_{ooj}-y_{oj}\right)^2+\left(z_{ooj}-z_{oj}\right)^2\right) \\ &+(1-\alpha)\left(\frac{1}{M}\right)\sum_{j=1}^{M}\left(\frac{1}{N_j}\right)\sum_{i=1}^{N_j}\left(\sqrt{\left(x_{ooj}-x_i\right)^2+\left(y_{ooj}-y_i\right)^2+\left(z_{ooj}-z_i\right)^2}-r\right)^2 \end{aligned} \quad (6)$$

$$G = \left(\frac{1}{N_j}\right)\sum_{i=1}^{N_j}\left(\sqrt{\left(x_{ooj}-x_i\right)^2+\left(y_{ooj}-y_i\right)^2+\left(z_{ooj}-z_i\right)^2}\right)$$

$$H = \left(\frac{1}{N_j}\right)\sum_{i=1}^{N_j}\left(a_i^2+b_i^2-c_i^2\right)$$

$$a_i = \sqrt{x_i^2+y_i^2-z_i^2}$$

$$b_i = \sqrt{\left(x_{ooj}-x_i\right)^2+\left(y_{ooj}-y_i\right)^2+\left(z_{ooj}-z_i\right)^2}$$

$$c_i = \sqrt{x_{ooj}^2+y_{ooj}^2-z_{ooj}^2}$$

where, M, N_j are the number of the spheres and the number of the points in the jth sphere, a is a weighted ratio of the function, ($xooj$, $yooj$, $zooj$) is the location of the sphere which is estimated in 2nd step, and (xoj, yoj, zoj) is the estimated location of the sphere from the 1st step. The 3-D position (xi, yi, zi) is shown in (3). r is the radius of the sphere. Note that the radius of all spheres is the same. The function F is the main function for the calibration. The function G is defined as the average of the length between the 3-D position (xi, yi, zi) and the center of the sphere ($xooj$, $yooj$, $zooj$). The function H is based on the Pythagorean theorem. We calibrate the camera and projector parameters by minimizing the function F with the constraints described below.

$$\begin{array}{ll} \text{Minimize} & F \\ \text{subject to} & G = r \\ & H \leq 0 \end{array} \quad (7)$$

The 1st constraint means that the average of the length between the 3-D position (xi, yi, zi) and the center of the sphere $(xooj,\ yooj,\ zooj)$ is equal to the sphere's radius. The second constraint means the average of the angle (which formed by the intersection of A with B; A is the line passing through the 3-D position $(xi,\ yi,\ zi)$ and the center of the sphere $(xooj, yooj, zooj)$, B is the line between the 3-D position (xi, yi, zi) and the origin) based on the Pythagorean theorem is an acute angle. In this function F, the calibration parameters are $u0$, $v0$, f, k, l, d, a, r, $xooj$, $yooj$, $zooj$. They are calibrated minimizing the function F according to the weighted ratio a.

3.4 Multiple-Viewpoint Extension

The present calibration method has advantage of the scalability for multiple-rangefinder calibration. The present calibration can be applied to multiple-rangefinder calibration simultaneously. For the multiple-viewpoint calibration, the relation between the coordinates in each range finder is calibrated. The relation represents the external parameter, which consists of the rotation parameter and the parallel transfer for the coordinate transformation. In general the transformation (from a 3-D position (x_i^k, y_i^k, z_i^k) in k coordinate to standard 3-D position (x_i^s, y_i^s, z_i^s) in coordinate s) can be represented by following equation,

$$\begin{pmatrix} x_i^s \\ y_i^s \\ z_i^s \end{pmatrix} = R^k \begin{pmatrix} x_i^k \\ y_i^k \\ z_i^k \end{pmatrix} + \begin{pmatrix} t_x^k \\ t_y^k \\ t_z^k \end{pmatrix} \tag{8}$$

Here, (t_x^k, t_y^k, t_z^k) is parallel transfer from the coordinate k to the standard coordinate s and R^k is the general rotation matrix which is represented by the rotation angle $(\theta_x, \theta_y, \theta_z)$ (from coordinate k to the standard coordinate s). The detail of the coordinate transformation can be found in [13]. When a sphere is captured by multiple rangefinders, the sphere's locations in various directions are uniformed in world coordinate. Multiple spheres which are captured by multiple rangefinders in each position are respectively same. In the multiple-rangefinder calibration, we utilize the relation between sphere's positions. We defined the objective function F^+ based on the objective function F.

$$\begin{aligned} F^+ = {} & \alpha\left(\frac{1}{M}\right)\sum_{j=1}^{M}\left(\frac{1}{N_j}\right)\sum_{i=1}^{N_j}\left(\frac{1}{L}\right)\sum_{k=1}^{L}\left(\left(x_{ooj}^k - x_{oj}^k\right)^2 + \left(y_{ooj}^k - y_{oj}^k\right)^2 + \left(z_{ooj}^k - z_{oj}^k\right)^2\right) \\ & + (1-\alpha)\left(\frac{1}{M}\right)\sum_{j=1}^{M}\left(\frac{1}{N_j}\right)\sum_{i=1}^{N_j}\left(\frac{1}{L}\right)\sum_{k=1}^{L}\left(\sqrt{\left(x_{ooj} - x_i\right)^2 + \left(y_{ooj} - y_i\right)^2 + \left(z_{ooj} - z_i\right)^2} - r\right)^2 \end{aligned} \tag{9}$$

where, k represents the index for k th rangefinder. We use the equation rotation as a constraint for the multiple-viewpoint calibration further to the constraints of G, H. We obtain the calibration parameters by minimizing the objective function F^+.

$$\begin{array}{ll} \text{Minimize} & F^+ \\ \text{subject to} & G = r \\ & H \leq 0 \end{array} \tag{10}$$

4 Calibration Result

The proposed method has been implemented and tested on real data. We used ipopt [14], which is a nonlinear optimization programming, for minimizing the functions using the interior point method

4.1 Single-Rangefinder Calibration Result

The measurement distance is about from 300 to 600 mm. The width and height of the measurement volume is 400×400 mm. The radius of the target sphere is around 17.5 mm. We acquire 2-D images of 10 spheres, and scan them based on light section method. We employed light-section-based image sensor [15] for the camera. Figure 4 shows the results of 2-D images and sheet light scanning images

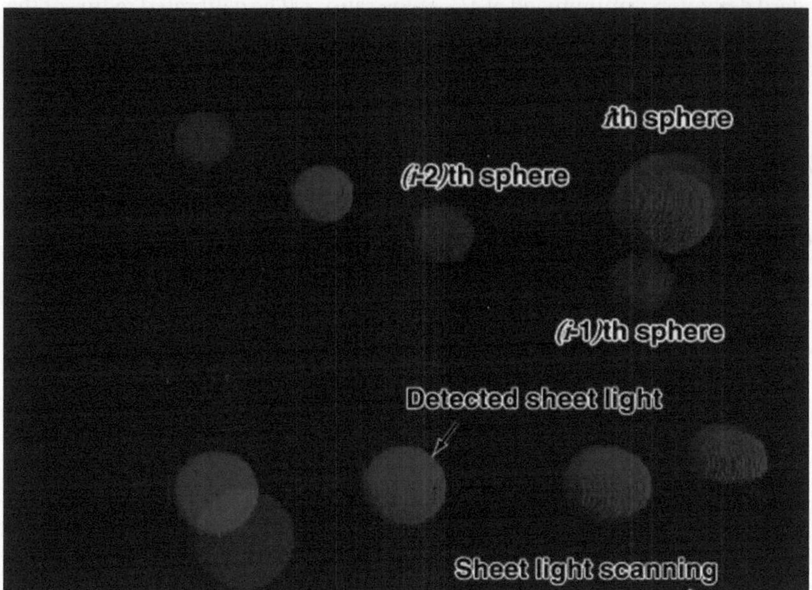

Fig. 4 Combined image of measured spheres

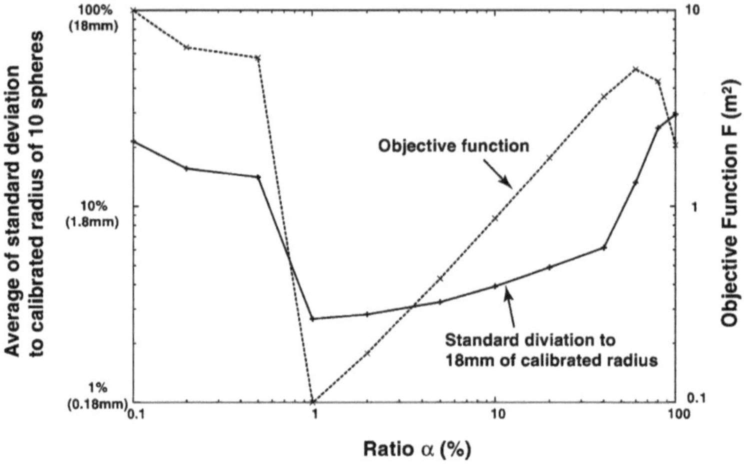

Fig. 5 Convergence result of the average of mean radius error and objective function according to the ratio α

with VGA (640 × 480 pixel) resolution. The figure is the combined image of each sphere image. The spheres are not obtained simultaneously. The lines represent the detected sheet light. Figure 5 shows the calibrated radius error according to the ratio α. The X-axis corresponds to the weighted ratio α and the Y-axis represents the standard deviation to the calibrated radius of sphere and the value of the objective function. The dashed line represents the value of the objective function F according to the ratio. The solid line represents the standard deviation to the calibrated radius. The standard deviation is minimized at 1% of the ratio α. The calibrated radius of the sphere is 18mm. We achieved 0.5 mm of the relative error to the radius (18 mm). The relative error of each sphere, the center of the sphere obtained by the camera image, and the center of the sphere obtained by the proposed method are summarized in Table 1. Figure 6 shows the histogram of the measured radius in each position. Figure 7 shows a sample of the reconstructed 3-D image of sphere by the proposed

Table 1 Calibrated result

Error[1] %(mm)	Center 1[2]	Center 2[3]
5.1(0.92)	(161.9, −113.2, 406.3)	(162.2, −111.7,409.4)
2.1(0.38)	(96.9, −115.1, 373.7)	(98.3, −114.2, 377.5)
2.2(0.4)	(8.8, −121.4, 396.0)	(9.4, −117.083,385.5)
2.2(0.4)	(−73.9, −119.1, 381.0)	(−74.6, −118.6, 380.5)
2.4(0.43)	(−51.7, −116.5, 317.1)	(−52.6, −118.3, 322.8)
1.5(0.27)	(138.7, −30.1, 461.1)	(140.0, −32.2, 466.5)
2.0(0.36)	(35.8, 0, 503.6)	(35.8, −4.9, 500.0)
2.6(0.47)	(−34.2, 19.8, 513.3)	(−35.3, 18.2, 512.2)
2.2(0.4)	(−106.2, 50.4, 515.9)	(−106.6, 50.8, 516.9)
1.8(0.32)	(100.347, 0, 306.49)	(99.7, 6.7, 302.4)

[1] Relative error to calibrated radius of sphere.
[2] (x*oj*, *yoj*, *zoj*) obtained by camera image.
[3] (*xooj*, *yooj*, *zooj*) obtained by proposed calibration.

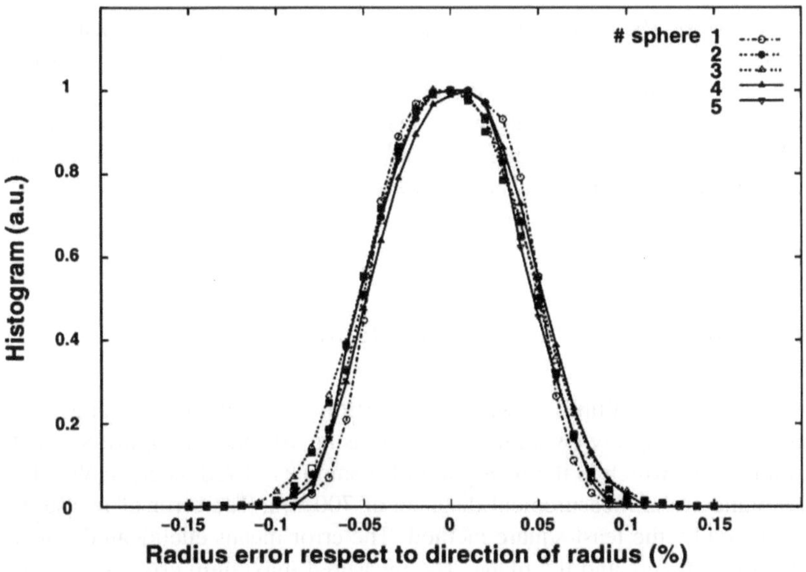

Fig. 6 Histogram of each radius error

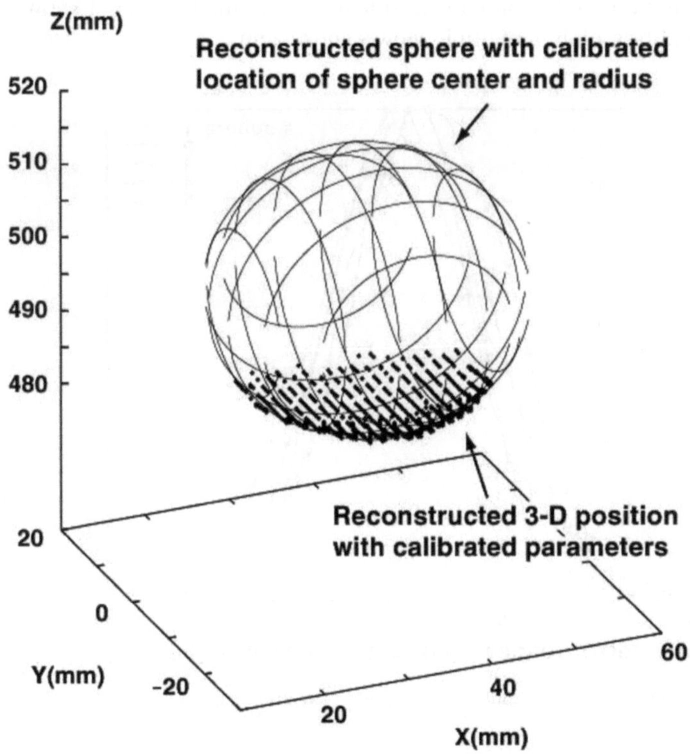

Fig. 7 One of the reconstructed result with calibrated parameters

calibration in case of 1% of the ratio α. The dots represent the reconstructed 3-D position using the calibrated parameters. The solid lines represent the reconstructed spheres. We also measured a plane at the measurement distance of 1000 mm. The error of measured plane is estimated by the least-square method. The error means euclidean distance between measured points and the plane. The measured maximum error is 2.1 mm. The result achieved the high-accuracy range-finding similar to the non-self calibration result [8].

4.2 Multiple-Rangefinder Calibration Result

We also carried out the multiple-viewpoint calibration. We calibrate two rangefinders with proposed multiple-viewpoint calibration using 10 spheres. Figure 8 shows the calibration result which is the histogram of 5 samples of radius error. We also measured a plane at the measurement distance of 700 mm. The error of measured plane is estimated by the least-square method. The error means euclidean distance between measured points and the plane. The measured maximum error is 1.9 mm. Figure 8 shows the relation between the range error and the measurement distance. Our result achieved the high-accuracy range finding similar to our previous multiple-viewpoints rangefinding result [16] employing typical linear camera calibration. We achieved the low calibration cost (only using spheres where the place is unknown) and high-accuracy calibration in a multiple viewpoint system.

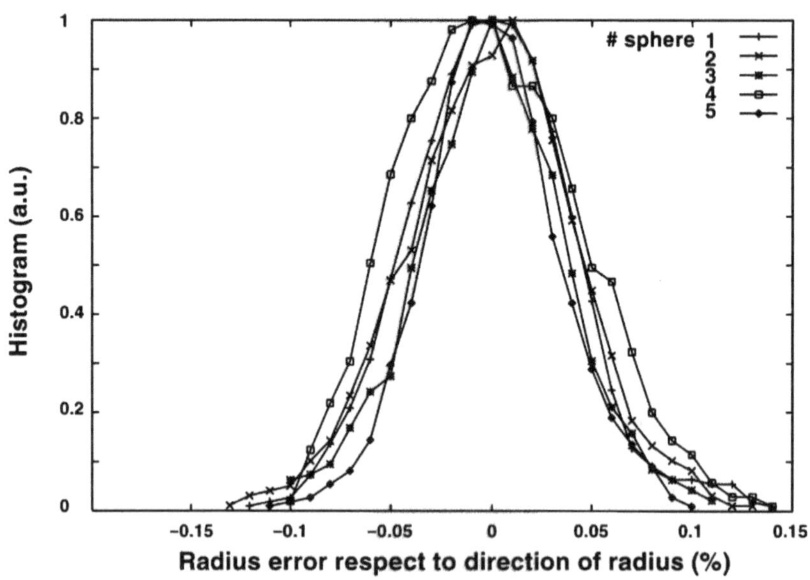

Fig. 8 Histogram of each radius error in multiple-viewpoint measurements

5 Conclusion

This paper shows camera and projector calibration method based on light section method using spheres aimed for high-accuracy calibration with low calibration cost. The proposed calibration method fits the scanning data based on light-section method to the sphere shapes. We propose the objective function with both of the locations and the shapes of spheres for the calibration. We measured a plane with the maximum error of 2.1 mm around 1000 mm. The range error is similar to the other non-self calibration result. And the method is feasible for multiple-viewpoint rangefinder calibration. We also proposed and evaluated the multiple-viewpoint rangefinder calibration. Our method has an advantage of only using spheres where the place is unknown with high accuracy. We achieved the high-accuracy range finding.

References

1. S. J. Maybank and O. D. Faugeras (1992) A theory of self-calibration of a moving camera. Journal of Computer Vision, 8(2):123–151
2. R. I. Hartley (1994) An Algorithm for Self Calibration from Several Views. In: Proc. IEEE Computer Vision and Pattern Recognition, pp 21–23
3. E. E. Hemayed (2003) A survey of camera self-calibration. In: Proc. IEEE international Conference on Advanced Video and Signal Based Surveillance, pp 351–357
4. K. Deguchi (1982) A Survey of Recent Camera Calibration Techniques. IPSJ SIG Notes. 93(25)
5. R. Y. Tsai (1987) A Cersatile Camera Calibration Technique for High-Accuracy 3D Machine Vision Metrology Using Off-the-Shelf TV Cameras and Lenses. IEEE journal of Robotics and Automation, RA-3(4):323–344
6. Z. Zhang (2000) A flexible new technique for camera calibration. IEEE Transactions on Pattern Analysis and Machine Intelligence, 22(11):1330–1334
7. J. Heikkilä (2000) Geometric camera calibration using circular control points. IEEE Transaction Pattern Analysis and Machine Intelligence, 22(10):1066–1077
8. W. Sun and J. R. Cooperstock (2006) An empirical evaluation of factors influencing camera calibration accuracy using three publicly available techniques. Journal of Machine Vision and Applications, 17(1):51–67
9. T. Kanade, P. Rander and P. J. Narayan (1997) Virtualized reality: constructing virtual worlds from real scenes. IEEE Multimedia, 4(1):34–47
10. R. I. Hartley and A. Zisserman (2004) Multiple View Geometry in Computer Vision. Cambridge University Press, Cambridge
11. M. Agrawal and L. S. Davis (2003) Complete camera calibration using spheres: A semi-definite programming approach. In: Proc. 9th IEEE International Conference on Computer Vision, pp 782–791
12. H. Zhang, K.Y. K. Wong and G. Zhang (2007) Camera calibration from images of spheres. IEEE Transaction on Pattern Analysis and Machine Intelligence, 29(3):499–502
13. D. A. Forsyth and J. Ponce (2002) Computer Vision: A modern Approach. Prentice Hall, Englewood Cliffs
14. Computational infrastructure for operations research. http://www.coin-or.org

15. Y. Oike, M. Ikeda and K. Asada (2004) Design and Implementation of Real-Time 3-D Image Sensor With 640x480 Pixel Resolution. IEEE Journal of Solid-State Circuits, 39(4):622–628
16. Y. Yachide, Y. Oike, M. Ikeda and K. Asada (2005) Real-time 3-D measurement system based on light-section method using smart image sensor. In: Proc. IEEE international Conference on Image Processing, pp. 1008–1011

An Adaptive Wiener Filter for Automatic Speech Recognition in a Car Environment with Non-Stationary Noise

Ziming Qi and Tom Moir

Abstract In real-time environments an Automatic Speech Recognition (ASR) system in a car has to receive the driver's voice only whilst suppressing background noise. Previous research by the authors presented a real-time Voice Activity Detector (VAD) which operates within a geometrical zone defined around the head of the desired speaker. The VAD only works well for a single sound source. During the period of desired speech the VAD does not work well if the unwanted speech is incoming at the same time as desired speech. As a solution to the above issue, this paper presents a real-time Wiener filter for noise reduction, improving the SNR by significant amounts in a car environment [1]. Experiments show that a simple template matching automatic speech recognition system obtained a 100% hit- rate [2].

Keywords Acoustic signal processing · adaptive wiener filtering · automatic speech recognition · automobile environment

1 Introduction

One of the most challenging and important problems in Intelligent Transport Systems (ITS) is to keep the driver's eyes on the road and his hands on the wheel. Automatic Speech recognition (ASR) offers one such solution to this problem. Speech control in car is a safe solution e.g. to enter a street name in a Global Positioning System (GPS) navigation system by speech is better than to do it by hand. However, speech recognition in a car has the inherent problem of acquiring speech signals in noisy environments. There are two types of additive noise in a car cabin: stationary and non-stationary. Stationary noise in car is from the engine (though it

Ziming Qi
Applied Technology Institute, Unitec New Zealand, Auckland, New Zealand,
e-mail: tqi@unitec.ac.nz

Tom Moir
School of Engineering and Advanced Technology, Massey University at Auckland, Auckland, New Zealand

varies with speed), road, wind, air-conditioner etc. Non-stationary noise is from the car stereo, navigation guide, traffic information guide, bumps, wipers, indicators, conversational noise and noise when passing a car running in the opposite direction [3]. Therefore noise reduction methods for speech enhancement in a car have been investigated for such applications. The Griffiths-Jim acoustic beamformer is a major technology in reducing stationary or non-stationary noise in car cabin [4]. Three microphones beamforming was used to detect the desired and undesired periods of speech by defining a geometrical "active zone". With three microphones this word boundary detector can retrieve the desired speech embedded with noise from varieties of noisy backgrounds. Simulation experiments have shown that the VAD algorithm is an effective speech detecting method that performs to an average of 80% success rate [5].

In real-time environments an Automatic Speech Recognition (ASR) system in a car has to receive the driver's voice only whilst suppressing background noise. Previous work [6] by the authors uses a real-time Voice Activity Detector (VAD) which operates within a geometrical zone defined around the head of the desired speaker. However, VAD only works well for a single sound source. During the period of desired speech the VAD does not work well if the unwanted speech is incoming at the same time as the desired speech. As a solution to the above issue, this paper presents a real-time Wiener filter for noise reduction, improving the SNR by slightly better than 28 dB in a noisy car environment [1].

The considered materials are derived from [1, 2] where they were first presented and discussed.

2 Wiener Filter

The Wiener filter was introduced by Norbert Wiener in 1949 [7] and independently for the discrete-time case by Kolmogorov [8]. Wiener-Kolmogorov filters have the following assumptions: (a) signal and (additive) noise are stochastic processes with known spectral characteristics or known autocorrelation and cross-correlation, (b) the performance criterion minimizes the mean-square error. An optimal filter can be found from a solution based on scalar or multivariable methods [9]. The goal of the Wiener filter is to filter out noise that has corrupted a signal by statistical means [10]. A Wiener filter block diagram as described by Haykin [11] is shown in Fig. 1. From Fig. 1, a microphone signal y (an M-dimensional vector) is filtered by the Wiener filter **W** (an M × M filter matrix) and the output z (an M-dimensional vector) has to estimate the desired signal with some residual error.

$$e = d - z = d - W^T y \tag{1}$$

Figure 1 shows that when $z = d$ then $e = 0$. This means that when $e = 0$, z is the estimated value of d. Therefore, when a desired signal comes with noise (white or coloured), a selected **W** matrix is available for estimating the desired signal. Thus the goal of speech enhancement is to compute the **W** matrix. Wiener filtering

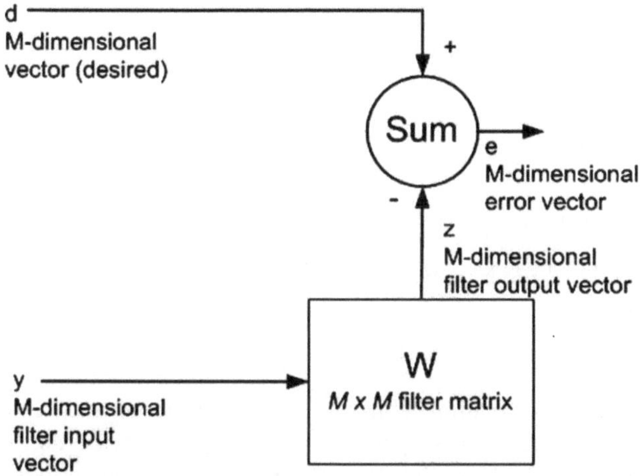

Fig. 1 Wiener filter

is commonly used in the time domain or the frequency domain to reduce noise for single-channel and multi-channel (multivariable) signals in speech or image enhancement. In order to compute and update the **W** matrix, the standard Wiener filter minimizes the mean-square error $E[e^T(k)e(k)]$ between the filter output signal and desired speech output [12]. This method is a so called adaptive filter if the signal and noise model have time-varying statistics [11, 13, 14, 15]. Doclo & Moonen [16] introduced a Wiener filter with N microphones, with each microphone signal defined as $y_n[k], n = 0, 1, \ldots, N-1$, at some sample k, the speech as $x_n[k]$ and noise component $v_n[k]$ all received at the nth microphone. Thus the mixture

$$y_n[k] = x_n[k] + v_n[k], n = 0, 1, \ldots N-1 \tag{2}$$

In single-microphone speech enhancement, the number of microphones is $N = 1$ such that the (2) simplifies to

$$y_0[k] = x_0[k] + v_0[k] \tag{3}$$

Let the filter $w_n[k], n = 0, 1, \ldots, N-1$ have length L

$$w_n[k] = [w_n^0[k] \quad w_n^1[k] \quad \ldots \quad w_n^{L-1}[k]]^T \tag{4}$$

and consider the L-dimensional data vectors $y_n[k]$, the M-dimensional stacked filter $w[k]$ (with $M = LN$), and the M-dimensional stacked data vector $y[k]$, defined as

$$y_n[k] = [y_n[k] \quad y_n[k-1] \quad \ldots \quad y_n[k-L+1]]^T \tag{5}$$

Then we have

$$w[k] = [w_0^T[k] \quad w_1^T[k] \quad \ldots \quad w_{N-1}^T[k]]^T \tag{6}$$

$$y[k] = [y_0^T[k] \quad y_1^T[k] \quad \ldots \quad y_{N-1}^T[k]]^T \tag{7}$$

Therefore, the output signal is the convolution

$$z[k] = \sum_{n=0}^{N-1} w_n^T [k] y_n [k] = w^T [k] y [k] \tag{8}$$

And also,

$$e[k] = d[k] - W[k]^T y[k] \tag{9}$$

In order to compute W[k], Doclo & Moonen [16] use the solution of the Wiener-Hopf equation

$$W[k] = R_{yy}^{-1} [k] \left(R_{yy} [k] - R_{vv} [k] \right) \tag{10}$$

where $R_{yy} [k]$ is the $M x M$ auto-correlation matrix of the input signal and it is estimated during speech plus additive noise periods. $R_{vv} [k]$ is the $M x M$ auto-correlation matrix of the input signal which is estimated during noise-alone periods.

From (8), when the **W** matrix is known, z[k] is estimated. Therefore, when the **W** matrix is pre-computed, a block diagram can be illustrated as in Fig. 2.

Therefore, a schematic for a single microphone was built as showed in Fig. 3. Microphone signal y (an M-dimensional vector) is the input signal and z (an M-dimensional vector) is the filtered signal, which is an estimate of the desired signal.

For a single microphone, we have one filter. In Fig. 3, the W matrix can be compute by collecting the input M-dimensional vector in the single microphone y under noise periods and the same microphone vector y under signal + noise periods. Therefore, $R_{yy} [k]$ is only updated at signal + noise periods and $R_{vv} [k]$ is only updated during noise alone periods. From (4) and (6), we have

$$w_0[k] = [w_0^0[k] \quad w_0^1[k]]^T \tag{11}$$

From (5) and (7), we have

$$y_0[k] = [y_0[k] \quad y_0[k-1]]^T \tag{12}$$

From (8), we have

$$z[k] = w^T [k] y [k] \tag{13}$$

During noise-alone periods, the noise input vector

$$Y(k) = [y(k), y(k-1), \ldots y(k-M+1)]^T \tag{14}$$

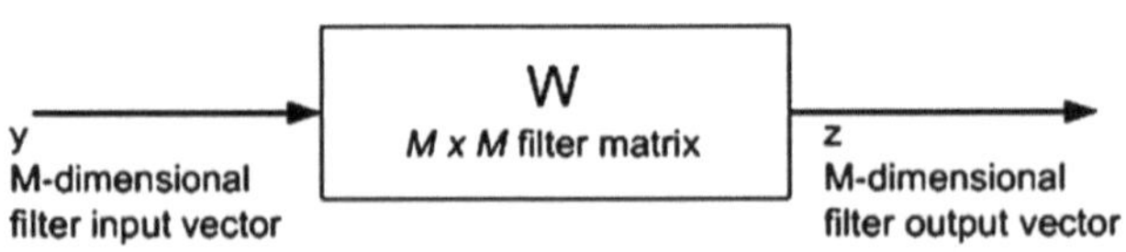

Fig. 2 A Wiener filter with pre-computed W matrix

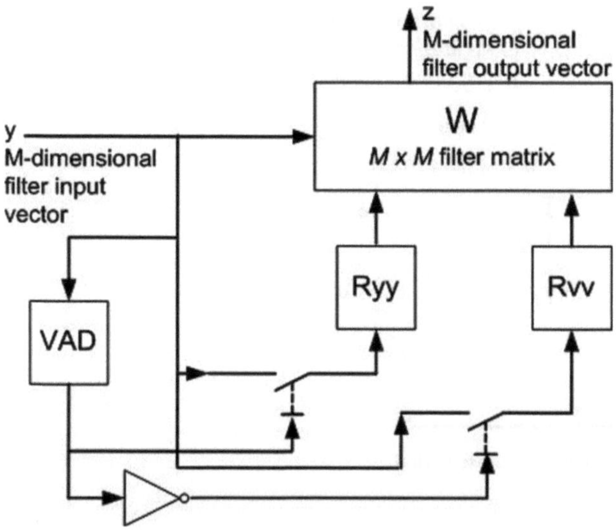

Fig. 3 W Matrix Calculation for single microphone case

where n defines the order of the noise process. The Toeplitz measurement correlation matrix is

$$R_{vv} = E\left[y(k)y^T(k)\right] \tag{15}$$

$$R_{vv} = \begin{bmatrix} r_{vv}(0) & r_{vv}(1) & \dots & r_{vv}(M-1) \\ r_{vv}(-1) & r_{vv}(0) & \dots & r_{vv}(M-2) \\ \dots & \dots & \dots & \dots \\ r_{vv}(-M+1) & r_{vv}(-M+2) & \dots & r_{vv}(0) \end{bmatrix} \tag{16}$$

The term $r_{vv}(0)$ is always real valued and

$$r_{vv}(k,k-n) = E\left[y(k)y^*(k-n)\right] \tag{17}$$

During speech and noise periods, a noise + signal vector as primary input is defined as

$$D(k) = [d(k), d(k-1), \dots d(k-M+1)]^T \tag{18}$$

where n defines the order of the noise process and the Toeplitz signal + noise correlation matrix is

$$R_{yy} = E\left[d(k)d^*(k)\right] \tag{19}$$

$$R_{yy} = \begin{bmatrix} r_{yy}(0) & r_{yy}(1) & \dots & r_{yy}(M-1) \\ r_{yy}(-1) & r_{yy}(0) & \dots & r_{yy}(M-2) \\ \dots & \dots & \dots & \dots \\ r_{yy}(-M+1) & r_{yy}(-M+2) & \dots & r_{yy}(0) \end{bmatrix} \tag{20}$$

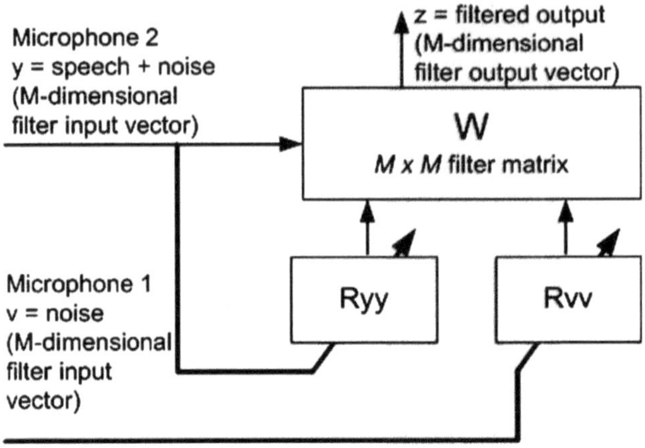

Fig. 4 W matrix updated during speech or noise period

The term $r_{yy}(0)$ is always real valued and

$$r_{vv}(k,k-n) = E\left[y(k)y^{*}(k-n)\right] \tag{21}$$

Experiments showed good results when the arrangement of Fig. 3 was used. That is when the noise source comes from a permanent location and the noise source model is not changing with time. The block diagram in Fig. 3 can be used in 1-microphone application e.g. mobile phone applications. However, for a real-time environment e.g. in a car the noise sources do not come from a permanent location

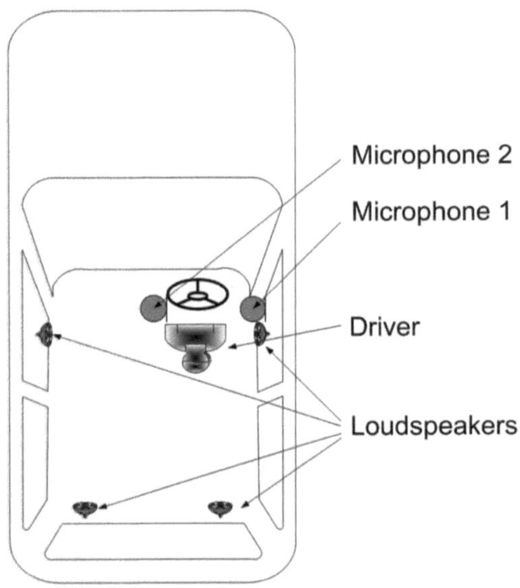

Fig. 5 In-car test plan

and the noise source model is non-stationary. If the noise characteristic changes between noise and speech periods then the **W** matrix will be in error.

Therefore, a new Wiener filter is built as in Fig. 4 with test conditions as shown in Fig. 5. Since Microphone 1 is close to the noise source, Microphone 1 is set as the noise input. Microphone 2 acts as a speech and noise input. Microphones 1 and 2 record desired speech from an equal distance whilst an unwanted speech or noise source is close to Microphone 1.

Essentially this technique is similar to the dual microphone method of Widrow et al. [17] except here we will not use least-mean squares(LMS) but instead update the Wiener filter directly by estimation of the signal + noise and noise covariance matrices and by direct solution of the Wiener-Hopf (10).

3 Automatic Speech Recognition

The enhanced signal from the noise canceller is fed to a speech recognizer. At the automated speech recognizer (ASR) we compute features that represent the spectral-domain content of the speech. These features are computed with a "frame", which is the number is of samples within a defined period of, for example 10 ms. A neural network can be used to classify a set of these features into phonetic-based categories at each frame. A search is used to match the neural-network output scores to the target words (the words that are assumed to be in the input speech), in order to determine the word that was most likely uttered. It is also possible to analyse the results by looking at the confidence we have in the top-scoring word. The word can be rejected as out-of-vocabulary if the confidence falls below a pre-determined threshold. Using the *M square* **W** matrix in the Wiener filter, the samples of input speech are filtered and outputted as a *1D* array with size M. This *1D* array is considered as a frame, which would be used in a typical neural network. For the **W** matrix, the larger M means more computing resource cost. In the meantime, the Frame size in a neural network is considered by the accuracy of speech recognition. Normally a smaller frame size means larger error.

Since filtered signals from a Wiener filter has as much as 25 dB SNR, word boundaries are easily identified by energy density. In a Wiener filter, the incoming voice is sampled at an 11025 Hz sample rate. The W matrix is 100×100 and is updated by the 100 samples from Microphone 1 and another 100 samples from Microphone 2. These samples from Microphone 2 are also filtered by the W matrix which is from the output as the filtered signal. Since the spectrogram is essentially a "fingerprint" of a person's voice, it shows particular features specific to the subject themselves. Therefore, dictionary fingerprint and sampled fingerprint are built on 100 samples. Based on the voice fingerprinting, different subjects will have different parallel bands as well as other features such as delays and time shifts. In this paper, the spectrograms of two different subjects (dictionary fingerprint $(d_1, d_2, \ldots d_n)$ and sampled word fingerprint $(w_1, w_2, \ldots w_n)$) are compared from each other with Cross-correlation. The dictionary fingerprint and the word fingerprint have to be

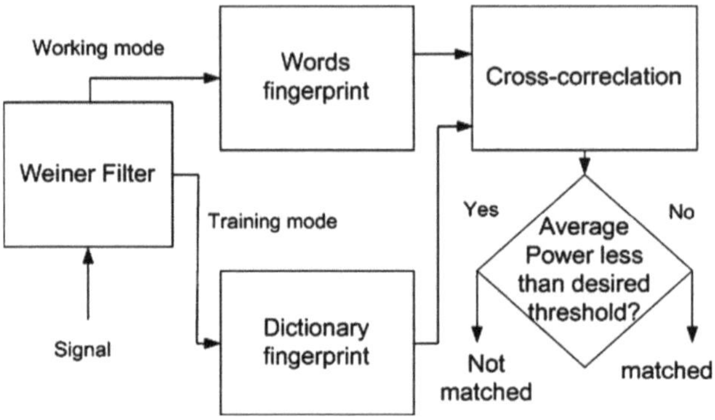

Fig. 6 ASR block diagram

amplitude normalized. This processing step simply amplifies dictionary fingerprint and the incoming fingerprint by positive or negative gain so that they have same level of peak. When we calculate the cross-correlation of the dictionary fingerprint and the word fingerprint, the average **RMS** (Root Mean Square) power of the result of cross-correlation shows difference between the same word or not. A process flowchart is showed in Fig. 6.

4 Experiments

Case study one

A simulation was built to test these assumptions using a sample rate of 11025 Hz as in Fig. 7. The anti-alias filters had a cut-off frequency of 5 kHz.

In Fig. 8, pre-recorded files from Microphone 1 or 2 were recorded at a sample rate of 11025 Hz. The numbers of elements in the array for these pre-recorded files have length M. As the method of Fig. 3 suggests, during active speech periods the

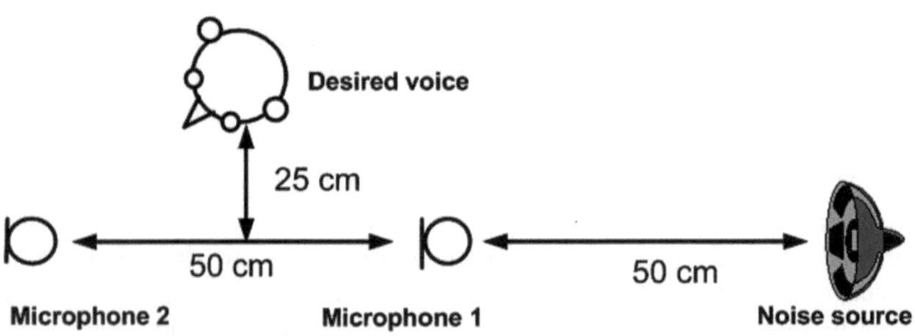

Fig. 7 Experimental setup

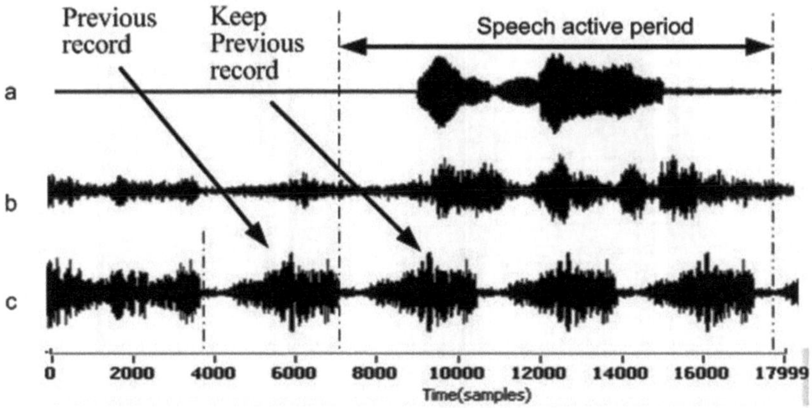

Fig. 8 The W matrix is updated during a noise period only or a speech activity period. (**a**) Wiener filter output (**b**) Waveform of Signal + noise from Microphone 2 (**c**) Waveform of noise from Microphone 1

elements of the array from Microphone 1 is not updated and keeps the previous record of noise periods, as (c) in Fig. 8. Therefore, the Wiener filter matrix **W** is updated during the noise periods only, or during active speech periods. The Wiener filter output is showed in (a) whilst Microphone 2 (b) is incoming speech + noise (input y). Since the noise statistics change between noise periods and speech periods then the **W** matrix cannot be updated correctly.

In Fig. 9, using the method shown in Fig. 4, the Wiener filter output is showed in Fig. 9(a). Figure 9(b) is microphone 2 - incoming speech + noise (input y). Microphone 1 shown in Fig. 9(c) is changed during the desired speech period. Comparing Figs. 8 and 9, in order to achieve a cleaner estimate from the original speech, the method of Fig. 4 is clearly an improved method.

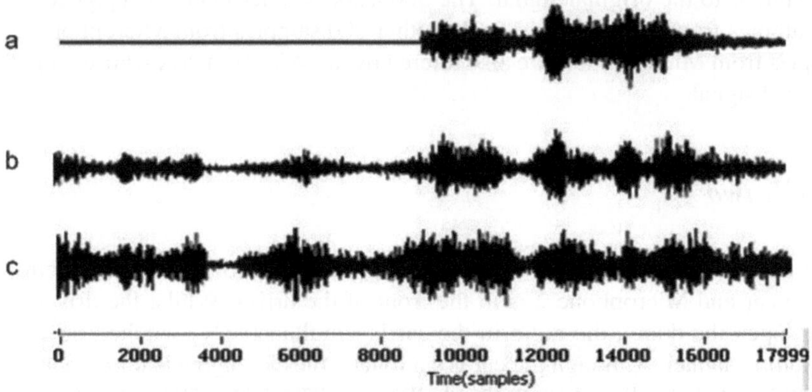

Fig. 9 The W matrix is updated in real-time. (**a**) Wiener filter output (**b**) Waveform of signal + noise from Microphone (**c**) Waveform of noise from Microphone 1

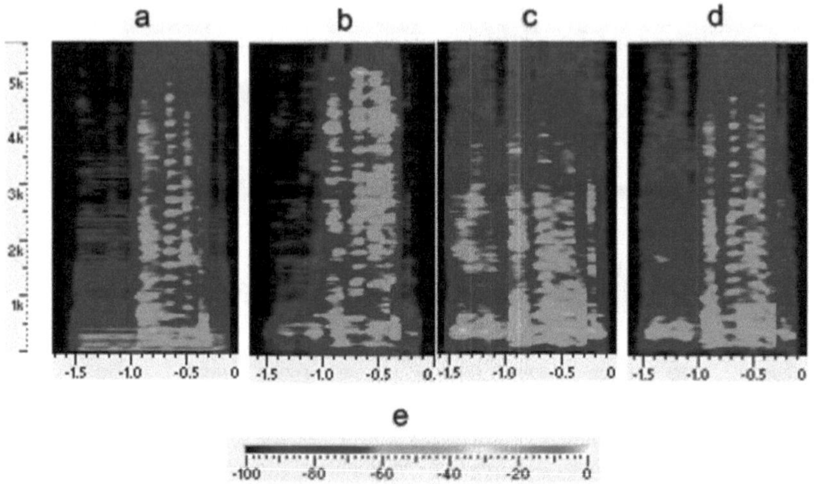

Fig. 10 Spectrograms of the filtering process. The horizontal axis represents time (in seconds) and the vertical axis is frequency (in Hz) (**a**) clean speech "open the door" (**b**) Filtered speech (**c**) recording of Microphone1 (**d**) recording of Microphone2 (**e**) Intensity scale in dB

Using waveform statistics, the results show this filter reduces the unwanted speech by approximately 30 dB.

Figure 10 shows the spectrograms of the original speech phrase "open the door" (Fig. 10(a)). From Microphone 1 in Fig. 10(b), the spectrogram shows the desired speech phase "open the door" + simultaneous unwanted speech babble from a Chinese radio channel. Fig. 10(d) shows the spectrogram at Microphone 2, the desired speech phase "open the door" + simultaneous unwanted speech from a Chinese radio channel. Figure 10(c) is the spectrogram of the filtered signal using 500 samples from microphone 2 and filtered by a 500×500 Weiner filter Matrix. Comparing with Fig. 10 (a) and (b), it is clear that spectrogram of "open the door" of the filtered signal is similar to the original signal. The 500×500 Weiner matrix is updated by the 500 samples from Microphone 1 and another 500 samples from Microphone 2. The samples from Microphone 2 are also filtered by the Wiener matrix whose output is the filtered signal.

Case study two

An in-car test was performed with layout as in Fig. 5. Microphone 1 is in the central area of the car and Microphone 2 is in the front of the driver. Whilst the driver is speaking "open the door", the radio in the car is simultaneously switched on to a Chinese radio channel. Although the car is stationary, the car engine is idling during the test making the overall problem quite challenging. The noise volume was typical of a car radio - but not deafeningly loud. The absolute dB level was not measured but rather dB is just used here as a relative measure of signal or noise strength.

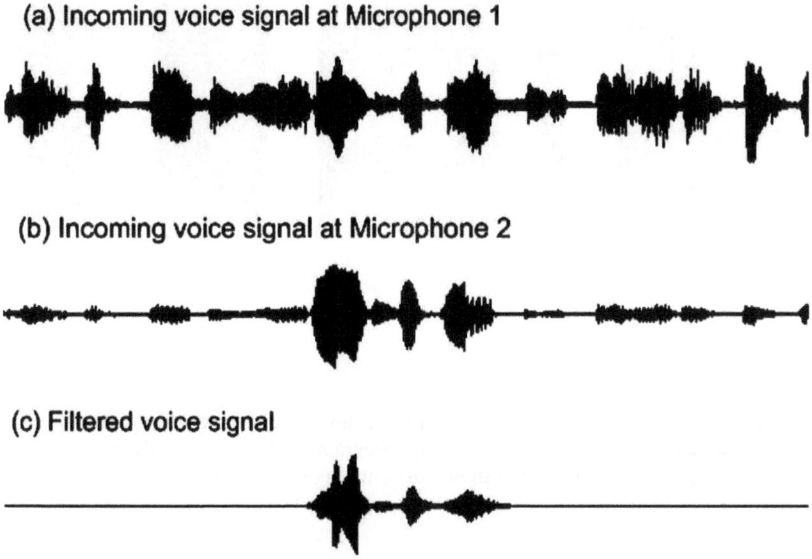

Fig. 11 Test results in a stationary car with engine and car radio on

Test results are shown as Fig. 11. Figure 11(a) is the incoming signal at Microphone 1, which is receiving the voice from all radio loud-speakers as well as the driver's voice as positioned in Fig. 10. Figure 11(b) is the incoming signal at Microphone 2, which is receiving the voice from all radio loud-speakers as well as simultaneously the driver's voice. Figure 11(c) is the filtered voice signal, which is keeping the driver's voice whilst reducing the voice from the radio loud-speakers. Table 1 shows the waveform statistics using Cool Edit Pro regarding the average power during the driver's speech period + speech from the radio period and also the speech only from the radio period. In Table 1, SNR is computed as follows:

$$\text{SNR} = \text{Average Signal Power - Average Noise Power (in dB)} \tag{22}$$

The results show Signal-to-Noise Ratio (SNR) is improved by 28.26 dB between the filtered signal and the noisy signal received at Microphone 2. This result was performed at an 11025 Hz sample rate. The W matrix is 1000×1000 and is updated by the 1000 samples from Microphone 1 and another 1000 samples from

Table 1 SNR Analysis

	Average power in only speech from radio period (dB)	Average power in driver's speech + speech from radio period (dB)	SNR (dB)
Filtered signal	−60.17	−17.44	42.73
Signal at Microphone 2	−31.91	−17.39	14.52
SNR improved			28.21

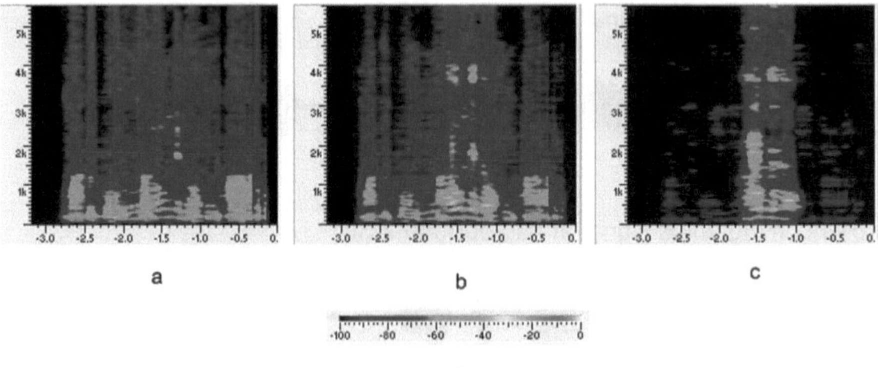

Fig. 12 The spectrograms of the waveforms in Fig. 11. The horizontal axis represents time (in second) and the vertical axis is frequency (in Hz). (**a**) Spectrogram at Microphone1. (**b**) Spectrogram at Microphone2. (**c**) Spectrogram of filtered signal. (**d**) Intensity scale in dB

Microphone 2. These samples from Microphone 2 are also filtered by the W matrix which forms the output as the filtered signal as in Fig. 11(c).

Figure 12 shows the spectrograms of the waveforms of Fig. 11. The horizontal axis represents time (in seconds) and the vertical axis is frequency (in Hz). (a) is the spectrogram at Microphone1. (b) is the spectrogram at Microphone2. (c) is the spectrogram of the filtered signal. (d) shows the intensity scale in dB.

Case study Three

An in-car test was performed with layout as in Fig. 5 Microphone 1 is in the central area of the car, in such a position where it will in all probability pick up the radio more than the drivers voice. Microphone 2 is in the front of the driver, so it is in a position where it will more likely pick up the drivers voice than the radio signal.This process was performed at an 11025 Hz sample rate. The W matrix is 100×100 and is updated by the 100 samples from Microphone 1 and another 100 samples from Microphone 2. These samples from Microphone 2 are also filtered by the W matrix which forms the output as the filtered signal.

A LabVIEW program in a laptop is designed to create and update the Wiener matrix. The speech signal is filtered by this Wiener matrix and then fed to a cross-correlation algorithm for a measure of similarity with one of the fingerprints.

In order to test the repeatability and speaker-dependence, two groups of data are collected: a male driver's speech "left" and "right"; a female driver's speech "left" and "right". Whilst the driver is speaking, the radio in the car is simultaneously switched on to a radio channel. The voice from the car radio is different from time to time as is the case with a normal radio transmission.

Figure 13 shows a sample of spectrograms of the waveforms of a female driver's speech "right" and "left" when the in-car radio was on and the engine was idling.

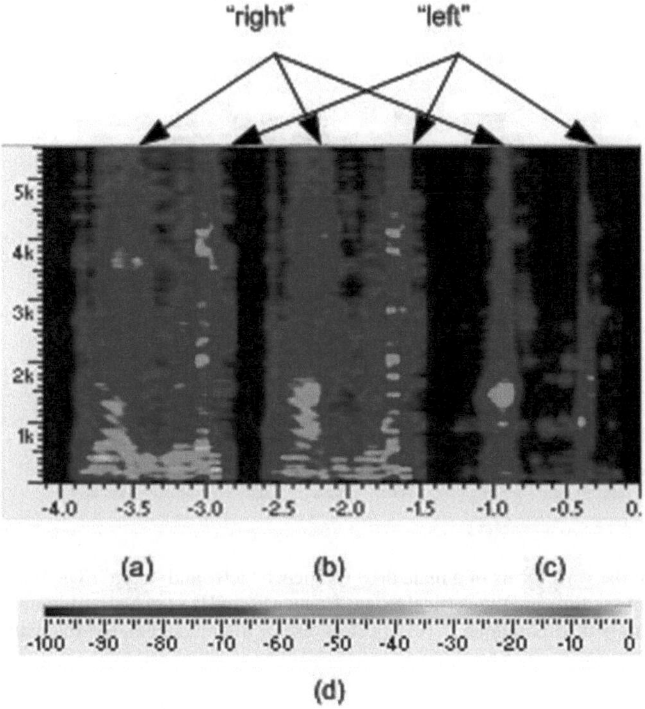

Fig. 13 The spectrograms of the waveforms of a female driver's speech "right" and "left". The horizontal axis represents time (in second) and the vertical axis is frequency (in Hz). (**a**) Spectrogram at Microphone 1. (**b**) Spectrogram at Microphone 2. (**c**) Spectrogram of filtered signal. (**d**) Intensity scale in dB

Figure 13(3) shows the driver's speech "right" and "left" was clearly separated from that of the radio.

Figure 14 shows a sample of spectrograms of the waveforms of a male driver's speech saying "left" and "right" when in-car radio was on and the engine was idling. Figure 14(3) shows the driver's speech "left" and "right" was clearly separated from that of the radio noise. There are 6 "samples" of a Female driver's speech "right" and "left" and 6 "samples" of a male driver's speech "left" and "right" in this test.

More than 100 experimental records of speech saying "left" and "right" have been performed to calculate the ASR success rate. We present 5 of these results at random for clarity.

Firstly we used a recording to create a fingerprint of a female driver's speech saying "right". Then we collected 5 tests of the female driver's speech "left" and 5 samples of the female driver's speech "right". Fingerprint "right" was cross-correlated with 5 records of "left" and 5 records of "right". Then the average power was calculated. The results are shown in Table 1. If a threshold is set at the pre-set value (between −45.48dB and −43.41dB), then the incoming signal "right" will be recognized with a hit-rate of 100%.

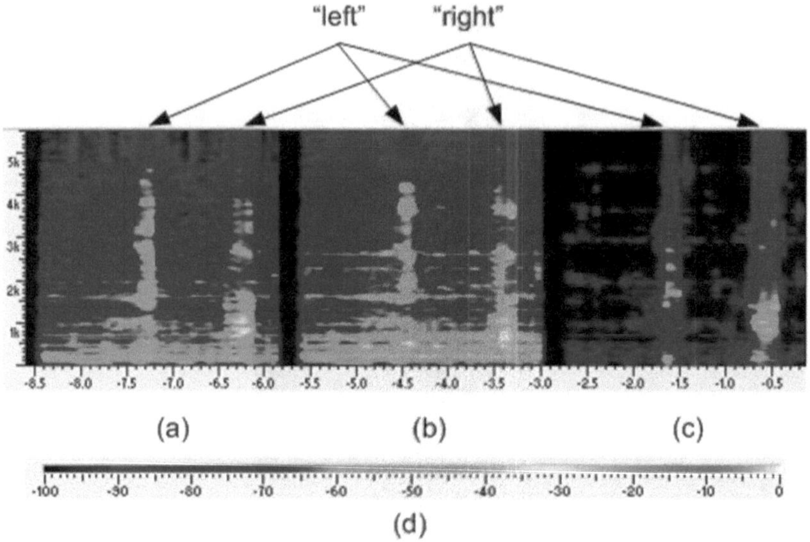

Fig. 14 The spectrograms of the waveforms of a male driver's speech "left" and "right". The horizontal axis represents time (in second) and the vertical axis is frequency (in Hz). (**a**) Spectrogram at Microphone 1. (**b**) Spectrogram at Microphone 2. (**c**) Spectrogram of filtered signal. (**d**) Intensity scale in dB

We trained and created a fingerprint of a male driver's speech "right". Then we collected 5 records of the male driver's speech "left" and 5 records of the male driver's speech "right". Fingerprint "left" was cross-correlated with 5 records of "left" and 5 records of "right". Then the average power was calculated. The results are showed in Table 2. As shown in Table 2, if a threshold is set at the pre-set value (between −48.81dB and −56.89) then the incoming signal "right" will be recognized with a hit-rate of 100%.

We trained and created a fingerprint of a female driver's speech "right". Then we collected 5 records of a female driver's speech "left" and 5 records of the female driver's speech "right". The results as shown in Table 3 show that there are no successes since there no threshold can be selected to recognize the female driver speech "right".

Table 2 Average power in dB result of cross-correlation between fingerprint "right" and 5 records of "right" or 5 records of "left" from a female driver

	"left"(dB)	"right"(dB)
Sample 1	−49.38	−39.52
Sample 2	−47.82	−40.68
Sample 3	−45.48	−41.49
Sample 4	−51.10	−41.52
Sample 5	−47.99	−43.41

Table 3 Average power in dB at result of cross-correlation between fingerprint "right" and 5 records of "right" or 5 records of "left" from a male driver

	"left"(dB)	"right"(dB)
Sample 1	−59.65	−40.56
Sample 2	−67.65	−40.81
Sample 3	−62.99	−46.44
Sample 4	−56.89	−42.27
Sample 5	−63.97	−48.81

We trained and created a fingerprint of a male driver's speech "right". Then we collected 5 records of a female driver's speech "left" and 5 records of the female driver's speech "right". The results as Table 3 showed that there are no successes since no threshold can be selected to recognize the male driver speech "right".

The above testes have been done for a limited number of records. However, as with the cases of Tables 2, 3, more than 100 records have been performed for each threshold with 100% success rate for these cases (as in Table 2 and 3). For the cases in Tables 4, 5, when we used the thresholds as in Tables 2, 3, a success rate of 40–60% in 100 records was found. These result shows a high hit-rate rate speaker-dependent ASR in a noisy environment.

Table 4 Average power in dB at result of cross-correlation between fingerprint "right" from a female driver and 5 records of "right" or 5 records of "left" from a male driver

	"left"(dB)	"right"(dB)
Sample 1	−53.04	−51.85
Sample 2	−63.75	−56.23
Sample 3	−47.84	−55.78
Sample 4	−57.14	−55.65
Sample 5	−67.72	−53.73

Table 5 Average power in dB at result of cross-correlation between fingerprint "right" from a male driver and 5 records of "right" or 5 records of "left" from a female driver

	"left"(dB)	"right"(dB)
Sample 1	−60.40	−55.00
Sample 2	−45.75	−48.05
Sample 3	−56.23	−53.85
Sample 4	−57.10	−50.95
Sample 5	−49.04	−42.64

5 Conclusion

Previous research suggested that a Wiener filter **W** matrix can be updated by knowledge of noise periods and signal + noise periods. The **W** matrix always acts as a filter on the noisy speech command signal. Therefore a VAD is needed to distinguish between these noise periods or signal + noise periods. This paper uses a method which eliminates entirely the need of a VAD switch using a continuously adapting Wiener filter. The experiments show that a Wiener filter can be updated during noise periods and signal + noise periods, improving the SNR by slightly better than 28 dB. As an experiment, this paper introduces a novel design in real-time environments for recognizing small vocabulary control commands. In order to receive the driver's voice whilst suppressing background noise e.g. voice from a radio, a simple and novel speech recognition system is designed using LabView. Experiments show this voice recognizer has high accuracy and is capability of speaker-dependence in tests in a car when the car radio was on and the engine was idling. The filtered signals were clearly recognisable during informal listening tests and sounded vastly improved to the original. However, as with all of these kinds of automated 'smart-car' technologies, (in contrast with some other similar technologies) it is not necessary for the enhanced signal to sound better to the human ear, but only needs to be good enough to provide a Boolean on or off command.

References

1. Qi Z, Moir T (2007) An Adaptive Wiener Filter for an Automotive Application with Non-Stationary Noise. In: Proc. 2nd International Conference on Sensing Technology. Palmerston North, New Zealand, pp. 124–128.
2. Qi Z, Moir T (2007) A Design of Automotive Voice Recognizer Using LabVIEW. In: Proc. the Fourteenth Electronics New Zealand Conference. Wellington, New Zealand, pp. 303–308.
3. Shozakai M., Nakamura S, Shikano K (1998) Robust speech recognition in car environments. In: Proc. Acoustics, Speech, and Signal Processing, Seattle, USA, pp. 269–272 vol.1.
4. Cho Y, Ko H (2004) Speech enhancement for robust speech recognition in car environments using Griffiths-Jim ANC based on two-paired microphones. In: Proc. Consumer Electronics, 2004 IEEE International Symposium on, Reading, UK, pp. 123–127.
5. Chen W, Moir T (1999) Adaptive noise cancellation for nonstationary real data background noise using three microphones. Electronics Letters, 35(23): pp. 1991–1992.
6. Qi Z, Moir T (2005) An Automotive three-microphone Voice Activity Detector and noise canceller. In: Proc. Intelligent Sensors, Sensor Networks and Information Processing Conference, Melbourne, Australia, pp. 139–143.
7. Wiener N (1949) Extrapolation, Interpolation, and Smoothing of Stationary Time Series. Wiley, New York.
8. Kolmogorov A (1941) Interpolation and extrapolation of stationary random sequences. Mathematics series, 5: pp. 3–14.
9. Barrett J, Moir T (1987) A Unified Approach to Multivariable Discrete-Time Filtering Based on the Wiener-Theory. Kybernetika 1987. 23: pp. 177–197.
10. Brown R, Hwang P (1996) Introduction to Random Signals and Applied Kalman Filtering. 3 ed. John Wiley & Sons, New York.
11. Haykin S (2002) Adaptive Filter Theory. 4 ed., Prentice Hall, Englewood Cliffs.
12. Doclo S, Moonen M (2005) On the output SNR of the speech-distortion weighted multichannel Wiener filter. Signal Processing Letters, IEEE, 12(12): pp. 809–811.
13. Dabeer O, Masry E (2002) Analysis of mean-square error and transient speed of the LMS adaptive algorithm. Information Theory, IEEE Transactions on, 2002. 48(7): pp. 1873–1894.
14. Griffiths L (1978) An adaptive lattice structure for noise-cancelling applications. In: Proc. Acoustics, Speech, and Signal Processing, IEEE International Conference on ICASSP '78. Tulsa, OK, pp. 87–90.

15. Resende L, Romano J, Bellanger M (2004) Split Wiener filtering with application in adaptive systems. Signal Processing, IEEE Transactions on [see also Acoustics, Speech, and Signal Processing, IEEE Transactions on], 2004. 52(3): pp. 636–644.
16. Doclo S, Moonen M (2002) GSVD-based optimal filtering for single and multimicrophone speech enhancement. Signal Processing, IEEE Transactions on [see also Acoustics, Speech, and Signal Processing, IEEE Transactions on], 2002. 50(9): pp. 2230–2244.
17. Widrow B, Glover J, McCool J, Kaunitz J, Williams C, Hearn R, Zeidler J, Dong J, Goodlin R (1975) Adaptive noise cancelling: principles and applications. In: Proc. the IEEE Vol 63 No 12 pp. 1692–1716.

3D-Shape Recognition Based Tactile Sensor

Somrak Petchartee, Gareth Monkman and Anan Suebsomran

Abstract a surface recognition algorithm capable of determining contact surfaces types by means of tactile sensor fusion is proposed. The authors present a recognition processes for 3-dimensional deformations in a 2-dimensional parametric domain. Tactile information is extracted by physical contact with a grasped object through a sensing medium. Information is obtained directly at the interface between the object and the sensing device and relates to three-dimensional position and orientation of the object in the presence of noise. The technique called "eigenvalue trajectory analysis", is introduced and adopted for specifying the margin of classification and classification thresholds. The authors demonstrate mathematically that this approach, which complements existing work, offers significant computational advantages when applied to challenging contact scenarios such as dynamic recognition of contact deformations.

Keywords Contact classification · eigenvalue perturbation · tactile sensor · eigenvalue trajectory

1 Introduction

The authors present an efficient algorithm for recognizing contact situations under robotic prehension using deformable two-finger sensors. The deformation is reformulated using a Quadric parametric atlas to reduce the computationally demanding high number of degrees of freedom. Using the Quadric surface interpolation, the data captured by the tactile elements are used to construct an eigenvalue trajectory

Somrak Petchartee
Ferderal Arms Force University Munich, Germany, e-mail: somrak.petchartee@gmail.com

Gareth Monkman
University of Applied Sciences Regensburg, Germany,
e-mail: gareth.monkman@e-technik.fh-regensburg.de

Anan Suebsomran
King Mongkut's University of Technology, North Bangkok, Thailand, e-mail: asr@kmutnb.ac.th

from each unique matrix relating to measurements. Comprehensive details about the prototype sensor, data acquisition and processing principles together with incorporated mathematical derivations for part recognition are provided in a previously published work [9]. Here, additional objects are introduced to synthesize the classification performances. Most of the existing methods in contact recognition do not explicitly address problems due to the contact area and surface geometry. Due to the numerous degrees of freedom involved in these types of deformation, the complexity of the problem tends to be intractable for real-time computation.

A number of approaches have been put forward for the processing of tactile sensor data in order to yield useful characterizations of contact surfaces for such applications. There exist some research works on tactile shape recognition based on planar shapes using neural networks. The neural network for classification generally requires that planar shapes be presented in a fixed position, orientation and dimension. To avoid this complicated configuration, a pre-processing stage is required to normalise the shape of the input data after all the prehended object data has been obtained. Examples using the radius vector method for shape representation have shown good results with neural network classifiers [6, 7, 12]. Another example [8] uses the angles and lengths of planar polygons to represent the differences among the grasped objects. The object's data are acquired by robotic tactile sensors, and a neural network is then used to recognise the shape. Ng's method [7, 8] is straightforward and seems to work well on simple shapes. It is also rotation, translation, and scale invariant, but he uses only fixed-sized shapes in his experiments. In using planar shapes for shape recognition, the apparent problem is how to distinguish between objects having the same planar shape but difference surface curvatures. Also, the neural network is unable to recognise the partial angle and length sequences of the planar shapes. There also exist other techniques for contact identification based on tactile sensing. For example, Ibrayev and Jai [3, 4] proposed the recognition of low-degree polynomial curves based on minimal tactile data. In their application, Euclidean differential and semi-differential invariants were derived for quadratic and special cubic curves. Those invariants, independent of translation and rotation, were evaluated over the differential geometry at up to three points on a curve. The contact locations were then found on the curve, thereby localizing it relative to a touch sensor. Unfortunately, he did not present any implementation methods and experimental results. Most of the proposed methods do not consider the variation of the image of the contact surface obtained by the tactile sensor on the imposed force.

2 Contact Classification

Considering a general 3-D surface expressed in the contact point as

$$f(x,y,z) = 0,$$

the general surface function can be approximated locally at the contact point as the following second order polynomial equation:

$$ax^2 + by^2 + cz^2 + 2dxy + 2eyz + 2fxz + 2gx + 2hy + 2jz + k = 0 \quad (1)$$

Equation (1) can be rewritten in a quadratic form of a matrix equation: $P^T.Q.P = 0$, where

$$Q = \begin{pmatrix} a & d & \underline{f} & g \\ d & b & \underline{e} & h \\ \underline{f} & \underline{e} & \underline{c} & j \\ g & h & j & k \end{pmatrix} \quad \text{and} \quad P = \begin{pmatrix} x \\ y \\ z \\ 1 \end{pmatrix}.$$

The properties of the surfaces represented by Q can be translated, rotated and scaled. Given a 4×4 transformation matrix M of the form developed, the transformed Quadric surface Q^* is:

$$Q^* = (M^{-1})^T.Q.M^{-1} \quad (2)$$

The general transformation matrices (M) are of the Denavit-Hartenberg type combining both translation and rotation. One property of Quadric surfaces is that the matrices are symmetrical. They can always be diagonalized, and their eigenvectors can be chosen to form an orthonormal basis with respect to the canonical dot product. For example, if we form a matrix A with their eigenvectors as the columns, then we will obtain the diagonal matrix as vAv^T. Rayleigh's principle [10, 11] states that the smallest eigenvalue $\lambda_{\min}$ coinciding with the minimum of

$$R(v) := \frac{< v, Av >}{< v, v >}, < v, v > \text{ is the inner product.}$$

The minimizer of R is an eigenvector with the smallest eigenvalue $\lambda_{\min}$. Likewise, the largest eigenvalue and its corresponding eigenvector can be found by maximizing R. Let the eigenvalues of A be ordered as

$$(\lambda_{\min})\lambda_1 \leq \lambda_2 \leq \lambda_3 \leq \ldots \lambda_{n-1} \leq \lambda_n(\lambda_{\max}).$$

Then, $\lambda_1 x^* x \leq x^* A x \leq \lambda_n x * x$ for all $A \in M_n$ is obtained, where n is the matrix dimension, $\lambda_{\max} = \lambda_n = \max\limits_{x \neq 0} \frac{x^* A x}{x^* x}$ and$\lambda_{\min} = \lambda_1 = \min\limits_{x \neq 0} \frac{x^* A x}{x^* x}$. At this point, the aim is to prove that the smallest eigenvalue can facilitate contact identification.

$$\text{In our case, } v = \begin{bmatrix} x \\ y \\ z \\ 1 \end{bmatrix} \text{ and } A = \begin{bmatrix} a & d & 0 & g \\ d & b & 0 & h \\ 0 & 0 & 0 & j \\ g & h & j & k \end{bmatrix}, \; Av = \begin{bmatrix} ax + dy + g \\ dx + by + h \\ j \\ gx + hy + jz + k \end{bmatrix},$$

$$< v, Av > = \begin{bmatrix} x \\ y \\ z \\ 1 \end{bmatrix} \begin{bmatrix} ax + dy + g \\ dx + by + h \\ j \\ gx + hy + jz + k \end{bmatrix}.$$

Then:

$$< v, Av > = ax^2 + dxy + gx + dxy + by^2 + hy + jz + gx + hy + jz + k$$
$$\text{and } < v, v > = x^2 + y^2 + z^2 + 1,$$

which can finally be formulated as:

$$R(v) = \frac{< v, Av >}{< v, v >} = \frac{ax^2 + by^2 + 2dxy + 2gx + 2hy + 2jz + k}{x^2 + y^2 + z^2 + 1}. \tag{3}$$

The next step is to minimize $R(v)$ and obtain $\lambda_{\min}$, in order to deal with the question of how the Quadric parameter can be used for contact classification.

In considering the problem of minimizing a polynomial fraction on $R(v)$, it is known to be difficult even for degree-2 polynomials. Hanzon and Jibetean [2] and Jibetean and Laurent [5] have proposed a method for computing tight upper bounds based on perturbing the original polynomial and using semidefinite programming. Their works are on global optimization methods for multivariate polynomials and rational functions, and their method can be applied to our problem. According to their theory, it involves the following problem:

$$\inf_{x \in R^n} \frac{p(x)}{q(x)} \text{ with } p(x), q(x) \in R[x]. \tag{4}$$

Regarding the terminology, they use the infimum (inf) instead of the more common minimum (min) simply to stress that the optimal value may not be attained exactly in $R(v)$ but may only be approached asymptotically. Note that there are no other differences between the formulations involving inf and min.

Nevertheless, it is important to have a procedure which computes, in principle, the global optimum. In [2, 5], the main interest is in designing algorithms which guarantee finding the global solution. The authors' method is based on the relation between positive polynomials and the sum of squares of the polynomials. In fact, in this approach the effort is directed towards finding a real number α such that $p(x) - \alpha q(x) \geq 0, \forall x \in R^n$, where $p(x)$ and $q(x)$ are the given polynomials. Obviously the largest α satisfying the condition is the infimum of $p(x)/q(x)$.

Moreover, the authors of [2, 5] rewrote the rational optimization problem into a semi-define optimization problem (SDP) by denoting $F(x) = p(x) - \alpha q(x)$ which is known to have good computational complexity. Actually, in general they obtained the SDP as the relaxation of the original problem, which gives a lower bound for the solution of the original problem.

If the total degree of F is even where the total degree is d (d; power of polynomial), they can find a matrix Q such that $F(x) = z^T Qz$, $z = [1, x_1, x_2, \ldots, x_n, x_1 x_2, \ldots, x_n^d]$. The value z contains all monomials of the variables $x_1, \ldots, x_n$ having its degree less than or equal to d. Obviously, if such Q exists, then it is a symmetrical matrix. In conclusion, equation (4) can be rewritten as:

$$F(x) = z^T Qz = p(x) - \alpha q(x) \geq 0. \tag{5}$$

The matrix Q can always be constructed. If $Q\geq 0$, then $F(x)\geq 0$, $\forall x\in R^n$. A positive semidefinite matrix, or $Q\geq 0$, is a matrix all of whose eigenvalues are nonnegative. The Q matrix is called a positive semidefinite matrix if $x^T Qx\geq 0$. Which means for every points x in the vector space introduced, $x^T Qx$ is a positive value.

Here we rewrite the rational optimization problem as a certain constrained polynomial optimization problem for which exact or numerical methods can be subsequently used for solving it.

Referring to (3), if the definitions: $x_1 = x, x_2 = y$ and $x_3 = z$ are used, then:

$$\inf_{x\in R^3}\left(\frac{ax_1^2+bx_2^2+2dx_1x_2+2gx_1+2hx_2+2jx_3+k}{x_1^2+x_2^2+x_2^3+1}\right).$$

Using the method in [2, 5], the problem can be arranged into

$$ax_1^2+bx_2^2+2dx_1x_2+2gx_1+2hx_2+2jx_3+k-\alpha(x_1^2+x_2^2+x_3^3+1)=z^T Qz,$$

where $z=\begin{bmatrix}x_1 & x_2 & x_3 & 1\end{bmatrix}$. Then, it can be re-arranged as:

$$(a-\alpha)x_1^2+(b-\alpha)x_2^2-\alpha x_3^3+2dx_1x_2+2gx_1+2hx_2+2jx_3+k-1=z^T Qz. \quad (6)$$

Based on (6), it can be observed that one solution is $x^T Qx=0$ which is the definition of a Quadric surface.

$$Q=\begin{bmatrix}a-\alpha & d & 0 & g\\ d & b-\alpha & 0 & h\\ 0 & 0 & -\alpha & j\\ g & h & j & k-\alpha\end{bmatrix} \quad (7)$$

Introducing the Quadric parameter into Rayleigh's quotient for minimizing $R(v)$ based the method in [2, 5] leads to the result $\alpha=0$, at least in one case. The conclusion of these proofs can confirm that the smallest eigenvalue of the Quadric parameter is identical with the Quadric parameter itself. This means the smallest eigenvalue of Quadric parameter yields the identical characteristic of the Quadric shape. The one-to-one correspondence can be achieved by assigning the approximate trajectory threshold to the specific contact surface. It is then reasonable to use the smallest eigenvalue to classify contact surfaces.

In the same way it is possible to prove how transformations on Quadric surfaces are unaffected by the eigenvalue trajectory. If (2) is substituted into the Quadric equation; $P^T.Q.P=0$. Then it becomes $P^T.(M^{-1})^T.Q.(M^{-1}).P=0$. After rearrangement of the equation, expression (8) is obtained.

$$(M^{-1}.P)^T.Q.(M^{-1}.P)=0 \quad (8)$$

If the variables of the transformation matrix are introduced as follows:

$$M^{-1}=\begin{bmatrix}m_{11} & m_{12} & m_{13} & m_{14}\\ m_{21} & m_{22} & m_{23} & m_{24}\\ m_{31} & m_{32} & m_{33} & m_{34}\\ m_{41} & m_{42} & m_{43} & m_{44}\end{bmatrix}.$$

Using the previous procedure, there exists

$$v = M^{-1}.P = \begin{bmatrix} (m_{11}+m_{21}+m_{31}+m_{41})x \\ (m_{12}+m_{22}+m_{32}+m_{42})y \\ (m_{13}+m_{23}+m_{33}+m_{43})z \\ (m_{14}+m_{24}+m_{34}+m_{44}) \end{bmatrix} = \begin{bmatrix} Ax \\ By \\ Cz \\ D \end{bmatrix},$$

where $A = m_{11}+m_{21}+m_{31}+m_{41}$, $B = m_{12}+m_{22}+m_{32}+m_{42}$, $C = m_{13}+m_{23}+m_{33}+m_{43}$ and $D = m_{14}+m_{24}+m_{34}+m_{44}$.

The components of (3) are now listed below:

$$< v, Av > = \begin{bmatrix} Ax \\ By \\ Cz \\ D \end{bmatrix} \begin{bmatrix} aAx+dBy+gD \\ dAx+bBy+hD \\ jD \\ gAx+hBy+jCz+kD \end{bmatrix},$$

$$\begin{aligned} < v, Av > &= aA^2x^2 + bB^2y^2 + 2dABxy + 2gADx + 2hBDy + 2jCDz + kD^2, \\ < v, v > &= A^2x^2 + B^2y^2 + C^2z^2 + D^2 \end{aligned}$$

Finally, R(v) can be formulated as:

$$R(v) = \frac{aA^2x^2 + bB^2y^2 + 2dABxy + 2gADx + 2hBDy + 2jCDz + kD^2}{A^2x^2 + B^2y^2 + C^2z^2 + D^2}. \tag{9}$$

Using the methods given in [2, 5], then the problem can be arranged into:

$$\begin{aligned} (a-\alpha)A^2x_1^2 + (b-\alpha)B^2x_2^2 + 2dABx_1x_2 + 2gADx_1 + 2hBDx_2 + 2jCDx_3 \\ +(k-\alpha)D^2 - \alpha C^2x_3^2 = z^TQz. \end{aligned} \tag{10}$$

Based on (10), the same value of Q as in (7) is obtained. Equation (10) is similar to (6) but the differences here are only scalar values appearing in the coordination components. In conclusion, the vector which represented the coordination system has changed from $\begin{bmatrix} x\ y\ z\ 1 \end{bmatrix}^T$ to $\begin{bmatrix} Ax\ By\ Cz\ D \end{bmatrix}^T$. However, any transformations on the Quadric surfaces are unaffected on the eigenvalue trajectory.

3 Tactile Sensor and Application

Tactile data contains information about magnitudes, distributions and locations of forces. It also provides information about the contact area and the pressure distribution over it. With resistive tactile sensors, changes in electrical resistance are detected by an element made from electrically conductive foam. The soft, deformable surface detects continuous pressure with excellent sensitivity and resolution (Fig. 1).

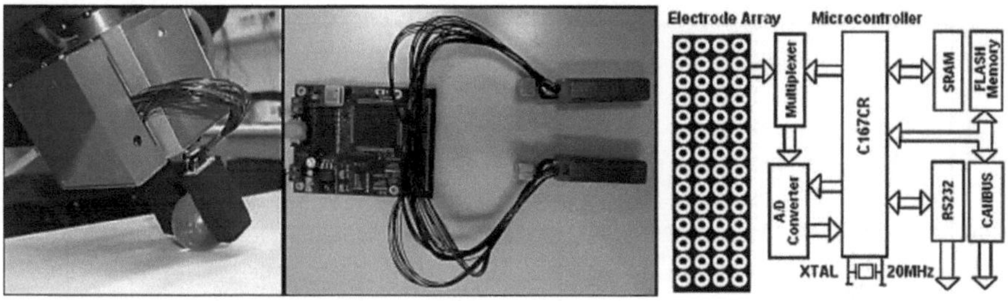

Fig. 1 Tactile sensor

The electrical resistance measured between two electrodes on the same side of the conductive foam (one tactile element) is derived from electrical conductivity through a number of simultaneously conducting paths. The tactile sensors have been developed with the following specifications: One finger consists of two 16×4 cells, two 16×2 cells, and one 6×2 cells, making up the total 408 cells for the two fingers. The width of the fingers is 20 mm, their length is 55 mm excluding an aluminum core and they have a thickness of 12 mm.

A 32-bit micro-controller, sampling circuits and memory are installed in the control unit. The micro-controller also has an analogue-multiplexer for accessing all 408 tactile elements. Stack memory is implemented inside the micro-controller by software for data manipulation. Initially, the micro-controller briefly obtains data from the tactile elements independently. Secondly, the tactile data are collected by the stack memory and the micro-controller accesses the remaining tactile elements again by controlling the multiplexer chip. When all tactile elements have been read by the micro-controller, then the micro-controller calculates several parameters which describe the features of the contact points. "Higher-order processing" is implemented on an external computer. In this design, few wires are required for connections between the control unit and the computer as the sensors only deliver a small number of parameters. This configuration reduces the amount of data traffic and computational overhead.

This prototype sensor can be used to provide intrinsic information of interest relating to grasped objects, such as geometrical features situated on the object surface. Information along the sensor surface can be obtained by a discrete number of tactile elements to enhance the measurement resolution. Feature heights are detectable down to 3 mm. Each tactile element can be calibrated to act as a force sensing element. The effective pressure detection threshold of a tactile finger is about 50 g/mm^2, giving a normal indentation of skin surface of 0.1 mm (about 3.3% strain).

Figure 2 shows variations in the contact resistance when forces are applied to the tactile sensor surface. In the experiment, a 3-mm $\times$ 20-mm $\times$ 55-mm foam sheet was placed on a flat surface. This was subsequently subject to depression by the indenter. Forces of 2, 4, 6 and 8 Newton applied to the foam surface yielded electrical resistances of 650, 250, 100 and 50 KΩ respectively, as shown graphically in the force-resistance curve of Fig. 2. The response was monotonic, although not perfectly

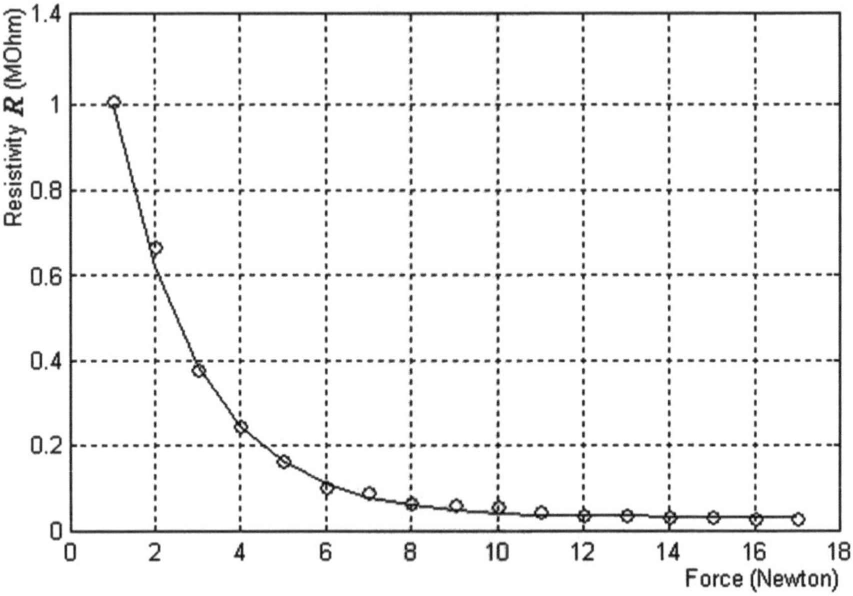

Fig. 2 The relationship between force and electrical resistance

linear, when the forces were small (between 0 and 4 N). The measured values show that such a tactile sensor has a high sensitivity within this range and a lower sensitivity to increasing forces outside of this range. The advantage of a smooth and continuous curve lies in the ease of mathematical fitting and the applicability of other computational processes. The authors found a well-defined mathematical relationship between the range of applied force and the values returned by the tactile element. From the graphical trend line shown in Fig. 2 it can be seen that the relationship between applied force and the returned sensor value obeys the formula:

$$R = (1.579e^{-0.449F} - 8.365 \times 10^{11} \times e^{-32.34F})\Omega. \tag{11}$$

where R is the electrical resistances returned by the tactile element and F is the force (Newton) applied to the tactile surface. The coefficients, 1.579 and 8.365×10^{11}, have the units Ω, while 0.449 and 32.34 have the units Newton^{-1}.

Five different shapes of objects have been used to test for the robot's ability to recognize object types. The robot makes contact with these objects, and the data from the tactile sensor array is stored and analyzed. Later, one of the five objects is grasped again but with different magnitudes of forces and with different positions and rotations. The ability to distinguish between object types is then calculated.

The test objects used were: an oval object with two major axes of 14 and 11.7 mm; a cylindrical object of 6.0 mm diameter and 20 mm length; a cube with the dimensions $10 \times 15.9 \times 10$ mm; a sphere with a 9.5 mm diameter; and dumbbell-shaped object 13 mm in length and 10 mm in diameter.

Following initiation of the aforementioned experiment, physical contact with the grasped objects typically produces only surface deformations. Values of pressure are extracted by means of given thresholds because the shape recognition algorithm only requires a pressure distribution. At this instant all tactile element output values are extracted and the deformation of tactile surfaces are interpolated and compiled in a property matrix. In the experiment, the training procedure is done to check classification performance, as shown in the following Pseudo-code. The Quadric surface properties are modified by multpilying with the transformation matrix for translation, rotation and scaling operations.

Algorithm 1. Contact-Classify ($T = (N, D_{N\times4\times4}, I_{4\times4})$)

```
Initialization: Objnum, i
while Objnum ≤ N
  for scale = 1; step = 0.1; scale ≤ 4
    for tranXY = −2; step = 0.1; tranXY ≤ 2
      for rotXYZ = 0; step = π/8; scale ≤ 2π
        tmpMatrix = Transform(D(Objnum), scale, tranXY, rotXYZ)
        traject(i) = MinEigen(tmpMatrix)
        Increase i
      end for
    end for
  end for
  Increase Objnum
end while
[â, b̂, ĉ, d̂, ...] = FindThreshold(traject)
T = CheckLevel(â, b̂, ĉ, d̂, ..., MinEigen(I))
```

The program rebuilds the eigenvalue trajectory under different transformation matrixes several times until the number of required Quadric parameters are obtained. In the experiment, the training procedure is done as outlined in the Pseudo-code in Algorithm 1. The inputs consist of N Quadric surfaces (D) which are randomly selected by the number I. The Quadric surfaces are transformed by translation, rotation and scaling matrices. The smallest eigenvalues are calculated and stored in the variable (*traject*) from which the trajectory is generated. The threshold of classifications ($\hat{a}, \hat{b}, \hat{c}, \hat{d}, \ldots$) are then determined. The smallest eigenvalue of the selected surface I is compared to this threshold of classifications. The algorithm returns the output in the variable T as an index. For a 4 by 4 matrix, there are four eigenvalues, the smallest one being in the order of 10^{-3}, and the largest one in the order of 10^2. Thus, the eigenvalues in the first three columns, in ascending order, are not used while the one in the last column is used because it is the smallest eigenvalue. This experiment applies graphical techniques to study the behavior of eigenvalues after the matrix elements change. This change normally requires numerical analysis and perturbation theory, but the technique called "eigenvalue trajectory analysis", illustrated in Fig. 3, is more applicable here.

This graph shows the smallest eigenvalue of the covariance matrix of the Quadric surface property independent of translations in all two axes (along the x-axis and the

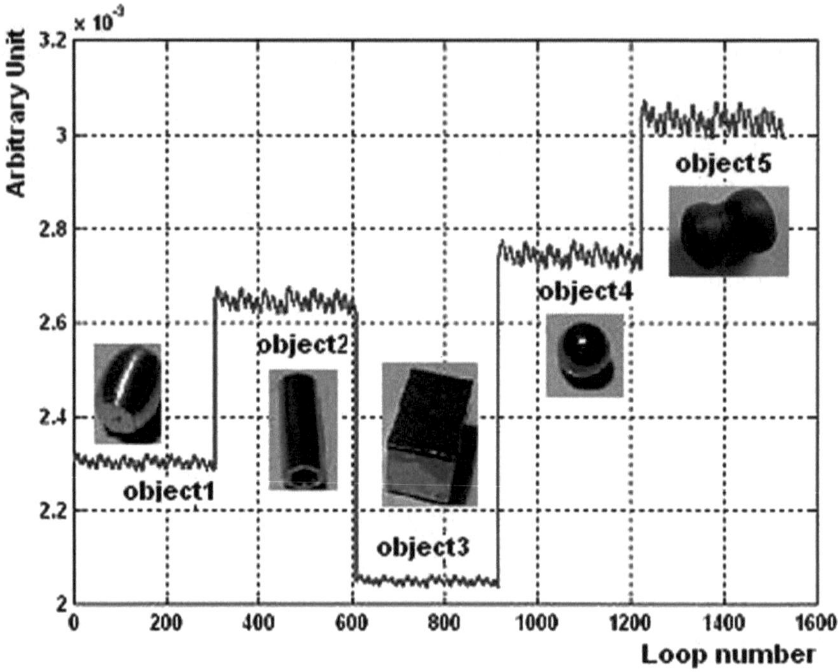

Fig. 3 Eigenvalue trajectories, for different objects, derived from Quadric parameters

y-axis), of rotations around any axis (around the x-axis, the y-axis and the z-axis), and of scalable values. After the trajectory of the eigenvalue has been derived, it can be used to classify the contact surface of the object by matching the eigenvalue levels of the surface property matrix with the object prototype.

There are many different numerical methods for solving the characteristic equation for the eigenvalues λ. In summary, a relatively straightforward algorithm extracts the eigenvalues by solving a n-degree polynomial, and then derives the eigenvector space for each eigenvalue. An important tool for describing the eigenvalues of square matrices is the characteristic polynomial. Conversely, n polynomial equations have exactly n roots.

In considering the machine hardware, each arithmetic operation is generally affected by round-off errors because the hardware can only represent a subset of real numbers (floating-point numbers). As an illustration, after an addition, subtraction, multiplication or division, the true value of an operation cannot be represented exactly as a floating-point number. Instead, it must be approximated by a nearby floating-point number before it can be stored in the memory. The difference between this approximation and its true value counterpart is denoted as the round-off error. Thus, an arithmetic operation is said to round correctly according to the machine precision.

In considering the software, the most common problems resulting from the round-off errors occur either when two quantities very close to each other are subtracted, or when a number is divided by another number which is close to zero. Other sorts of round-off errors can also occur when there are operations which attempt to convert a high-order number (64-bit number) to a low-order number (16-bit number) or when integers are converted to decimal numbers (such as in C). This means that round-off errors in finding eigenvalues mainly depend on the type of the arithmetic operations used.

Eigenvalues of large matrices should not be computed using the characteristic polynomial because computing the polynomial becomes expensive in itself and the exact (symbolic) roots of a high-degree polynomial can be difficult to compute. For instance, the Abel–Ruffini theorem implies that the roots of high-degree polynomials cannot be expressed simply using nth roots [1]. Moreover, although effective numerical algorithms for approximating the roots of polynomials exist, small errors in the eigenvalues can lead to large errors in the eigenvectors. In fact, eigenvalues using the characteristic polynomial give unexpected results in these tests. General algorithms for finding eigenvectors and eigenvalues are usually iterative methods, but only a few iterative methods can provide round-off errors small enough to be useful for the purposes of this work.

The easiest method is the power method in which a random vector is chosen and used it to compute a series of increasing power matrices. The aim of the power method is to find only the largest eigenvalue. The problem with this method is that if the data in matrix A has errors, the square of A^m will exacerbate them leading to higher round-off errors. Although the modified method enables estimation of the smallest eigenvalue, these two limitations are inevitable.

Even though the level enabling classification of objects is only in the order of 10^{-3}, the computing method used to find an eigenvalue needs a very small round-off error. Moreover, there are some methods used in finding the roots of polynomials via eigen-functions (such as those employed by Matlab). The algorithms simply involve computing the eigenvalues of the companion matrix. Although it is possible to prove that the results produced are the exact eigenvalues of a matrix within specific round-off errors of the given companion matrix, this does not mean that they are the exact roots of the polynomial with coefficients within the same round-off errors. This has been tested experimentally resulting in unusable results. Consequently, these methods may not be used for this type of contact identification.

On the other hand, popular methods such as the QR algorithm used in the LaPACK library (Linear Algebra Software Package), have good classification ability since a precise resolution within the order of 10^{-3} is possible with very small round-off errors.

4 Experimental Evaluation

The random noise associated with the interpolation in a real application can generate variations in eigenvalues. This requires an investigation into the eigenvalue trajectory under random noise conditions.

4.1 Effects of Random Noise on Eigenvalue Trajectory

This section investigates the sensitivity to noise in contact identification algorithms developed earlier. The effect of noise on the classification accuracy is considered when the incoming trajectories are corrupted by noisy data. The analysis of the simulation results shows varying abilities of the algorithms to cope with the noise perturbations. In most instances, high prediction robustness was achieved. Few simulations showed sensitivity to noise.

Figure 4 indicates that noise on the trajectory dampened some graph levels during experiments. Nonetheless, it did not reduce contact classification capability with respect to the overall error growth. Contact classification can still be achieved through level checking.

Noise was simulated using a random function with a normal distribution with values in the range [1, 20]. They were then added to every tactile data element. All of the eigenvalue trajectories were tested through 10 iterations with a range of noise levels. The noise levels for all tactile elements were randomly and simultaneously increased, yet only to the maximum values of 8% of the ADC's maximum (255), and the performance of algorithms was demonstrated as shown in Fig. 4. This leads to the conclusions that if the noise level is maintained below 8% it will not be statistically meaningful for, nor affect, classification.

By experiment it was also clear that classification capability is reduced if the random noise peaks are greater than 8% of the ADC's maximum value. Invalid

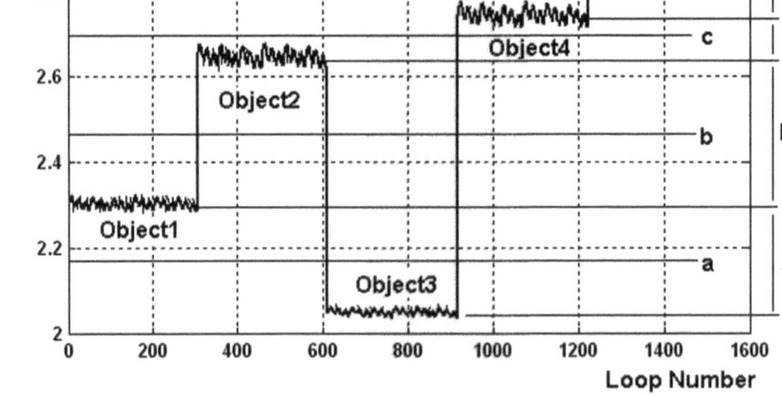

Fig. 4 Noise tolerances

classification was tested by increasing noise to a level higher than 8%, and consequentially, crossing levels of eigenvalue trajectory appeared.

By observation, the use of a noise filter on the tactile data reduces the effect of noise on the eigenvalue trajectory, and such a filtration must be performed before surface interpolation was applied.

4.2 Margin of Classification

The principal idea used to classify the contact is in the matched threshold of the eigenvalue trajectory. For example, in Fig. 4, the "a" level can be used to distinguish between object 1 and object 3; the "b" level can be used to distinguish between objects 1 and 2; and so on. As formerly mentioned, random noise has an effect on the eigenvalue trajectory. The windows of different sizes (A, B, C, D) have their uses in defining different thresholds for the eigenvalue trajectory, and classification by thresholds has to be adjusted dynamically.

In cases where the noise amplitude falls outside of the range [1, 20], this classification method cannot guarantee correct classification of object shapes. Another limitation of the algorithm is the differences in the object shapes and noise affected in surface interpolation. If the shapes of the test objects are similar, then the classification can still fail as random noise can reduce the window size to zero.

Also, some object shapes are very sensitive to noise which leads to them being incorrectly identified in the interpolation process. As displayed in Fig. 4, object 2 and object 4 yield little difference in terms of the smallest eigenvalues, which leads to a classification failure. In reality, the noise mixed with the tactile data also has an effect on the smallest eigenvalue. If the window size is too small, then the added noise will make the classification capability approach zero.

In terms of actual implementations using tactile sensors, the set of objects and other criteria also affect the optimal classification performance. It also seems reasonable to think that the degree of deviations or differences between any two object shapes should eventually entail classification performance between them, and experiments appear to have uncovered such a case.

According to Fig. 4, the thresholds of objects 1, 2, 3, 4 and 5 correspond with the oval, cylindrical, cube (box), ball shapes and dumbbell-shape object, respectively. Each object has a different eigenvalue in the eigenvalue trajectory with no particular increasing or decreasing order in terms of their levels. This can lead to miscalsssification in the case of two very close threshold values. Table 1 demonstrates the test results, whose tests were repeated ten times on each object shape. The classification results reveal correct recognitions as well as misrecognitions. These differences may indicate that there are indeed limitations on the ranges of classification due to the similarity of test objects, fitting performance, and random noise.

For example, the cylinder object can be misclassified as a ball object in 10% of cases. Yet, the box is less likely to be misclassified as another object, because their thresholds are not in neighbouring ranges. The experiment has also clearly shown

Table 1 Statistical error of classification

Object features	Percent of misclassify					Percent error (%)
	Oval	Cylinder	Box	Ball	Dumbbell-shape	
(1) Oval	0	0	0	0	0	0
(2) Cylinder	0	0	0	10%	0	10
(3) Box	0	0	0	0	0	0
(4) Ball	0	10%	0	0	0	10
(5) Dumbbell-shape	0	0	0	0	0	0

that classification capability reduces if the random noise peaks are greater than 8% of the ADC's maximum value. Invalid classification was achieved by increasing noise levels above 8%, and consequentially. As a result, contact classification cannot be achieved by a simple level checking.

5 Conclusions

In this paper, a novel and fast framework for contact recognition based on the eigenvalue trajectory of a 3-dimensionally deformed surface onto Quadric parameters has been presented. The technique called "eigenvalue trajectory analysis", is introduced and adopted for specifying the margin of classification and classification thresholds. It has been shown that the matrix from the parameter of Quadric surfaces by interpolation of tactile data may be formulated by eigenvalue decomposition and can reflect under all contact geometries. The smallest component of an eigenvalue can be used to estimate and identify object shapes without using any other references, whereas classification is used as the principal indication of surface identity. The shape reflectance parameter pertaining to (unique to) each surface may be recovered and identified. It has been shown that the reliability of the surface classification method and the accuracy of transformation are dependent of object shapes.

The quadric surfaces used still have a limited number of possibilities which can be accurately represented. The five non-singular quadrics are the ellipsoid, elliptic paraboloid, hyperbolic paraboloid, hyperboloid of one sheet, and hyperboloid of two sheets. The singular quadrics include cones, cylinders, swept hyperbolas and parabolas and the degenerates: planes. To improve classification the surface properties must be modified. One option is to parameterize the Quadric surface in higher space and obtain the higher surface discrimination whilst preserving the eigenvalue trajectory behavior.

It is called a quadric surface if the surface is the equation of second degree in three-dimensional Cartesian Coordinates. However, quadric surfaces are multi-dimensional spaces defined by coordinates $\{x_0\ x_1 \cdots x_D\}$, where the general quadric is defined by the algebraic equation, $\sum_{i,j=0}^{\varphi} Q_{i,j}x_ix_j + \sum_{i=0}^{\varphi} P_ix_i + R = 0$, in which Q is a

(φ+1)-dimensional matrix, $P$is a (φ+1)-dimensional vector and R a constant. For example, if a Quadric parameter is projected into a higher coordinate system, then the parameters Q and P will be:

$$Q = \begin{bmatrix} a & d & f & g & w_1 \\ d & b & e & h & w_2 \\ f & e & c & j & w_3 \\ g & h & j & k & w_4 \\ w_1 & w_2 & w_3 & w_4 & w_5 \end{bmatrix} \quad \text{and} \quad P = \begin{bmatrix} x\, y\, z\, w\, 1 \end{bmatrix}^T.$$

In this case, Q is still the symmetrical matrix and the smallest eigenvalue used for classification will be λ_5. w_1, w_2, w_3, w_4 and w_5 are the lists of parameters which must be found in order to project onto a higher dimensional space. Future research concerns formulating a set of such challenge problems.

References

1. Dehn E (1930) Algebraic Equations: An Introduction to the Theories of Lagrange and Galois. Columbia University Press New York.
2. Hanzon B, Jibetean D (2003) Global Minimization of a Multivariate Polynomial using Matrix Methods. Journal of Global Optimization, 27(1):1–23.
3. Ibrayev R, Jia Y-B (2004) Tactile recognition of algebraic shapes using differential invariants. In: Proc. IEEE International Conference on Robotics and Automation, pp 1548–1553.
4. Ibrayev R, Jia Y-B (2005) Semi-Differential Invariants for Tactile Recognition of Algebraic Curves. In: Proc. International Journal of Robotics Research, 24(11):951–969.
5. Jibetean D, Laurent M (2005) Semidefinite approximations for global unconstrained polynomial optimization. The SIAM Journal on Optimization, 16(2):490–514.
6. Lynch MR, Rayner PJ (1989) Optical Character Recognition Using a New Connectionist Model. In: Proc. IEE 3rd International Conference on Image Processing And Its Application, UK.
7. Ng E (1989) Robot Arm World Visualisation By Tactile Exploration. Project Report, E&E Engineering, Auckland University, New Zealand.
8. Ng EWK et al. (1991) Tactile robot shape recognition using geometrical angle/length sequences. In: Proc. IEEE International Joint Conference on Neural Networks, 1:307–312.
9. Petchartee S, Monkman G (2007) 3D-Shape Recognition based Tactile Sensor. In: Proc. International Conference on Sensing Technology, New Zealand.
10. Temple G (1928) The Theory of Rayleigh's Principle as Applied to Continuous Systems. In: Proc. the Royal Society of London. Series A, 119(782):276–293.
11. Temple G, Bickley W (1933) Rayleigh's Principle and Its Applications to Engineering. Oxford University Press, Oxford.
12. Yan H (1990) Closed Boundary Recognition using a Multi-layer Neural Network. In: Proc. the 1st Australian Conference on Neural Networks, Sydney University Electrical Engineering, Australia.

Development of A Low Cost Compact Low Profile Phase Array Antenna for RFID Applications

Nemai Chandra Karmakar, Sushim Mukul Roy and Muhammad Saqib Ikram

Abstract The 900 MHz frequency band has found many applications in RFID systems all over the world. A 3x2-element planar phased array antenna for RFID readers in a low cost compact package is developed. The antenna covers 860–960 GHz frequency band with more than 10 dB input return loss, 12 dBi broadside gain and up to 40° elevation beam scanning with a cascade of 4-bit reflection type phase shifters. Once implemented in the mass market, RFID smart antennas will contribute tremendously in the areas of RFID tag reading rates, collision mitigation, location finding of items and capacity improvement of the RFID system.

Keywords Radio frequency identification · smart antenna · RFID tag · barcode · anti-collision · RFID reader · switched beam array · ESPAR · beamforming · phased array antenna · phase shifter

1 Introduction

The Radio Frequency Identification (RFID) system is a wireless data transmission and reception technique for automatic identification, asset tracking, security surveillance and many other emerging applications. An RFID system consists of three major components: (i) a reader or integrator, which sends interrogation signals to an RFID **trans**mitter res**ponder** (transponder) or tag, which is to be identified; (ii) an RFID tag, which contains the identification code; and (iii) a middleware, which maintains the interface and the software protocol to encode and decode the

Nemai Chandra Karmakar
Department of Electrical and Computer Systems Engineering, Monash University, Bldg. 72, P.O. Box 35, Clayton, VIC Australia 3800, e-mail: nemai.karmakar@eng.monash.edu.au

Sushim Mukul Roy
Department of Electrical and Computer Systems Engineering, Monash University, Bldg. 72, P.O. Box 35, Clayton, VIC Australia 3800

Muhammad Saqib Ikram
Department of Electrical and Computer Systems Engineering, Monash University, Bldg. 72, P.O. Box 35, Clayton, VIC Australia 3800

S.C. Mukhopadhyay, G.S. Gupta (eds.), *Smart Sensors and Sensing Technology*,

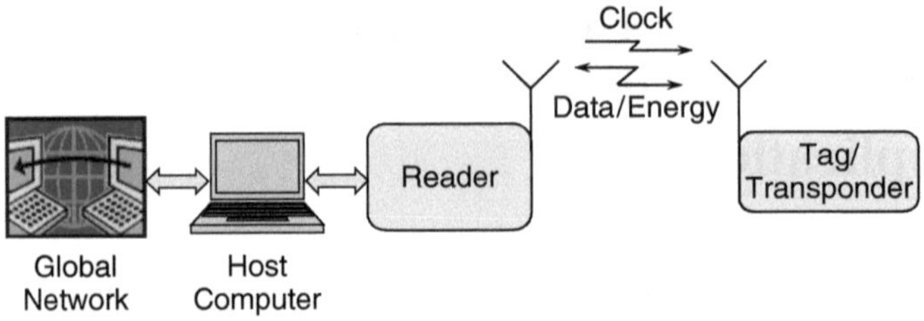

Fig. 1 Generic RFID system

identification data from the reader into a mainframe or a personal computer. In a further embodiment, the computer can be connected to the global network for remote real-time asset tracking and security surveillance. Figure 1 below illustrates a generic block diagram of the RFID system. At the dawn of the new millennium, as barcodes and other means for identification and asset tracking are becoming inadequate for recent demands, RFID technology has been facilitating logistics, supply chain management, asset tracking, security access control, intelligent transportation and many other areas at an accelerated pace. A recent Google search of the terminology "RFID" brought up thirty eight million hits. This large huge number of URLs represents the significant activities and applications of RFID in various sectors in either commercial domains or government agencies.

Due to the flexibility and numerous advantages of RFID systems compared to barcodes and other identification systems available so far, RFIDs are now becoming a major player in retail and government organizations. Patronization of the RFID technology by organizations such as Wal-Mart, K-Mart, the USA Department of Defense, Coles Myer in Australia and similar consortia in Europe and Asia has accelerated the progress of RFID technology significantly in the new millennium. As a result, significant momentum in the research and development of RFID technology has developed within a short period of time. The RFID market has surpassed the billion dollar mark recently [1], and this growth is exponential, with diverse emerging applications in sectors including medicine and health care, agriculture, livestock, logistics, postal deliveries, security and surveillance and retail chains. Today, RFID is being researched and investigated by both industry and academic scientists and engineers around the world. Recently, a consortium of the Canadian RFID industry has put a proposal to the Universities Commission on the education of fresh graduates with knowledge about RFID [2]. The Massachusetts Institute of Technology (MIT) has founded the AUTO-ID centre to standardize RFID, thus enabling faster introduction of RFID into the mainstream of retail chain identification and asset management [3, 4]. The synergies of implementing and promoting RFID technology in all sectors of business and day to day life have overcome the boundaries of country, organisation and discipline.

This work concerns development of a smart antenna for RFID readers. The smart antenna enhances the performance of automatic identification systems, asset tracking in real time and inventory control in warehouses. It is important to note that smart antennas for RFID readers are currently in the developmental stage. To the best of the author's knowledge, such smart antenna systems for RFID have not yet emerged as the mainstream enabling technology. Therefore, smart antenna implementation systems for RFID readers are either still on the drawing boards of designers or under investigation, mainly by various commercial research groups as patents, be they either conceptual or physical developments.

The chapter is organised as follows. An introduction to RFID is presented first. An in-house developed phased array antenna for 900 MHz band RFID readers is then presented. The phased array antenna comprises an L-shaped slot loaded half disk (semi-circular) patch antenna, a corporate feed network, a four-bit phase shifter array and control electronics. The design of each component from scratch is presented. The results of the prototype components are presented next followed by conclusion.

2 Phased Array Antenna for RFID Readers

Smart antennas for RFID readers are scanned beam antenna systems, which point beams to selective transponders within their main lobe radiation zones, thus reduce reading errors and collisions among tags. This technique exploits *spatial diversity,* and in many cases polarisation *diversity* of tags. The directed beam also reduces the effects of multipath fading [5]. This new approach to RFID antenna technology is being adopted by RFID manufacturers such as Omron Corporation, Japan [6], RFID Inc. [7] and RFSAW, USA [8]. In March 2006 Omron Corporation developed a new electronically controlled antenna technology in their UHF band RFID reader systems [9]. Recently, RFID Inc. [7] has introduced a 32-element compact package smart antenna with associated switches at 125 KHz. This new product advances RFID applications in factory automation, process controls and original equipment manufacturer (OEM) markets. However, the technical details of these smart antennas for RFID readers are not available to readers due to commercial in confidence considerations. This paper presents technical information on smart antennas for RFID readers and this is provided in subsequent sections.

Smart antennas are electronically steerable antennas, which are fixed with supporting platforms such as gantries. Electronic beam steering can be achieved by (i) switching on certain antenna elements and turning off the rest as in switched beam array antennas; (ii) changing the phase excitations of individual elements as in phased array antennas; and (iii) adaptively controlling the weight vectors of individual elements using array signal processing chips. Adaptive antennas can also be sub-classified as multiple input multiple output (MIMO) or multiple input single output (MISO). Finally, by combining switching and phase shifting the turned on subset of the antenna element, a switched beam phased array is formed. Figure 2 shows classifications of smart antennas for RFID readers. Following are detailed accounts of the development of a phased array smart antenna for RFID applications.

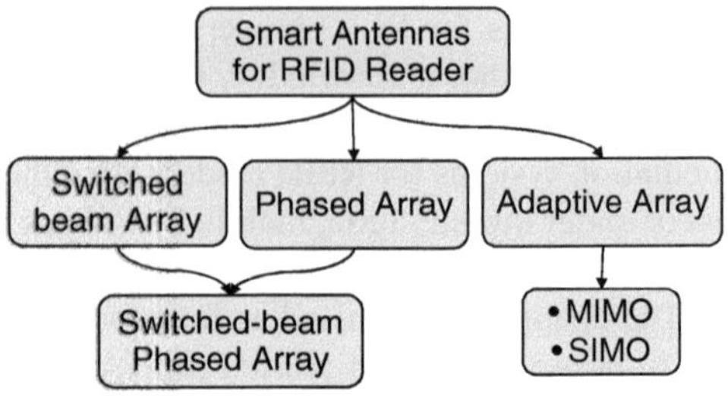

Fig. 2 Classification of smart antennas for RFID readers

3 Development of Phased Array Antenna at 900 MHz

So far various smart antenna systems for RFID readers have been discussed. In this section, the technical development of a planar phase scanned antenna developed by the Monash RFID Research Group led by the author is presented. The 900 MHz frequency band has many applications in RFID systems. The application areas are: 868–870 MHz for short range devices in CEPT countries including Australia and most European countries that recognize the U.S. participation in the CEPT Radio Amateur License (http://www.arrl.org/FandES/field/regulations/io/#cept, accessed 11 Sep. 2007); 902–928 MHz ISM band RFID for North America; 918–926 MHz RFID for Australia; and 950–965 MHz for Japan [10, 11]. Therefore, there is a demand for a smart reader antenna which can cover all these bands of operations. To fulfill this requirement the author has developed a 3×2-element planar phased array antenna operating at 900 MHz frequency band covering 100 MHz bandwidth. The objective is to make a low cost, low profile and light weight antenna for easy mounting and portable operation, aimed at mass markets worldwide. These research and development activities in RFID have also trained personnel for this potential RFID market. Table 1 shows the specification requirements of the smart antenna for RFID readers.

Table 1 Specification requirements of 900 MHz RFID reader phased scan array antenna

Frequency	910 MHz
1o dB RL Bandwidth	100 MHz
Beam scanning	3D azimuth and elevation beam scanning
Beamwidth	42°
Peak gain	12 dBi
Weight	Less tan 1 kg
Dimensions	3 cm × 50 cm × 50 cm

3.1 Array Antenna Design

To make the antenna compact and fully planar the most suitable candidate is the microstrip patch antenna, but the conventional microstrip patch antenna is a half-wavelength resonator. At 900 MHz, the antenna element is sizable compared to the wavelength and is not portable and easily mounted. The resonant frequency of a circular patch antenna at dominant TM_{110} mode is expressed [12] as:

$$f_r|_{110} = \frac{1.8412\,c}{2\pi a_e \sqrt{\varepsilon_r}} \tag{1}$$

Where, c = velocity of light in free space (3×10^{10} cm/s)
a_e = effective radius of the circular patch antenna (fringing field effect included)
ε_r = dielectric constant of the microwave laminate on which the patch is etched

Using the design (1), the circular patch antenna on a low dielectric constant microwave laminate for efficient radiation at 900 MHz has a diameter of approximately 19 cm, which is quite large for array application. Besides this obtaining a large bandwidth with of approximately 100 MHz bandwidth at 900 MHz resonant frequency is not a trivial task. This constitutes 11% bandwidth at 900 MHz. A comprehensive investigation for a suitable compact antenna configuration, which can meet the required bandwidth and antenna size without much compromising antenna radiation efficiency and gain is made. Finally, an L-slot loaded crescent antenna [13] is culled as the most suitable candidate for portability and ease of mounting.

Antenna Element Design. The antenna element is designed with full-wave electromagnetic solver CST Microwave Studio. Figure 3a shows the L-shaped slot loaded half disk antenna. The slot loading creates parasitic loading comprised of extra capacitance due to the slot and inductance due to the modified current path on the radiating half disk. This broad 10 dB return loss (RL)bandwidth of approximately 110 MHz is achieved as shown in Fig. 3b. The single antenna element is designed on FR4 substrate with a dielectric constant of 4.4 and loss tangent of 0.002. The thickness of the material is 0.787 mm. The radius of the half disk antenna is 8 cm. The L-shaped slot is added around the coaxial feed probe to increase the bandwidth to more than 100 MHz centered at 910 MHz. The total height of the antenna element is 25 mm. As shown in Fig. 3b the single antenna element produces a RL bandwidth of more than 100 MHz centered at 910 MHz with a maximum RL of 35 dB. Figure 3c shows the radiation patterns of the antenna element in the broadside direction at 910 MHz. As the figure shows, the antenna generates a beam of 70° beamwidth with a gain of approximately 8 dBi. The asymmetry in H-plan radiation pattern (green curve) is due to the anti-symmetric surface current on the antenna configuration in H-plane.

Beamforming Network Design. The next design is the beamforming network for the 3×2-element array antenna. Figure 4a shows the layout of the beamforming network developed using Agilent's Advanced Design System (ADS) 2005A. The beamforming network comprises a 1-to-6-way power divider, 4-bit reflection type phase shifters and the biasing control of the 4-bit phase shifter array. All components are designed on Taconic TLX0 of thickness of 0.787, dielectric constant of 2.45 and loss tangent of 0.0006.The details of the reflection type phase shifter can be found in previous work by the author and a colleague [14]. The phase shifters are comprised of a quadrature hybrid coupler in a very compact package with reactive loading. The reactive loading is comprised of high impedance transmission lines of

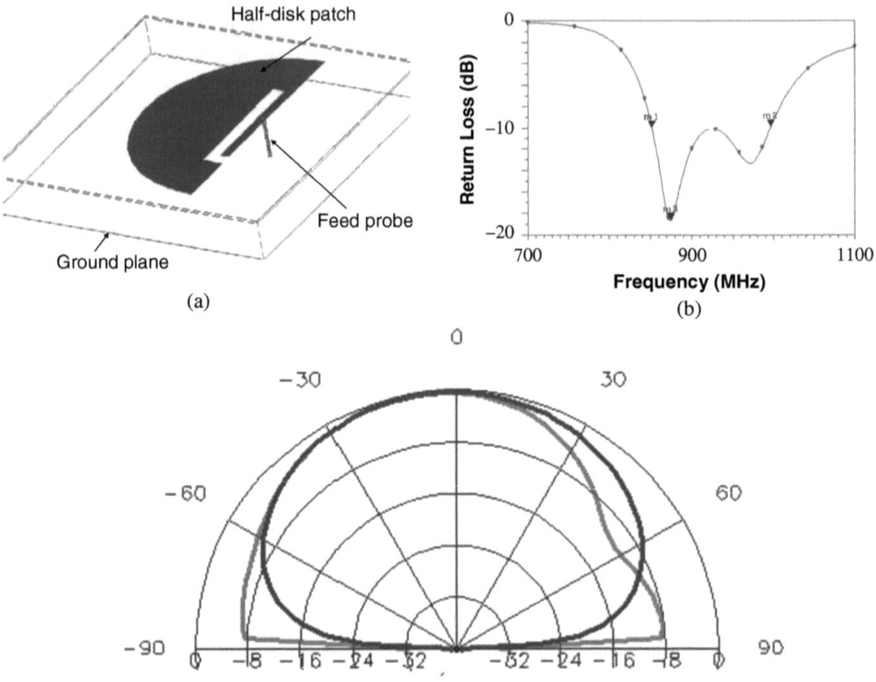

Fig. 3 CST Microwave Studio generated (**a**) layout, (**b**) input return loss vs frequency of the half disk patch antenna, and (**c**) CST simulated radiation patterns (blue — E-plane and green – H-plane) at 900 MHz. FR4 laminate: $\varepsilon_{r1} = 4.2$, thickness 1.6 mm and airgap height 23.4 mm

various lengths followed by Philips BA682 pin diodes. A diode compensation technique is used to operate the diode in the 900 MHz band. The diode is a very low cost band switching diode with high parasitic capacitance at its OFF state. Detailed information about the diode compensation technique can be found in [15]. The reflection type phase shifter is selected due to its large bandwidth and uniform configuration for all bits. The total beamforming network is delicately phase matched to each branch with zig-zag transmission lines. The feed network is a corporate type feed network with the first stage of a 2-way T-junction power divider followed by a three-way power divider. Thus 1-to-6-way equal power division is achieved. In this primary design amplitude tapering was not attempted. However, the readers are referred to [16].

Integrated Array Antenna. Mutual coupling greatly affects the array antenna's performance, especially when the antenna elements are brought too close together. The situation worsens when the heights of the antenna elements are considerable (in this case 2.5 cm). Metallic barriers have been used to reduce the mutual coupling effect between the antenna elements as shown in Fig. 4b. Thus the antenna aperture area is maintained at 50 cm × 50 cm. This area is about 70% of a conventional patch array antenna at that frequency. The antenna is housed in a microwave transparent

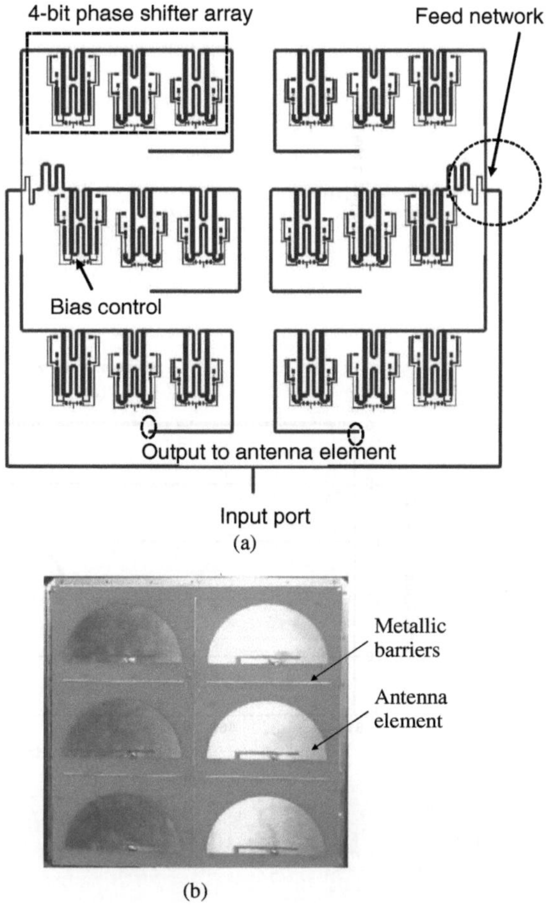

Fig. 4 (**a**) Beamforming network; and (**b**) photograph of the complete array antenna

plastic casing (not shown here). The total weight of the antenna is about 1.7 kg, making it suitable for easy transportation and mounting.

3.2 Results of Array Antenna

Figure 5a shows the s-parameters vs frequency for the 1-to-6-way power divider. The input return loss of the power divider is about 30 dB and the transmission loss for each output port is about −8 dB. The output ports are well phase matched at the center frequency of 910 MHz and fall linearly with the frequency. The maximum phase offset at the band edge is less than ±15°. The phase shifters have an average insertion loss of 3.5 dB in both ON and OFF states and the maximum phase deviation of about ±18° for the four bit phase shifter array. This quantization error in phased array antennas is quite common. Figure 5b shows the input return loss of

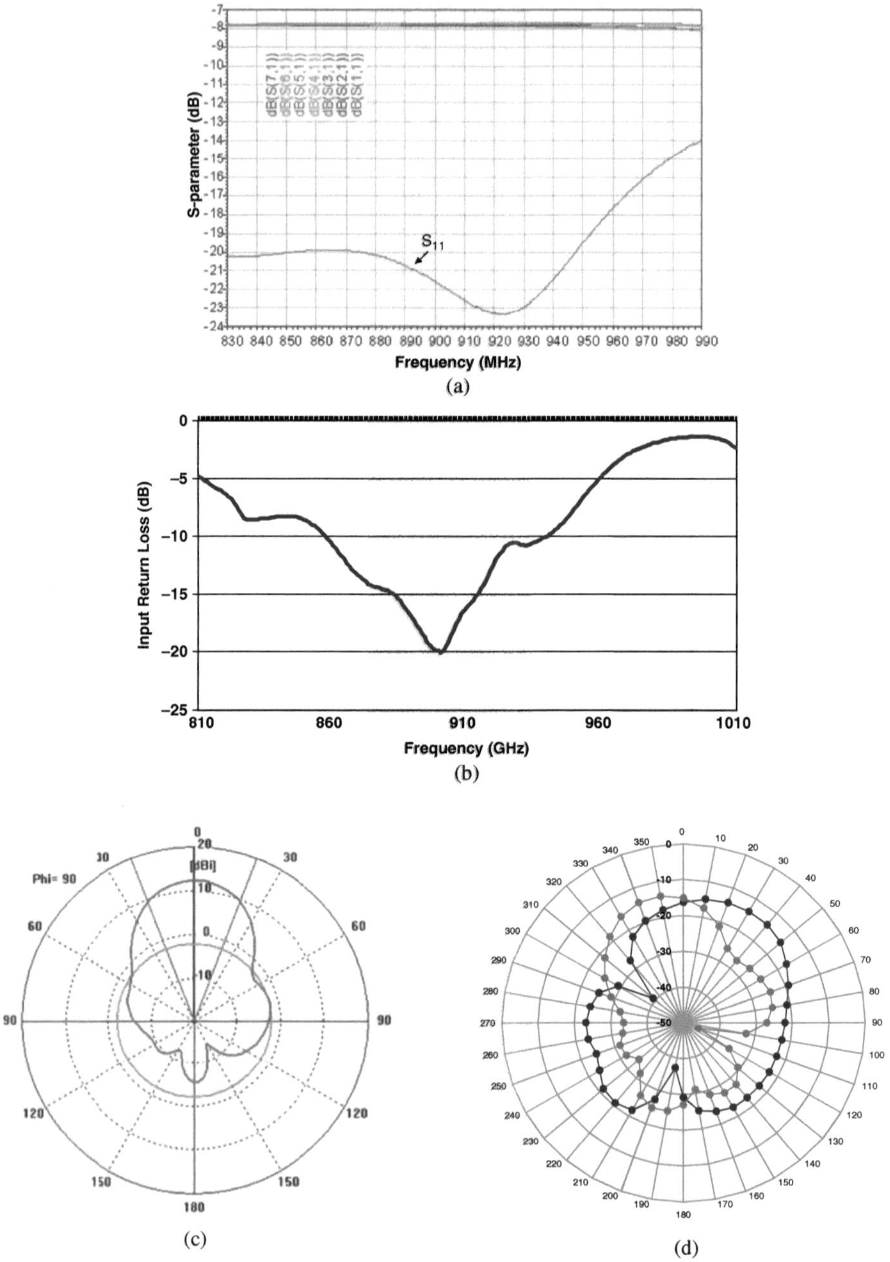

Fig. 5 (**a**) s-parameter vs frequency of 1-to-6-way power divider; (**b**) input return loss of complete antenna array; (**c**) CST Microwave Studio simulated broadside gain pattern of the array and (**d**) measured squinted radiation patterns of the array (Pink: +20° and blue: −30° beams)

the assembled array antenna. The antenna is very well matched with approximately 10 dB return loss in the prescribed bandwidth of 100 MHz centered at 910 MHz. The RL at 910 MHz is about 20 dB.

Figure 5c shows the radiation pattern of the array antenna in the broadside direction. As the figure shows, the antenna generates a beam of 42° beamwidth with a gain of 12 dBi. Figure 5d shows the beam patterns when they are scanned at +20° and −30° from the broadside direction. The results indicate that the antenna is capable of scanning main beams in 3D orientation. However, due to the compact design with only 6 antenna elements the degree of freedom is confined to about ±40° elevation scan angles from the broadside direction. With the number of antenna elements this scan range can be improved at the expense of more complex beamforming network, associated switching electronics and a larger antenna aperture.

4 Conclusion

RFID is an enabling technology which is transforming the identification, asset tracking, security surveillance, medicine and many other industries at an accelerated pace. This paper has presented a 3×2-element low cost, light weight and compact planar phase scanned antenna array for RFID readers. A detailed design guideline for the component level physical layer development of the smart antenna has been presented. The antenna is designed at the 900 MHz frequency band with 100 MHz bandwidth to cover a plethora of applications in this frequency band. The development certainly has a potential international market.

Acknowledgements The work is supported by Australian Research Council's Discovery Project Grant # DP665523: *Chipless RFID for Barcode Replacement.* The software supports from Agilent and CST are also acknowledged.

References

1. Das. R. & Harrop, P., "RFID Forecasts, Players & Opportunities 2006–2016," IDTechEx, London, UK,. Retrieved from http://www.idtechex.com/products/en/view.asp?productcategoryid=93 on 11 September 2007.
2. GTA. RFID Industry Group RFID Applications Training and RFID Deployment Lab Request Background Paper January, 2007.
3. McFarlane, D. & Sheffi, Y. "The Impact of Automatic Identification on Supply Chain Operations", International Journal of Logistics Management, 14(1), 407–424, 2003.
4. Karkkainen, M. & Ala-Risku, T. "Automatic Identification – Applications and Technologies", Logistics Research Network 8th Annual Conference, London UK, September 2003.
5. Ingram, M. A., "Smart reflection antenna system and method", US patent no. US 6,509,836, B1, January 21, 2003.

6. Nakamura, T. and Seddon, J., "Omron Develops World's First Antenna Technology That Boosts UHF RFID Tag Read Performance," Omron Corporation press releases, 2006. Retrieved from http://www.omron.com/news/n_270306.html in June 2007.
7. Profibus Device Net-Ethernet IP-Modbus-Remote I/O, Retrieved from http://www.rfidinc.com in June 2007.
8. RFID Analyst. Issue 26, March 2003. Retrieved from http://rfid.idtechex.com/documents/en/sla.asp? documentid=67 in September 2007.
9. Nakamura, T. & Seddon, J. (2006) Omron Develops World's First Antenna Technology That Boosts UHF RFID Tag Read Performance, Omron Corporation press releases. Retrieved from http://www.omron.com/news/n_270306.htmlC in June 2007.
10. Mendolia, G. Gupta, O. & Toit, C.F. RF ID tag reader utilizing a scanning antenna system and method, US Patent No. US 2005/0113138 A1, May 26, 2005.
11. Calmpitt, H.G. RFID Certification Textbook, 2nd Ed., Arlington Heights, IL, American RFID Solution LLC, 2006.
12. Balanis, C.A., A*ntenna Theory: Analysis and Design*, 2nd Ed. Wiley, New York, 1982.
13. Deshmukh, A. A. & Kumar, G., "Various slot loaded broadband and compact circular microstrip antennas" *Microwave and Optical Technology Letters*, vol. 48, no. 3, pp. 435–439, 2006.
14. Karmakar, N. C. and Bialkowski, M. E., "An L-Band 90° Hybrid-Coupled Phase Shifter Using UHF Band PIN Diodes", *Microwave and Optical Technology Letters*, vol. 21 no. 1, pp. 51–54, Jan., 1999.
15. Karmakar, N. C., *Antennas for Mobile Satellite Communications*, Unpublished doctoral dissertation, The University of Queensland, Australia, 1999.
16. Karmakar, N. C., "A Low Sidelobe Sub-Array of Portable VSATS" *Proc. 2003 IEEE International Antennas and Propagation Symposium and URSI North American Radio Science Meeting*, Columbus, Ohio, USA, June 22–27, 2003.

Implementation and Characterization of a Novel Acoustic Imaging through Beamformers for Automotive Applications

Aimé Lay-Ekuakille, Giuseppe Vendramin and Amerigo Trotta

Abstract Traffic congestion is a severe problem on European freeways. According to a study of the European Commission, its impact will increase even up to 10% in the year 2010. Since building new infrastructure is no longer an appropriate option in most (Western) countries, there are many approaches towards a more effective road usage and a more 'intelligent' way of increasing the capacity of the road network. Examples of advanced traffic control systems are, e.g., 'intelligent' speed limits, adaptive ramp metering, or dynamic routing. This paper focuses on a proposal of a vehicle acoustic imaging that allows driver to recognize obstacles and other cars around him in a safe way. The proposed front end electronic and sensing system are included in a specific architecture supervised by a microcontroller. The used algorithm for gathering data belongs to LCMV and LCMP beamforming methods.

Keywords Acoustic imaging · acoustic sensors and transducers · beamforming · microcontroller · instrumentation

1 Rationale

Precise close distance ranging using radar systems is a very interesting and demanding radar application. Technical problems, like multiple echo suppression, signal interference, resolution of closely spaced targets and robustness to clutter noise have to be solved. Additionally, the available bandwidth is limited by technical means and official regulations. Distance ranging has plenty of applications as well as in:

Aimé Lay-Ekuakille
D.I.I.- University of Salento (Italy), e-mail: aime.lay.ekuakille@unile.it

Giuseppe Vendramin
D.I.I.- University of Salento, Italy

Amerigo Trotta
D.E.E.- Polytechnic of Bari, Bari, Italy

S.C. Mukhopadhyay, G.S. Gupta (eds.), *Smart Sensors and Sensing Technology*,

- automotive for anticolliding system, in helping driver for parking in normal and severe conditions;
- blind aided support for identifying obstacles in closed and opened areas;
- antintruding;
- children mobility control;
- industrial manufacturing automation, sliding object verification, etc..

Many on-board systems use acoustic propagation, as radar, to detect obstacles and other cars. When they are low-cost, they use simple acoustic radar and cannot release an image, although acoustic, of the obstacle or barrier. The main issue that one can wave for this kind of topic is: sensor array processing. Arrays play a basic role in radar, radio astronomy, sonar, communications, directions-searching, seismology, medical diagnosis and care and, measurements. The implementation of arrays of sensors to achieve certain performance criteria encompasses trade-offs among the array geometry, the number of sensors, signal-to-noise, and signal-to-interference ratios, as well as a number of factors.

2 Array of Sensors and Beamforming

2.1 Fundamentals

Consider an array of sensors or 'Smart Antenna' collecting spatial samples of propagating wave fields. Signals may be present from any given direction and always in the presence of noise. The desired signal is from a particular spatial location. From a signal processing perspective, the main objective is to detect the signal arriving from a particular look direction and cancel out any interfering signals and noise. A beamforming processor can achieve this by electrically steering the Smart Antenna, rather than mechanically as done in years gone by. The advantages of beamforming are improvements in communication quality, throughput and efficiency and as a result, have applications in fields such as radar, sonar, seismology and wireless communications.

A sensor array receives incoming signal information. A beamformer processes the spatial samples collected to provide the required spatial filtering [1]. The sensor array may be arranged in a number of different configurations, two of which are the Uniform Linear Array (ULA) and Uniform Circular Array (UCA) in 2D space. The beamformer linearly combines the spatially sampled time series from each sensor to obtain a scalar output time series. Beamformers are grouped into three different classes, fixed, optimum and adaptive. Fixed beamformers are analogous to bandpass filters, they strive to pass spatially signals from a desired look direction and suppress all other signals arriving from all other angles. Furthermore, for practical purposes, interfering signals and noise are to be suppressed relative to the look direction. Optimum beamformers create an optimum response based on the signal statistics and subject to specific constraints.

The beamformer mean output power response is optimized subject to look direction and derivative constraints as is to contain negligible contributions due to noise and interfering signals. A gain, or look direction constraint preserves the desired signal in the particular look direction. Constraining the derivative of the power response with respect to the spatial angle protects the desired signal from directional mismatch. In some cases a minimum variance beamformer is derived from the general formulation by imposing a linear constraint [2]. Since the performance of MPDR (Minimum Power Distortionless Response) beamformers is not particularly robust to various changes in the environment, LCMV (Linear Constrained Minimum Variance) and LCMP (Linear Constrained Minimum Power) beamformes are used the overcome robustness limitations [3].

2.2 LCMV Beamformer

One such solution proposed by Frost [4] and later by Tseng [5] to the optimization problem is Linearly Constrained Minimum Variance Beamforming (LCMV). However, the proposed LCMV solution is inadequate for practical operations as the optimum beamformer requires an a priori level of statistical knowledge of the incoming signal characteristics. An adaptive implementation recursively (or adaptively) estimates the interference, and as time proceeds [6], incorporates updated knowledge into a dynamic filter design. Attenuation of the desired signal and interference alike potentially occurs if this knowledge is inaccurate. In the event of unknown signal statistics and more practically realizable, an adaptive algorithm is preferred.

Frost proposes the statistical autocorrelation can be estimated as time progresses. As a result, Frost's Constrained LMS adaptive algorithm is modified to incorporate the derivative constraints. A beamformer is a spatial filter [7] that uses the output of a sensor array to preserve the amplitude of a signal in a desired look direction relative to background noise and directional interference, Fig. 1.

Suppose an antenna array receives N narrowband signals, $s_1(t)$, $s_2(t)\ldots.s_N(t)$, each arriving from a distinct look direction θ_1, $\theta_2\ldots\theta_N$ the array output vector is given by

$$x(t) = a(\theta)s(t) + n(t) \tag{1}$$

where $n(t) = [n_1(t)\, n_2(t) \quad \cdots \quad n_n(t)]^T$ is the noise output of the nth sensor. $n(t)$ and $s(t)$ are assumed to be stationary, zero mean and uncorrelated with each other. Also incoming signals are to hold the narrowband condition. The beamformer output is formed by applying a complex weights vector, w, to the N incoming signals received from the antenna array and summing the result, as shown in Fig 2. The beamformer output, $y(t)$ to a signal approaching from direction is expressed as

$$y(t) = w^H x(t) \tag{2}$$

where
$$w = [w_1\, w_2 \quad \cdots \quad w_N]^T$$
$$x(t) = [x_1(t)\, x_2(t) \quad \cdots \quad x_N(t)]^T$$

Fig. 1 Beamforming basics

where $x(t)$ is defined according to (1) as a function of the received signal and the steering vector $a(\theta)$. The power response, F of the beamformer is defined as

$$F = |y(t)|^2 \tag{3}$$

Practical use of a beamformer processor usually requires real time operation. To achieve real time operation using the aforementioned LCMV processor, an excessively elaborate and costly hardware requirement would be needed. Furthermore, the non-linear transformations of the LCMV solution are of little use for a continuous adaptive algorithm. However, to simplify the optimization problem and also eliminate hardware limitations, a self-designing or adaptive algorithm is preferred. Unlike the LCMV beamformer, the adaptive beamformer weights are determined via a continuous adaptive process, as shown in Fig. 3.

Upon attempt of utilizing the constraints for an adaptive algorithm, it becomes evident that there is an inaccuracy on the beamformer response due to a lack of restrictions on the complex part of the derivative constraints. To censure convergence of the adaptive algorithm to the actual roots, a small, artificial complex component is added to a previous root estimate before the algorithm is started.

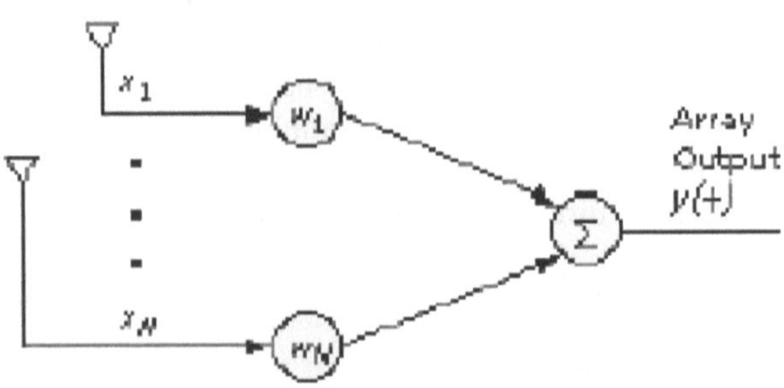

Fig. 2 Beamforming block diagram

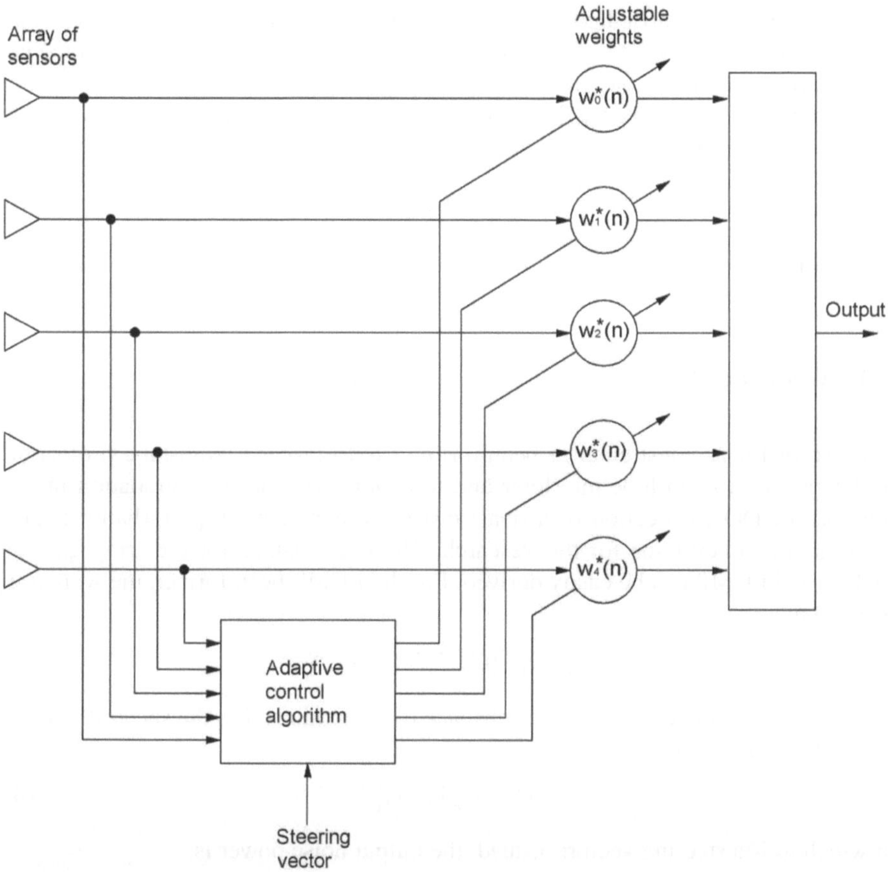

Fig. 3 Block diagram of an adaptive beamformer

2.3 LCMP Beamformer

An important class of beamformers are the so-called linearly constrained minimum power (LCMP) beamformers [8]. These beamformers determine the weights w so as to minimize the output power of the beamformer subject to a set of m linear constraints of the form

$$C^{\mathrm{T}}w = f \tag{4}$$

where C is the $N \times m$ constraint matrix, and f is the $m \times 1$ vector of constraint values. That is, the weights are the solution to

$$\min_{w} w^{\mathrm{T}} R_x w \text{ subject to } C^{\mathrm{T}}w = f \tag{5}$$

where $R_x = E\left\{x(t)x(t)^{\mathrm{T}}\right\}$ is the data covariance matrix. If $f = 1$, this is called the minimum power distortionless response (MPDR) beamformer. The LCMV (or LCMP) beamformer can be carried out by dividing the N-dimensional space into two subpaces, a constraint subpace and an orthogonal one. The first subspace is defined by the columns of C, an N x M_c matrix. The second subspace is defined by the columns of B, an N x $(N - M_c)$ matrix. The columns in B are linearly independent and each column is orthogonal to each column in C. Thus,

$$C^H B = 0 \tag{6}$$

where 0 is a $M_c x(N - M_c)$ matrix of zeros.

3 Development

The use of linear constraints in beamforming is to provide main-lobe protection in the case of a main-lobe interferer and to avoid performance degradation in the presence of DOA (direction of arrival) mismatch and/or array perturbations. The array gain is a key issue for this research. The relationships for the array gain of LCMV and LCMP can be easily derived. For the LCMP beamformer, the weight w is given by

$$w^H_{lcmp} = g^H \left[C^H S_x^{-1} C\right]^{-1} C^H S_x^{-1} \tag{7}$$

where g is the gain vector, C is the subspace of constraints, S is the spectral matrix. The output signal power is

$$P_s = \sigma_s^2 \left|w^H v_s\right|^2 \tag{8}$$

in which v_s is a steering vector; instead, the output noise power is

$$P_n = \sigma_n^2 \left|w^H \rho_n w\right|^2 \tag{9}$$

where σ^2 is the variance and ρ_n is the normalized spatial spectral matrix and includes both the white noise and any additional interference. Hence the gain is

$$A_0 = \frac{\left|w^H v_s\right|^2}{\left|w^H \rho_n w\right|} \tag{10}$$

If a distortionless constraint is included, then the numerator of the previous equation is unity and the array gain is

$$A_0 = \left|w^H \rho_n w\right|^{-1} \tag{11}$$

The output SNR_0 is

$$SNR_{output} = \left(\frac{\sigma_s^2}{\sigma_n^2}\right) A_{optimum} \tag{12}$$

The array gain for the LCMV beamformer, according to common literature is

$$A_{lcmv} = \frac{1}{g^H \left[C^H \rho_n^{-1} C \right]^{-1} g} \tag{13}$$

while to derive the array gain for LCMP it is suitable to use the optimum processor expression defined by Frost in [4] and put it in (13) in order to get the following expression:

$$A_{lcmp} = \left\{ g^H \left[C^H S_x^{-1} C \right]^{-1} C^H S_x^{-1} \rho_n S_x^{-1} C \left[C^H S_x^{-1} C \right]^{-1} g \right\}^{-1} \tag{14}$$

The array gain for the LCMP case is equal to the array gain for the LCMV case when the distortionless constraint is imposed and the signal is perfectly matched. Thus,

$$A_{lcmp} = \frac{1}{g^H \left[C^H \rho_n^{-1} C \right]^{-1} g} \tag{15}$$

One interpretation of the optimum processor is shown in Fig. 4. The beamformer first forms a set of M_c constraint beams and then combines them to form $Y(\omega)$. One can note the processor can be viewed as operating in a M_c-dimensional constraint subspace.

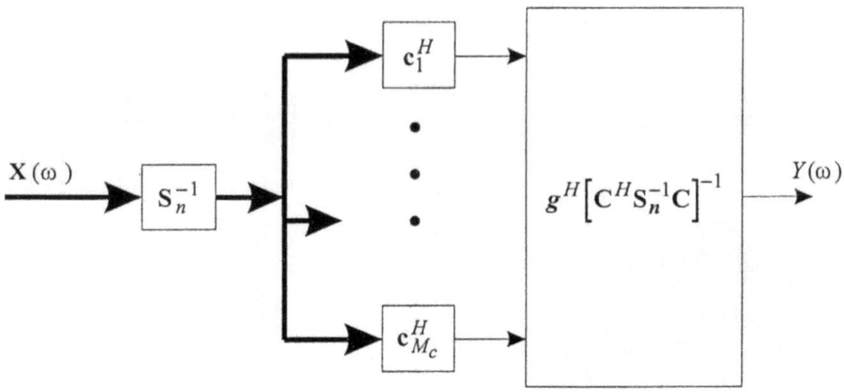

Fig. 4 Optimum constrained receiver

3.1 DOA Mismatching and Modelling

The application of beamforming must take into account the DOA (direction of arrival) of the system to be designed. The objective of the research is to design and utilize an array of sensors capable of retrieving an acoustic imaging of the item or

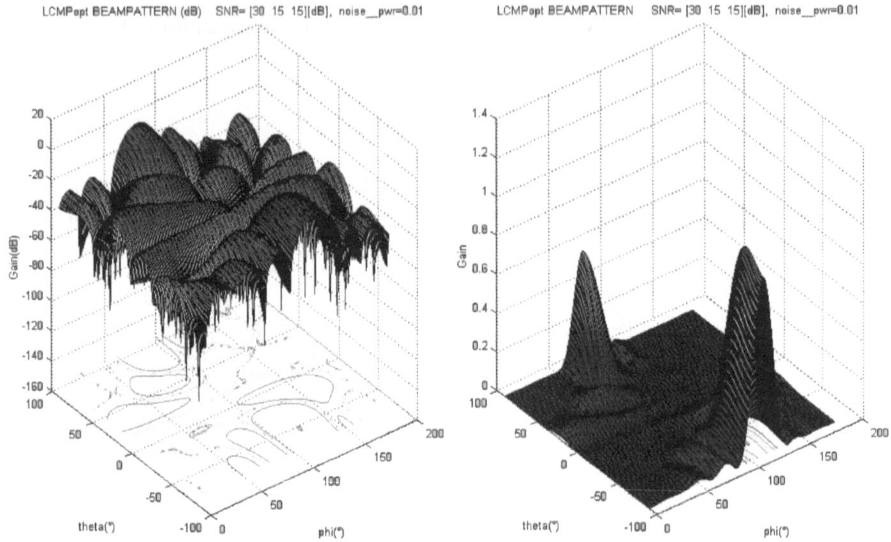

Fig. 5 LCMP beampattern

the obstacle towards which the car is approaching. The modelling has been enacted for different scenarios with variable SNR and INR. The main issue of this modelling is to verify in which scenarios errors is under the established threshold (5%) and results of Figs. 5, 6 and 7 are within that range if SNR is greater than 10 dB.

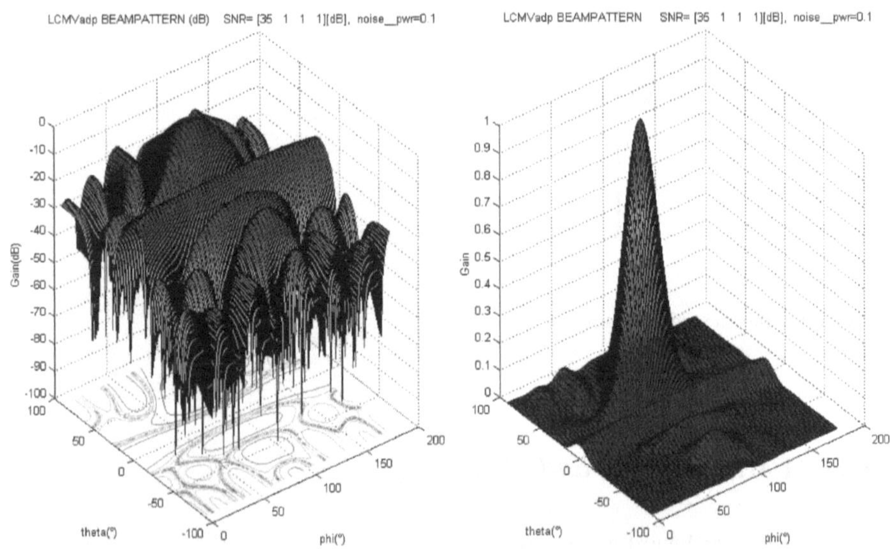

Fig. 6 LCMV beampattern

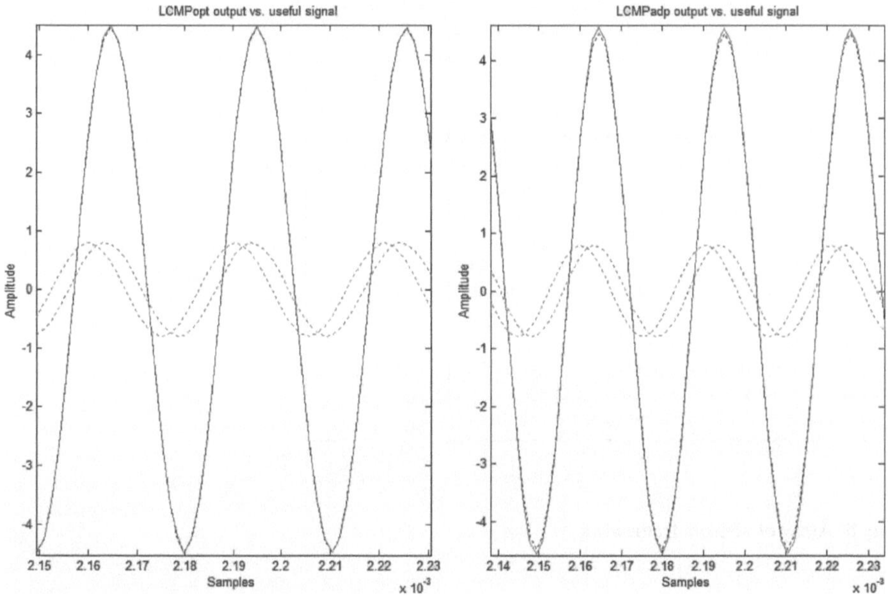

Fig. 7 LCMP output vs used signal

For the above values, the behaviour of adaptive algorithm is near to that of optimum algorithm. Hence the following specifications are used; attention must be paid in analyzing in detail the beampattern.

This yields to a consideration according to which its value that is overlapped towards constrained directions is:

$$(\theta 1, \varphi 1) = (-66, 110) \rightarrow 0\,\text{dB}$$
$$(\theta 2, \varphi 2) = (-22, 44) \rightarrow -70\,\text{dB}$$
$$(\theta 3, \varphi 3) = (44, 88) \rightarrow -50\,\text{dB}$$

While in the other DOA of signals, it is:

$$D(\theta 1, \varphi 1) = (-66, 110)[30\,\text{db}] \rightarrow 0\,\text{dB}$$
$$D(\theta 2, \varphi 2) = (-44, 22)[15\,\text{db}] \rightarrow -25\,\text{dB}$$
$$D(\theta 3, \varphi 3) = (-44, 132)[15\,\text{db}] \rightarrow -19\,\text{dB}$$

In this case the system attenuation in correspondence to interferences DOA is sufficient to eliminate them

Once the framework (Fig. 8) is ready, they can be mounted within the chassis of Fig. 9 according to the dimensions indicated in Fig. 9. A PC-4018S sensor is adopted as a sensing device.

The characteristics of sensor (Fig. 10) are suitable for the purposes of this paper (Fig. 11).

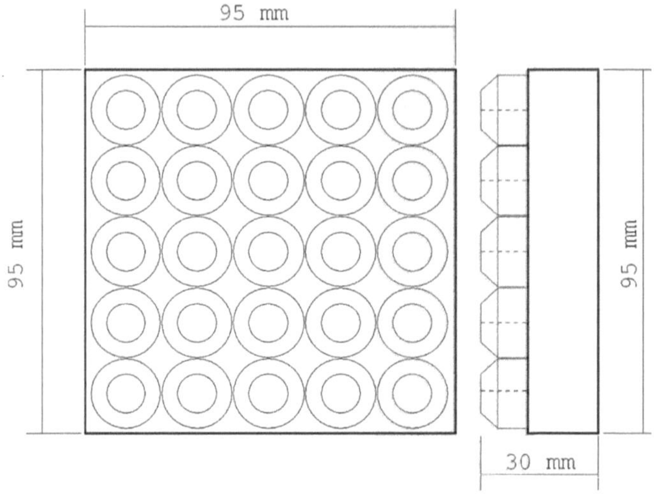

Fig. 8 Array of sensors framework

Fig. 9 Array of mounted sensors

3.2 Frond End Design

The overall architecture is shown in Fig. 12 and it is organized with the following main subunits: array of nxm (n = 5, m = 5) microphones, n buffers, an array for signal conditioning, a driving circuit, a beamformer, a dsPIC control. The front end electronic has been tested using Multisim design software.

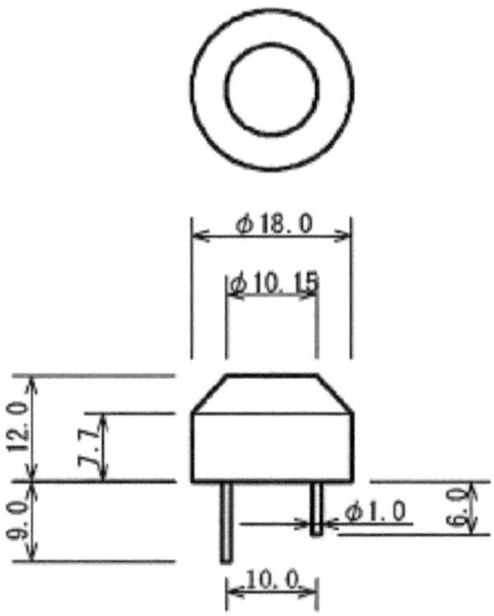

Fig. 10 PC40-18 S sensor details

Item / Type	Model	Center Frequency (kHz)	Sound Pressure (dB)	Sensitivity (dB)	−6dB Directivity Typical (deg)	Equivalent Circuit Cb (pF)	R (Ω)	L (mH)	Ca (pF)	Appearance
Conical	PC40-18S	40	>100	−80	80	2200	400	130	130	E

Fig. 11 PC-18S characteristics

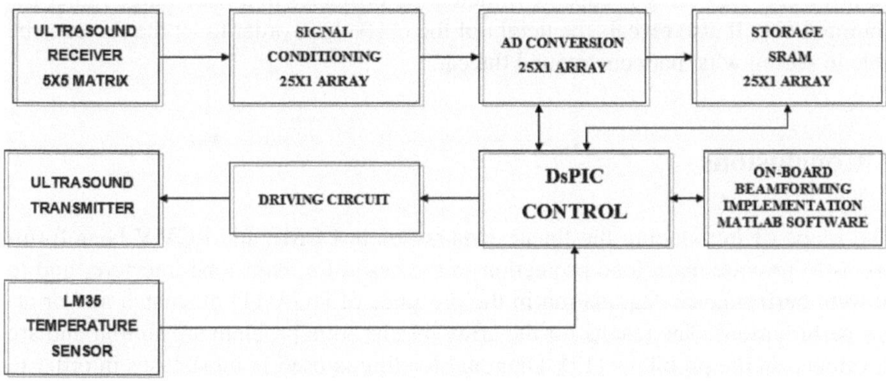

Fig. 12 General architecture

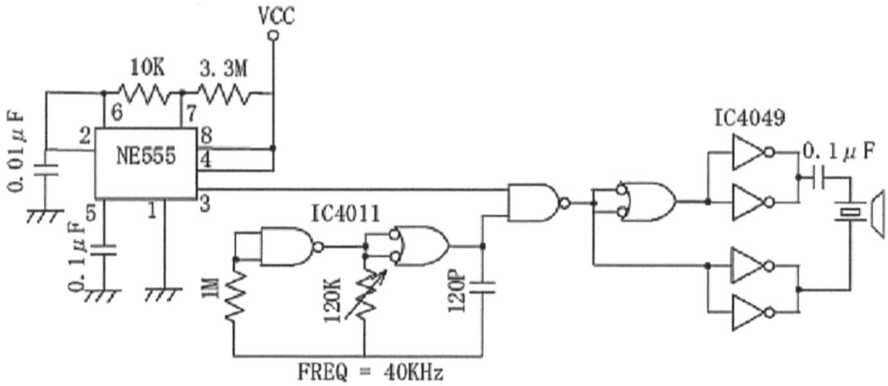

Fig. 13 PC40-18S piezoelectric transceiver

The choice of PC40-18S (Figs. 13, 14, 15) is compatible with the objective of this present work. It presents some interesting aspects amid different manifold characteristics. Figure 16, instead, shows the results of circuit electrically excited.

After the transducer, a receiving circuit is needed (Figs. 17 and 18) by using μPC4570, that is an ultra low-noise, wideband high slew-rate, dual operational amplifier. Input equivalent noise is three times better than the conventional 4558 type op-amps. The gain bandwidth products and the slew-rate are seven times better than 4558. In spite of fast AC performance, the μPC4570 is extremely stable under voltage-follower circuit conditions. Supply current is also improved compared with conventional wideband op-amps. The μPC4570 is an excellent choice for preamplifiers and active filters in audio, instrumentation, and communication circuits [9, 10].

A square waveform (40 kHz and 5V of amplitude) is needed as input of receiving circuit and the output is the same but with an increasing bias and amplitude as highlighted in Figs. 19 and 20.

As one can see in Fig. 21, the built array of 5×5 sensors is mounted on the rear part of a car. This array is capable of detecting obstacles in a wide range rather the common one. It also reveals the height of the obstacle in order to let the driver to be able in seeing what happens around the car.

4 Conclusions

The scope of introducing the linear constraints, in LCMP and LCMV beamformers, is to provide main-lobe protection in the case of a main-lobe interferer and to prevent performance degradation in the presence of DOA[11] mismatch and/or array perturbations. The results for the array perturbation problem are similar and are developed in the problems [12]. Diagonal loading is used in most cases in order to obtain satisfactory performance. It is noticed that the results for the linear array with discrete interference are characteristic of the general case [13].

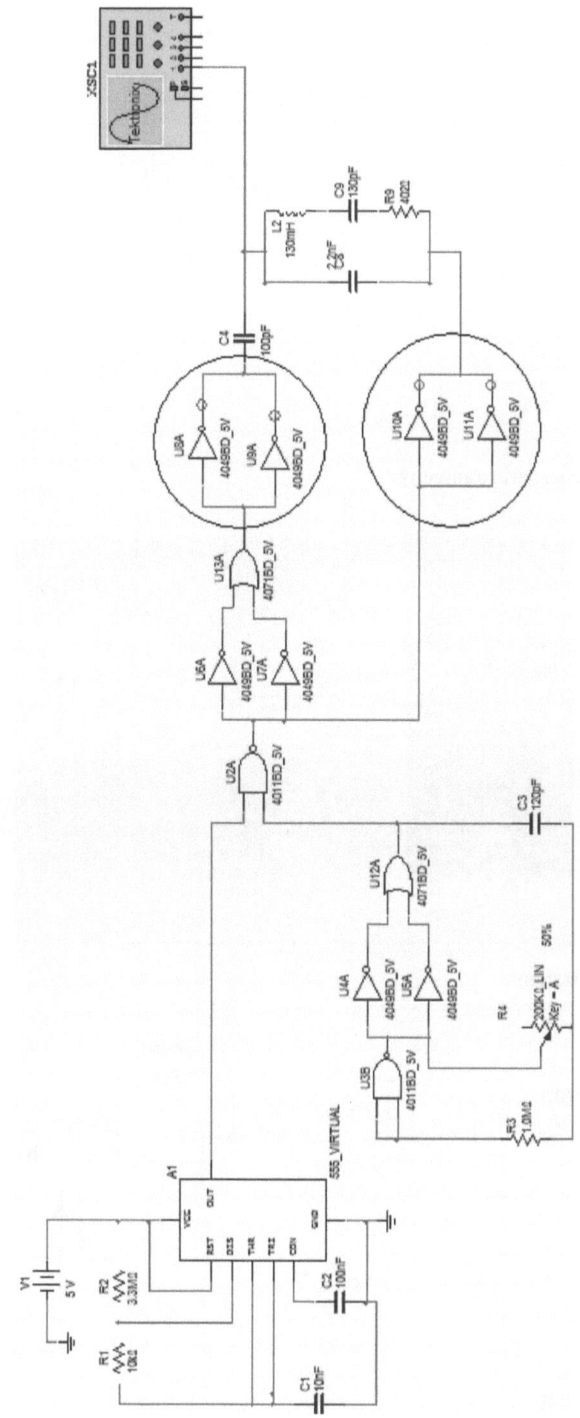

Fig. 14 Piezoelectric transceiver based on Multsim

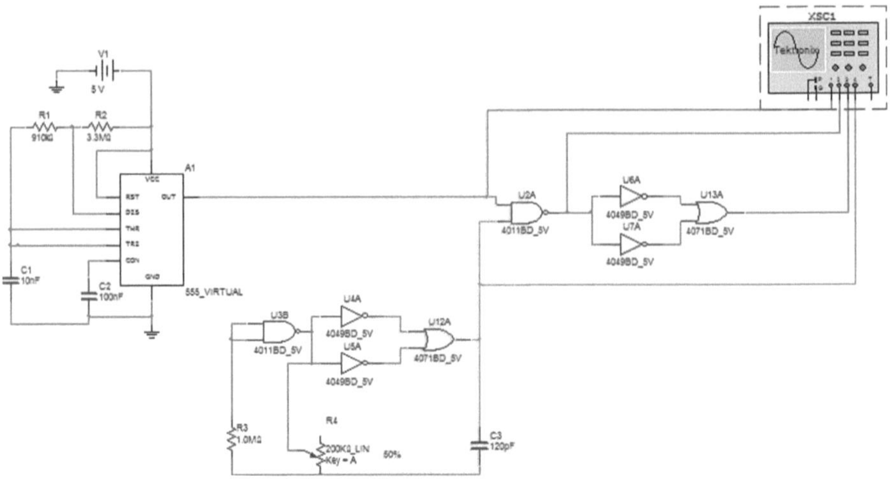

Fig. 15 Modified piezoelectric transceiver

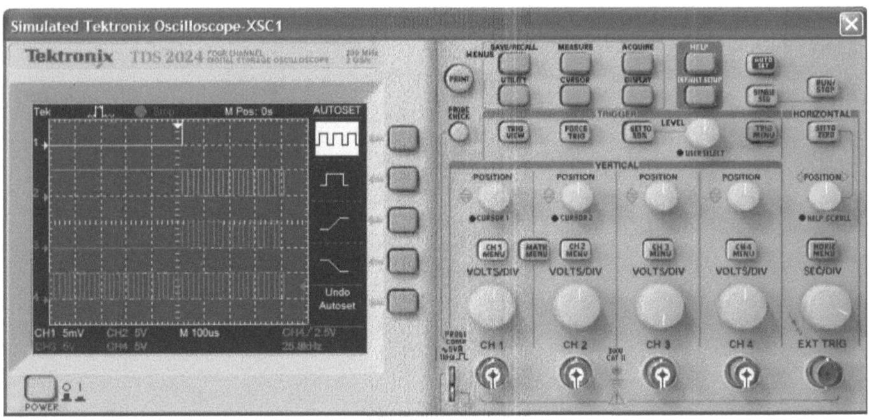

Fig. 16 Signal visualization

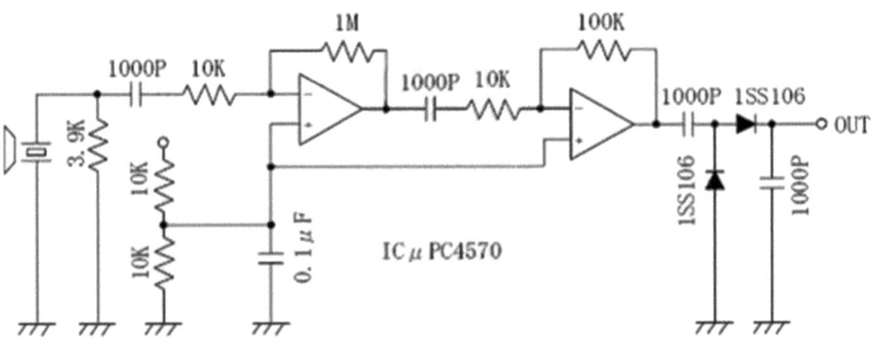

Fig. 17 Receiving circuit

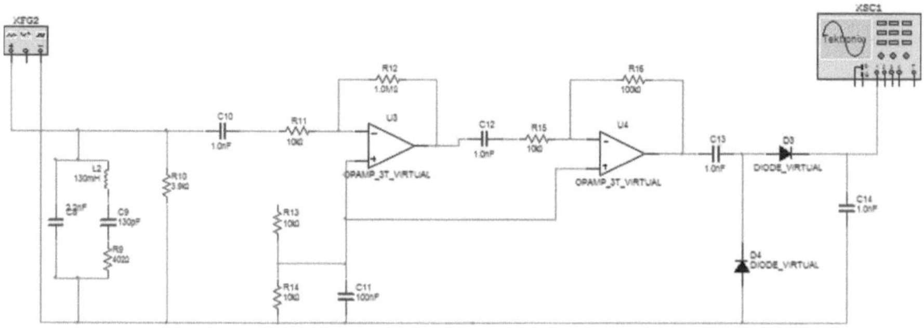

Fig. 18 Receiving circuit on multisim

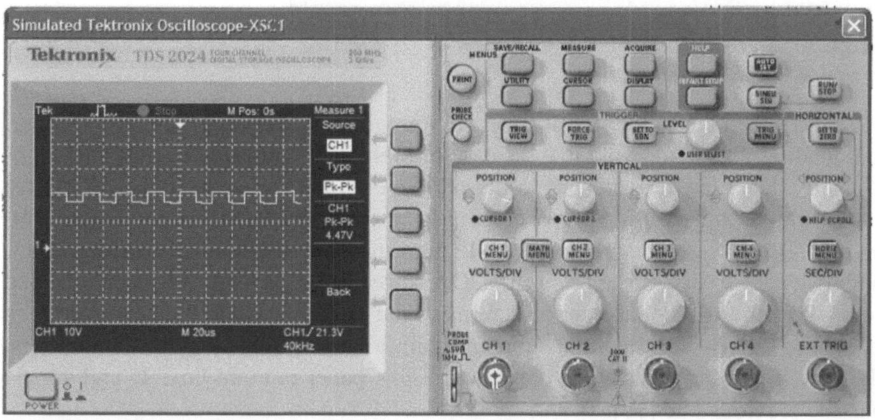

Fig. 19 Snapshot signal (a)

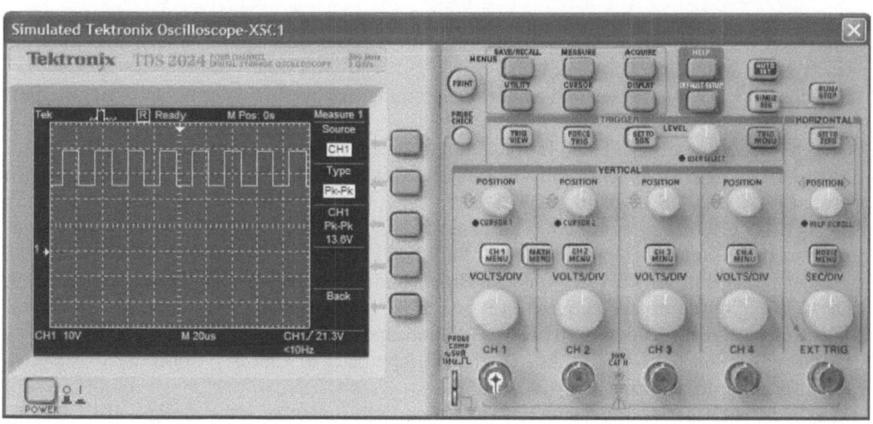

Fig. 20 Snapshot signal (b)

Fig. 21 Array general view

The key element in the beamformer performance is an appropriate level of diagonal loading. It is also found that the case of high SNR compared to INR led to significantly degraded performance in the presence of DOA mismatch [14]. A logical step would be to perform a preprocessing to estimate the DOA prior to beamforming [15]. By virtue of the previous LCMP algorithm, an interference direction of 1.1 radians has been established and the DOA (direction of arrival) has been changed from $-90°$ up to $+90°$ in order to determine the range in which the beamformer is capable to satisfy the constraints: unitary main-lobe amplitude and lower sidelobes. The new frond end electronics presented in this paper is more flexible and simple than the previous one [16].

References

1. Van Veen B.D, Buckley K.M (1988) Beamforming: A versatile approach to spatial filtering. IEEE ASSP Magazine, 5(2): 4–24
2. Darlington S (1958) Linear least–squares smoothing and prediction, with applications. Bell Syst. Tech. J. 37: 1221–1294
3. Brown R.G, Nilsson J.W (1962) Introduction to Linear Systems Analysis. Wiley, New York
4. Frost III O.L (1972) An algorithm for Linearly Constrained Adaptive Array Processing. In: Proc. 1972 IEEE, USA, vol 60, pp. 926–935
5. Tseng C.H (1992) Minimum Variance Beamforming with Phase-Independent Derivative Constraints. IEEE Trans. Antennas Propagat., 40(3): 285–294
6. Tseng C.H, Griffiths L.J (1993) Adaptive Broadband Beamforming Using Phase-Independent Derivative Constraint Techniques. In:Proc. 1993 International Conference on Acoustics, Speech and Signal Processing (ICASSAP), Minneapolis, MN, pp. 536–540
7. Griffiths L.J, Jim C.W (1982) An Alternative Approach to Linearly Constrained Adaptive Beamforming. IEEE Trans. On AP, 30: 27–34
8. Van Veen B.D (1988) Eigenstructure based partially adaptive array design. IEEE Trans. Antennas Propag., 36: 4–24

9. NEC (1998) Data sheet, Bipolar analog integrated circuit, Japan
10. Nippon Ceramic Ltd (1983) Ultrasonic Transducer, Japan
11. Bell K.L, Ephraim Y, Van Trees H.L (1996) Robust Adaptive Beamforming under Uncertainty in Source Direction-of-Arrival. In: Proc. IEEE Signal Processing Workshop on Statistical Signal and Array Processing, USA
12. Bell K.L, Van Trees H.L (1999) Adaptive and Non-Adaptive Beampattern Control Using Quadratic Beampattern Constraints. In: Proc. Conference Record of the Thirty-Third Asilomar on Signals, Systems, and Computers, USA
13. Bell K.L, Van Trees H.L (1998) Adaptive Beamforming for Spatially Spread Sources. In: Proc. Ninth IEEE SP Workshop on Statistical Signal and Array Processing, USA
14. Van Veen B, Roberts R (1987) A Framework for Beamforming Structures. IEEE Transactions on Acoustics, Speech, and Signal Processing, 35(4): 584–586
15. Van Veen B, Roberts R (1987) Partially Adaptive beamformer design via output power minimization, IEEE Transactions on Acoustics, Speech, and Signal Processing, 35: 1524–1532
16. Lay-Ekuakille A, Vendramin G, Trotta A (2007) Acoustic Sensing for Safety Automotive Applications. In: Proc The 2nd International Conference on Sensing Technology, New Zealand

Development of a Smart Home for Elder-People Based on Wireless Sensors

Anuroop Gaddam, Subhas Chandra Mukhopadhyay and Gourab Sen Gupta

Abstract A smart home specifically for the elder people has been planned to be developed using wireless sensors. The design intricacies and implementation details of a novel Wireless Sensors systems has been targeted to achieve that. The system is designed to support people who wish to live alone but, because of old age, ill health or disability, there is some risk in this, which worries their family or friends. The system works on the principle of using sensor units (SU) to monitor the appliance throughout a house and detect when certain desired electrical equipments are turned on. Rules are defined for appliances to turn on in certain time intervals. The rules are flexible and can be user-defined based on the daily activities of a person. Several levels of alarm conditions have been created based on combination of rules that are violated. Any number of sensor units may be installed in a house, one each to monitor an electrical appliance. A central controller unit (CCU) queries the sensor units and logs the data into a PC at a pre-defined rate. Communication between the SUs and the controller is using radio-frequency wireless media. The rules inference engine runs on the PC and whenever the situation requires, sends a text message to the care-givers or relatives. Since no vision sensors (camera or infra-red) are used, the system is non-invasive, respects privacy and has found wide acceptance. The system is completely customizable, allowing the user to select which appliances to monitor and define exactly what is classified as unusual behavior. Some other sensors such as bed sensors, gas sensor etc., are considered to make the system a complete system suitable to define a digital home.

Anuroop Gaddam
School of Engineering and Advanced Technology, Massey University, Palmerston North, New Zealand, e-mail: anuroop24@gmail.com

Subhas Chandra Mukhopadhyay
School of Engineering and Advanced Technology, Massey University, Palmerston North, New Zealand

Gourab Sen Gupta
School of Engineering and Advanced Technology, Massey University, Palmerston North, New Zealand

S.C. Mukhopadhyay, G.S. Gupta (eds.), *Smart Sensors and Sensing Technology*,

Keywords Smart home · wireless sensor network · radio frequency communication · non-intrusive monitoring system · current sensing · force sensor · bed sensor · hardware interfacing · elderly care

1 Introduction

There is growing trend of population of old people throughout the world, so new challenges are arising to provide a safe, secure and sound living environment. According to the current estimates, in 2005, 15.3% of the population in the developed regions of the world was aged 65 years and above [1]. In the developing regions it was 5.5%. What is alarming is the rate at which these numbers are predicted to rise. As shown in Fig. 1, the proportion of elderly will grow to 25.9 and 14.6% respectively for the developed and developing regions by 2050.

Quite often we are appalled by news headlines such as "Elderly man lay dead for days in his home" and "Woman found starved in flat". We are shocked that no one was looking after these people and imagine that they must have had an uncaring family. At the same time, however, as a society we value our rights to live independently and keep control of our own lives. Many elderly people dread the idea of being forced to live with their adult children, or in a rest home or other sheltered living arrangement. Yet at the same time they know there is a high risk

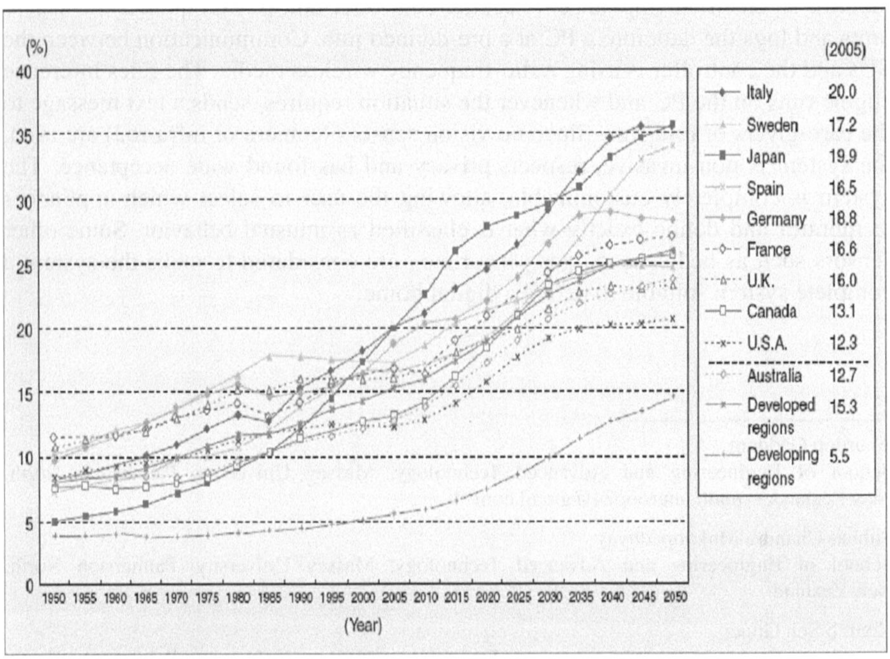

Fig. 1 The population of aged people in the world [1]

of injury or even death because of a fall or stroke. There are many people in our community who, because of old age, infirmity, memory loss or impaired judgment which can no longer be relied upon, are having pressure put on them by loving relatives to leave their home and give up their precious independence. Surely, with the technology of today, there is a better way for these people to resolve this problem. Today, the progress in science and technology offers miniaturization, speed, intelligence, sophistication and new materials at lower cost. In this new landscape, microtechnologies, information technologies and telecommunications are the key factors. With the population aging in most developing countries, there will be more and more elderly people living alone in future. Hence telemedicine and remote monitoring of home to make it "Digital Home" are gaining added importance and urgency. Our research was undertaken against this back drop.

Several research reports have documented the implementation of systems to monitor homes and patients [2, 3]. In [2], the in-house movements of elderly people are monitored by placing infrared sensors in each room of their homes. While such a method may not be as blatantly intrusive as using cameras, still it encroaches into the privacy of the person. In [3] the new possibilities for home care and monitoring are discussed using wireless micro-sensors which are minimally-invasive. Patient monitoring using personal area networks of wireless intelligent sensors is reported in [4]. The development of care support system to monitor the overall health of welfare facility residents who need constant care has been reported in [5]. The reported system is designed with wireless sensors, wireless repeaters and a host computer. The sensors, however, must be in physical contact with the monitored subject to record respiration and physical posture. These data are transmitted to the wireless repeater by the transceiver. The wireless repeaters, which are installed throughout the welfare facility, send data, including the repeater's ID, to the host computer. The ID is used to detect the resident's location in the welfare facility. When the resident is in an emergency situation, such as falling or in an inactive state for more that the allotted time, the host computer automatically alerts the situation to the care staff by an alarm sound and also by mobile phone.

Commercially available systems include panic buttons and camera based surveillance. The panic button approach requires a person being monitored to wear a "panic button". The signal is relayed to a base station which then communicates to the service provider to send help. The problem with this system is that it requires the monitored persons to be prepared to wear the sensor around their neck and remember what it is there for. It also requires them, when an emergency occurs, to push the button. This is not always possible because either the person fails to recognize that he is at risk, or because he is unconscious or physically incapacitated. However, the nature of the panic button does ensure an excellent response time. Most people loath camera based surveillance systems as they do not offer any privacy. The system detailed in this paper uses sensors to monitor electrical appliances in a house and detect when they are turned on. Every person has a reasonably fixed routine of daily activities to follow, albeit with occasional variations. For example, a kettle or toaster is turned on in the morning for making breakfast. Likewise, the TV is turned on in the evening. Based on the pattern of activities, rules are defined for appliances to turn on in certain time intervals. The rules are flexible and can be user-defined based on the activities of a person. Any number of sensor units may be installed in a house. In Sect. 2 we provide an overview of the system architecture. The various elements of the hardware sections of the Sensor Unit (SU) and the Central Controller unit (CCU) are presented in Sect. 3. The software, including the rules inference engine, is explained in Sect. 4. The prototype and the system test results are detailed in Sect. 5. The Sect. 6 describes the current and future work carried out to make it a more complete system. The paper concludes with a brief discussion and future work in Sect. 7.

2 System Architecture

Figure 2 shows the functional block diagram of the wireless sensors system based digital home.

Each electrical appliance that needs to be monitored is connected to a sensor unit (SU). The SU detects when the appliance is turned on, basically my measuring the current drawn by the appliance. The SU has been designed to detect a wide range of current, from small appliances such as a reading lamp to a room heater that draws a fairly large current. The SUs and the central controller unit (CCU) are equipped with RF (radio frequency) transceivers for two-way communication. The CCU polls the SUs at regular intervals and gathers the status of the appliances. The RF communication between the SUs and CCU is at 418MHz frequency. The maximum data transfer rate is 40 kbps. The CCU is connected to a personal computer (PC) using RS232 serial communication protocol. message service (SMS) facility of the cellular telephone network, to the mobile phone of a family member or a care giver who can then take appropriate action to contact the person and provide help if necessary.

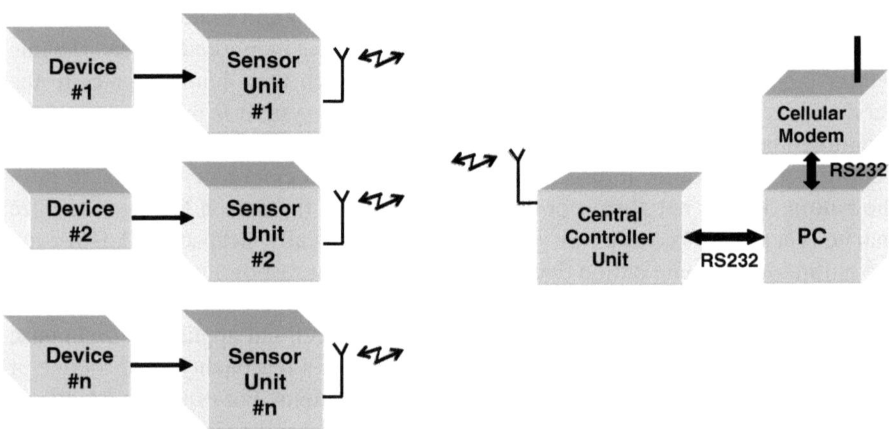

Fig. 2 Functional block diagram of the system hardware

3 Hardware Design

The hardware components of the system are the Sensor Units, the Central Controller unit, a PC and a cellular modem. In this section we describe in details the design of the custom built SUs and CCU. The PC and the cellular modems are standard off the shelf devices.

3.1 Design of the Sensor Unit

The various components of the SU are shown in Fig. 2. The five major hardware blocks in an SU are the power supply, RF transceiver module, micro-controller, current transformer circuitry and the LED display.

The power supply section steps down the 230V AC mains voltage using a transformer and converts it to a DC voltage using a bridge rectifier. 3-pin voltage regulator chips have been used to obtain 5 and 3.3 V DC voltages to power the RF transceiver, micro-controller and current sensing circuitry as shown in Fig. 3.

The Radio Frequency module used is a Radiometrix BiM-418 transceiver capable of half-duplex operation at 418 MHz frequency. The *TxSelect* and *RxSelect*

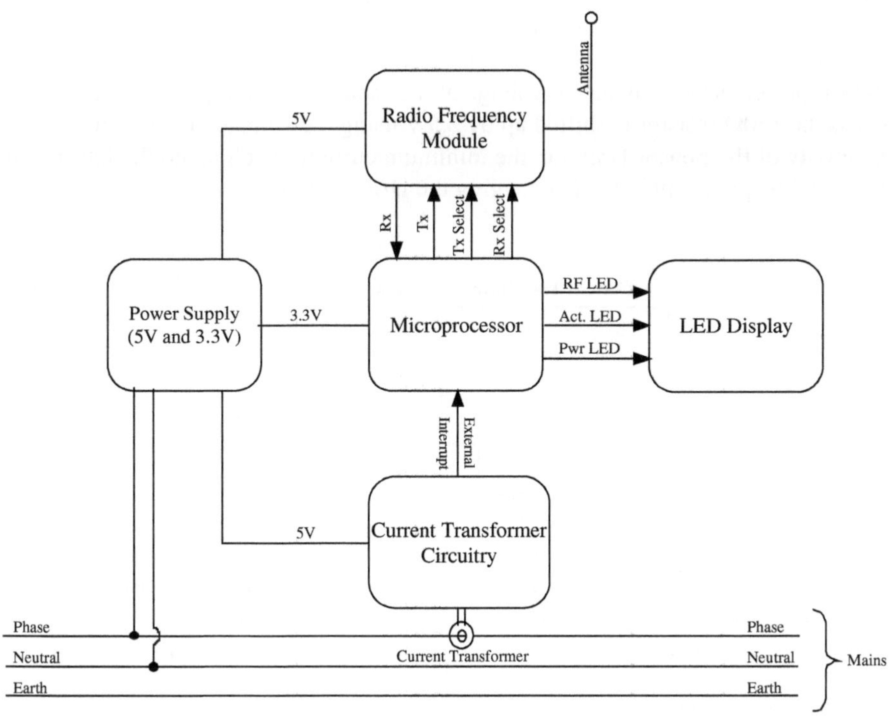

Fig. 3 Core components of the sensor unit (SU)

signals from the micro-controller configure the transmit or receive mode of the RF transceiver. The *Rx* and *Tx* are the receive and transmit data lines respectively. In our design, the data transfer rate is set to 38400 bauds.

The microcontroller used is the SiLab C8051F020 mixed signal microcontroller. The microcontroller development board was used to build the prototype as it provided ready connection terminals for analog and digital I/O.

There are three Light Emitting Diodes (LED) – one to indicate that power to the SU is on, another to show that the RF communication with the SU is currently in progress and the third to indicate that the appliance connected to the SU is active, i.e. turned on.

The current transformer and the associated circuitry are shown in Fig. 4.

A current transformer (CT) is used to detect current in the phase line of the AC mains connected to an appliance. When an appliance is turned on, the current drawn by it passes through the primary coil of the CT. This produces a proportional current in the secondary coil of the CT, which, flowing through a 120Ω resistor, generates an AC voltage of 50Hz frequency across it. This voltage is compared with the programmable voltage of the Digital-to-Analog converter (DAC) output of the microcontroller. The LM329 op-amp based comparator generates a transition at its output each time the CT output crosses the DAC threshold. In summary, when an appliance is turned on, the circuit generates a series of pulses, at a frequency of 50 Hz, at its out. These pulses, used as external interrupts to the microcontroller, are counted by the program running on the microcontroller.

The peak voltage across the 120Ω resistor is 2 V for a load current of 10 A, which is the maximum for our design. The comparator is run off a single +5 V supply. To prevent presentation of negative voltage at the input of the comparator, the voltage across the 120Ω resistor is shifted up by 2.5 V using the voltage divider circuit. The sensitivity of the Sensor Unit, i.e. the minimum current which it should detect, can be altered by programmatically changing the DAC output.

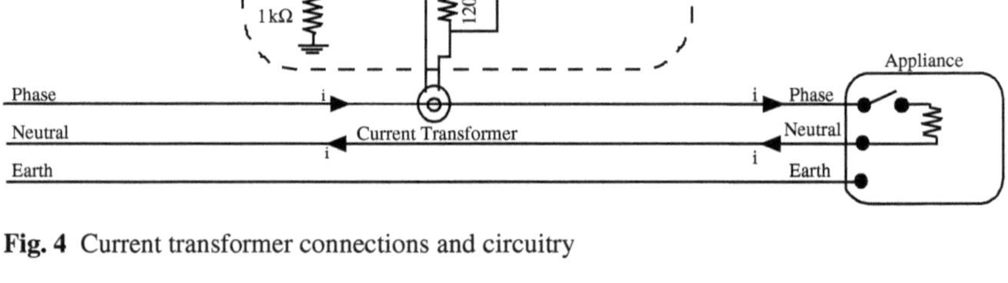

Fig. 4 Current transformer connections and circuitry

3.2 Detecting the Status of an Appliance

The task of the load detection software is to continuously evaluate whether the load is active or not. It does this by counting the number of pulses received from the CT circuitry on the external interrupt pin of the microcontroller over a predetermined period of time.

A timer has been configured to generate an interrupt every 500 ms. If 10 external interrupts occur within this period, the appliance is deemed to be active. Although ideally we would expect 25 pulses over half a second, this was sometimes not the case. The most obvious example of this is a regulated heater which controls the temperature by switching the load on and off. If this happens within the 500 ms monitoring window, the number of pulses counted is less than 25.

The flow chart of the algorithm for load detection is shown in Fig. 5.

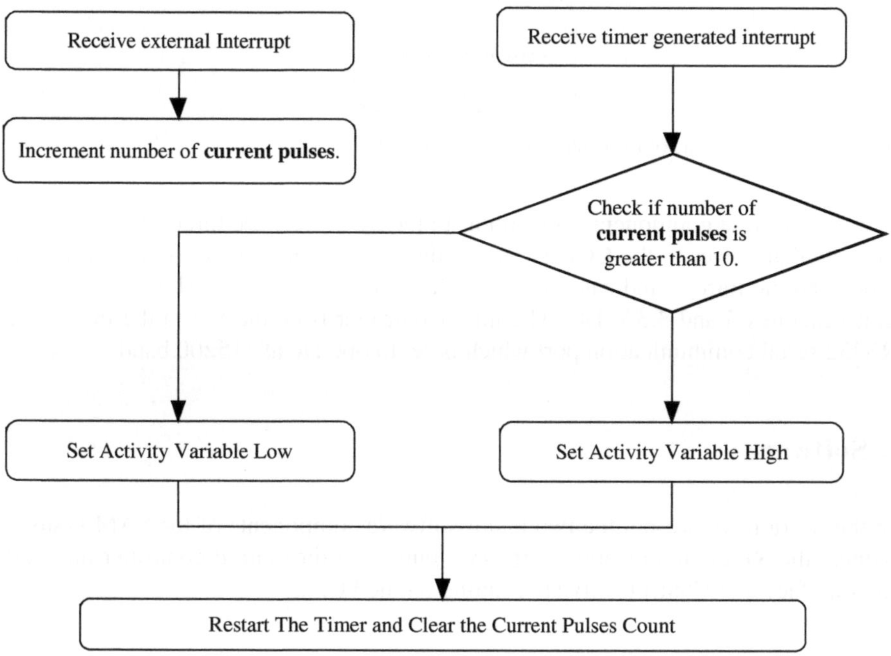

Fig. 5 Algorithm for load detection

3.3 Design of the Central Controller Unit

The various components of the CCU are shown in Fig. 6. The four major hardware blocks in the CCU are the power supply, RF transceiver module, micro-controller and the LED display.

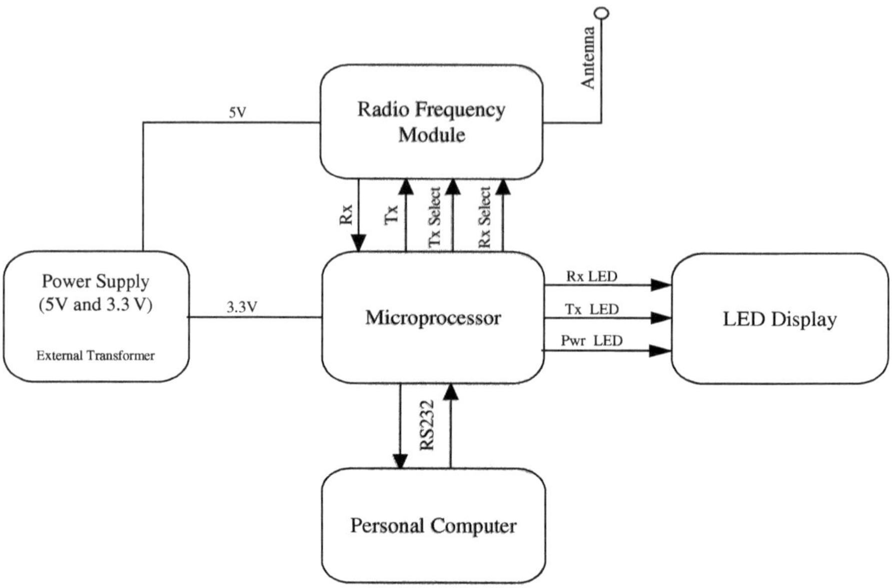

Fig. 6 Core components of the central controller unit (CCU)

Most of these components are similar to the SU. There are three LEDs – one to indicate that power to the CCU is on, another to show that the CCU is in transmit mode and the third to indicate that the CCU is in receive mode. The power supply unit generates 5 and 3.3 V DC. The microcontroller is connected to the PC on the RS232 serial communication port which is set to operate at 115200 bauds.

4 Software

In this section we present the two major software components of the SAM system, namely the RF communication software running on the central controller unit and the Interface and Control software running on the PC.

4.1 Radio Frequency Communication Protocol

As all the sensor units communicate with the central controller unit on the same frequency, it is important that a protocol be established to avoid data collisions. Data collisions occur when multiple devices try to communicate simultaneously on the same channel. The basic principle behind the protocol used in the SAM system is that sensors will respond only when told to do so. This means the controller is the only device that can transmit on the radio channel, unprompted. The controller is in fact the master which decides which SU should transmit

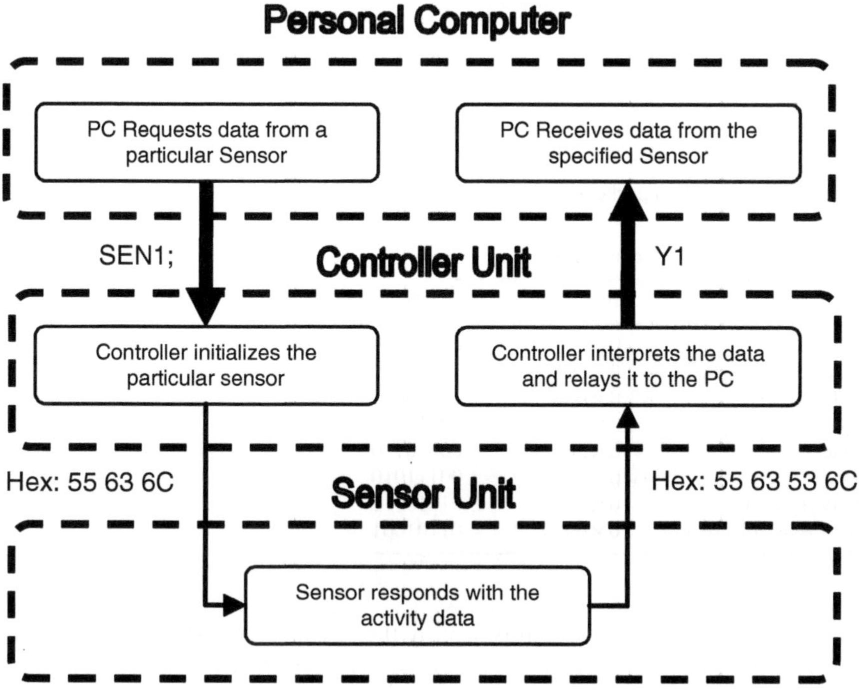

Fig. 7 RF communication protocol

Figure 7 explains the sequence of actions that take place when the SU and CCU communicate with each other.

The personal computer determines which sensor it requires data from, in this case it requires data from sensor one. It therefore sends the command "SEN1;" serially over the RS232 interface to the controller. The controller receives this command and converts it to an appropriate hexadecimal / binary packet for radio transmission. The packet is received by all the SUs but only the specified sensor unit will respond. This ensures that only one SU 'talks' at any time. The packet sent back by the SU contains the status byte indicating whether the appliance is active or inactive. The controller receives this response and transmits it to the personal computer, via the RS232 link. The computer then confirms that the correct sensor has responded and continues with the monitoring process, querying the next SU.

4.2 Radio Frequency Communication Packets

There are two packet formats that have been used, one that is transmitted by the controller and the other is the response from the sensor.

The format of the packet which is sent by the controller is shown in Fig. 8. It consists of three bytes - the start byte and the stop byte with the sensor unit ID (identity) byte sandwiched in between.

Start byte	ID byte	End byte

Fig. 8 RF communication packet sent by the controller

Start byte	ID byte	Activity byte	End byte

Fig. 9 RF communication packet sent by the sensor unit

Table 1 Bytes used for RF communication

Description	Hex value	Binary value
Start Byte	0×55	01010101
End Byte	0×6C	01101100
ID Byte: 1	0×63	01100011
ID Byte: 2	0×33	00110011
ID Byte: 3	0×36	00110110
ID Byte: 4	0×3A	00111010
Activity Byte: Active	0×53	01010011
Activity Byte: Inactive	0×69	01101001

The format of the packet which is sent by the sensor unit is shown in Fig. 9. It consists of four bytes - the start byte, sensor unit ID (identity) byte, activity byte and the stop byte.

All the bytes used in the packet are bit-balanced i.e. they have the same number of binary 0s and 1s. The bytes used in the RF packets are shown in Table 1.

4.3 Interface and Control Software

The interface and control software has been written in Visual Basic and runs on the PC. It allows the user to configure the sensor units; up to 6 six sensors may be selected and associated with different appliances. The Graphical User Interface (GUI) to accomplish this is shown in Fig. 10. Once the sensors have been set up, the monitoring can start. Figure 11 shows the monitoring screen.

Once the sensors have been configured the rule creation can begin. There are two methods of creating a rule. The user can use an elaborate rule creation wizard and be guided through the rule creation process with examples at each step. Or, the user can bring up an immediate window which allows for a much faster rule creation process. The rule creation wizard would be the recommended method initially, until the user has come to terms with the software. The GUI of the rule creation wizard is shown in Fig. 12.

The user can select a rule to be either a rule type one or rule type two. A rule type one checks whether an appliance is activated within a given time period. For example, has the television been used between 6 and 8 pm? A rule type two checks whether an appliance is still active after a predetermined time. For example, has the

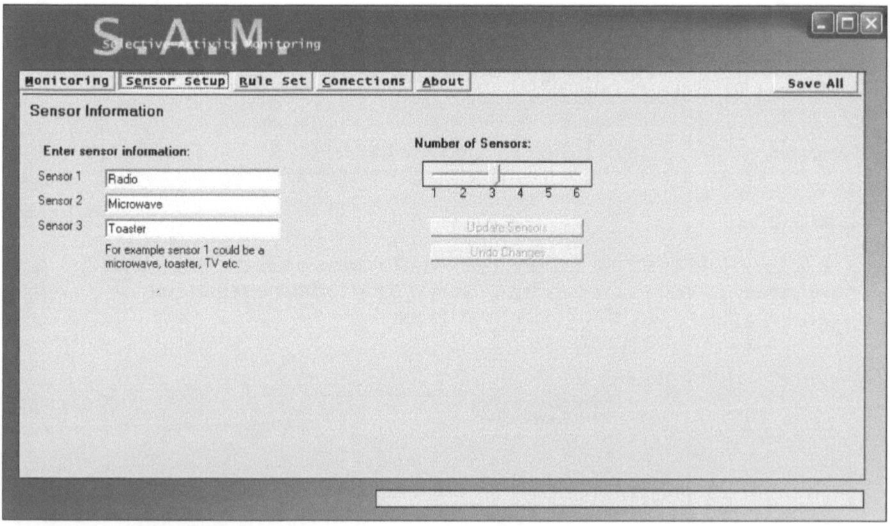

Fig. 10 GUI for senor setup

television been left active after 11 pm? The rules are based around a 24 hour cycle. The user can turn a rule off on particular days of the week. Figure 13 shows the immediate rule creation window for rule type 1.

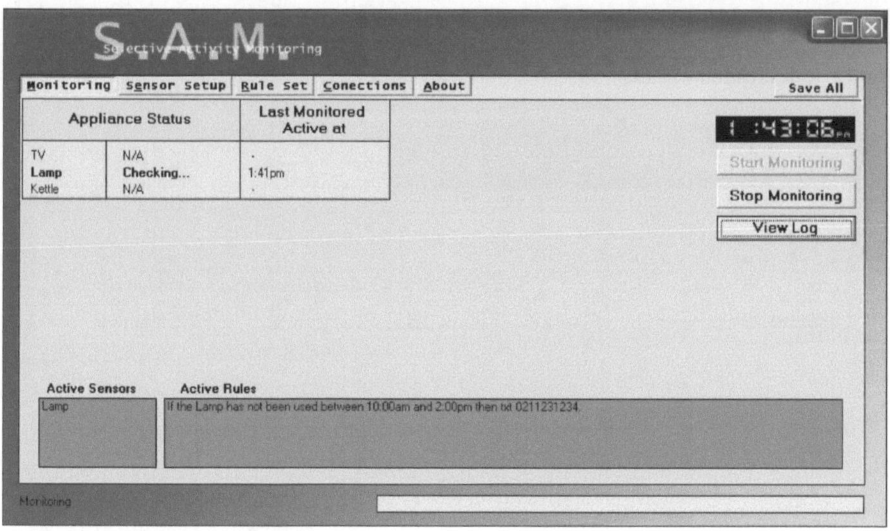

Fig. 11 Monitoring the appliances

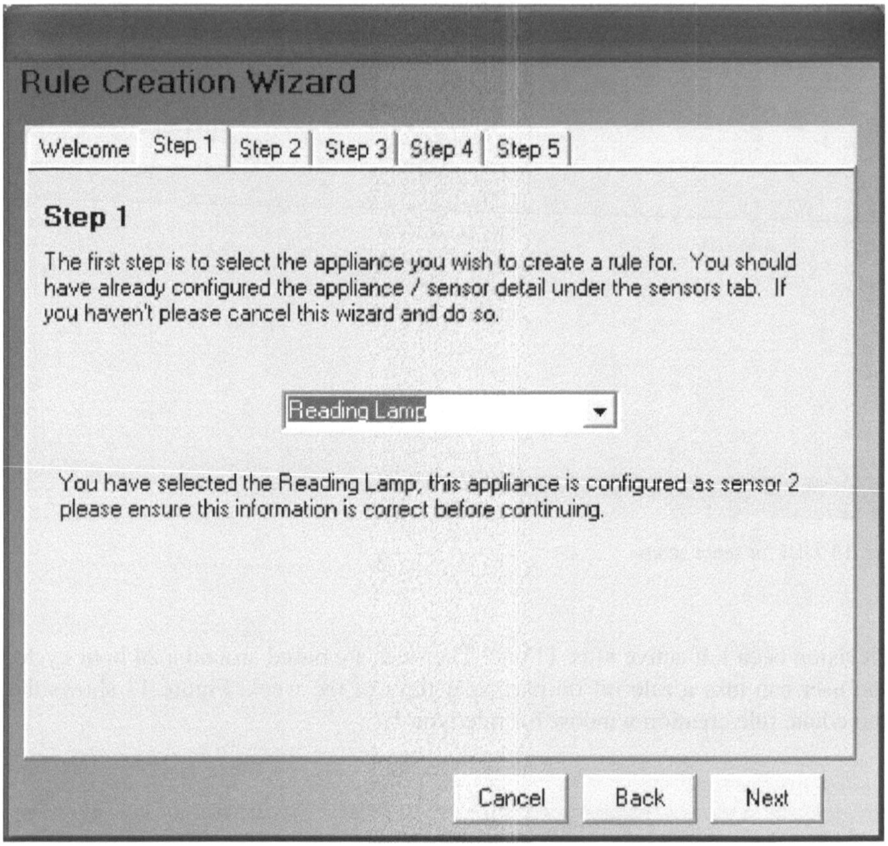

Fig. 12 Rule creation wizard

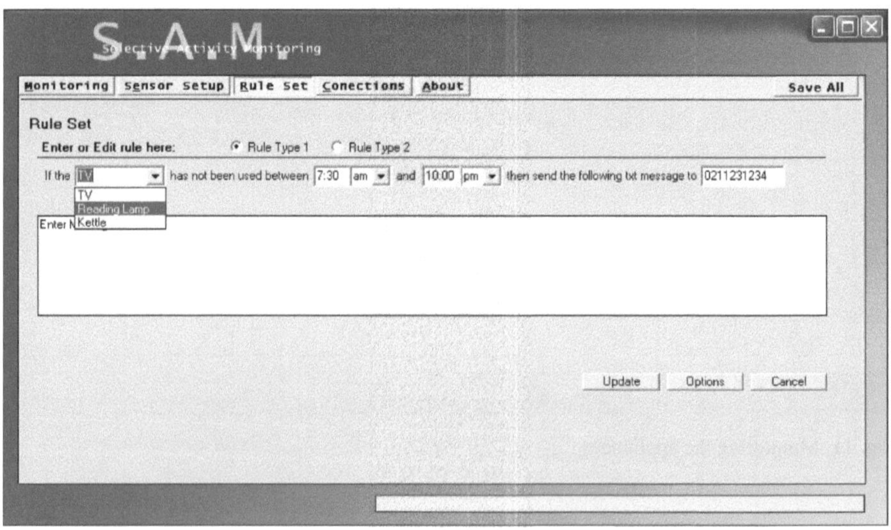

Fig. 13 Immediate rule creation window for rule type 1

5 System Prototype and Experimental Results

The system was fabricated and tested with three sensor units and one central controller unit in an average size home. The complete system is shown in Fig. 14. The sensor unit was successfully tested for various load currents, from 0.1 to 10 A. The communication was very reliable. A Wavecom Wismo cellular modem was used for sending text messages to a mobile phone. Figure 15 shows the internals of the fabricated sensor unit.

As part of the system testing, a data logger was developed and integrated into the SAM control software. It logs every activity which can be viewed later. Figure 16 shows the activity log viewer.

Fig. 14 Complete fabricated system

6 Current and Future Work

It has been realized that there is a need of incorporating a bed-sensor in the Selective Activity Monitoring system towards to smart digital home monitoring system. The main aim of the kind of sensor system is to monitor the bed activity of an old person or a patient who is living all alone in a house and need an immediate medical attention in case of an emergency or an abnormal situation.

The abnormal situation include, if a person who is supposed to have a habit of waking up from bed at early morning but sleeps till late morning on a particular day, or goes to bed at a particular time at the night but goes to bed very late or something happen during the night and or any other abnormal situation. The system can incorporate many such situations which usually happen in day to day life. If something of that sort happens the system will trigger a alarm signal which in turn alerts the concern person for immediate help. So the Bed Monitoring Wireless Sensor incorporated with the SAM device monitors the abnormalities in the person's daily activity.

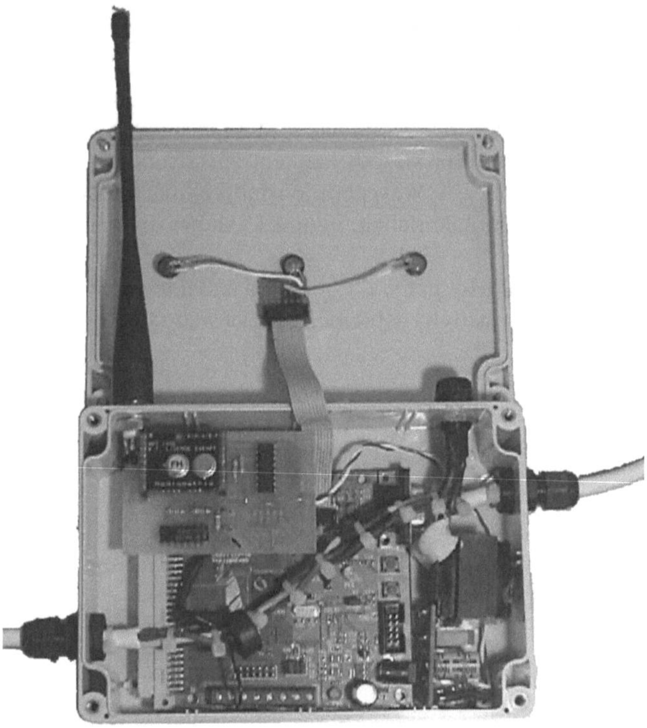

Fig. 15 The sensor unit

The bed monitoring system uses the Tekscan's FlexiForce Sensors [7]. The FlexiForce sensors as shown in Fig. 17 use a resistive-based technology. By application of an external force to the "active sensing area" of the sensor results in a change in the resistance of the sensing element which is inverse proportion to the force applied on the sensor. These sensors are constructed of two layers of substrate composed of polyester film.

On each layer, silver is applied for electrical contact which is then followed by a layer of pressure sensitive ink. Adhesive is then used to laminate the two layers of substrate together to form the sensor. The silver circle on top of the pressure-sensitive ink defines the "active sensing area." Silver extends from the sensing area to the connectors at the other end of the sensor forming the conductive leads.

Some of the characteristics of sensors, which can affect the results are: operating range of temperature is from −9° to 60°C, repeatability is $+/-2.5\%$ of full scale (conditioned sensor, 80% force applied, linearity is $<+/-5\%$, hysteresis is $<4.5\%$ of full scale.

The FlexiForce sensor used in the bed monitoring system is powered up by a +5 V supply as shown in Fig. 18. Since the basic operation of the FlexiForce Load sensor is similar to the variable resistor, the other end pin is used as an inverting input for LM311N, a low input current single voltage comparator [8]. A 10 K ohm potentiometer is used at the non-inverting terminal of the voltage comparator to

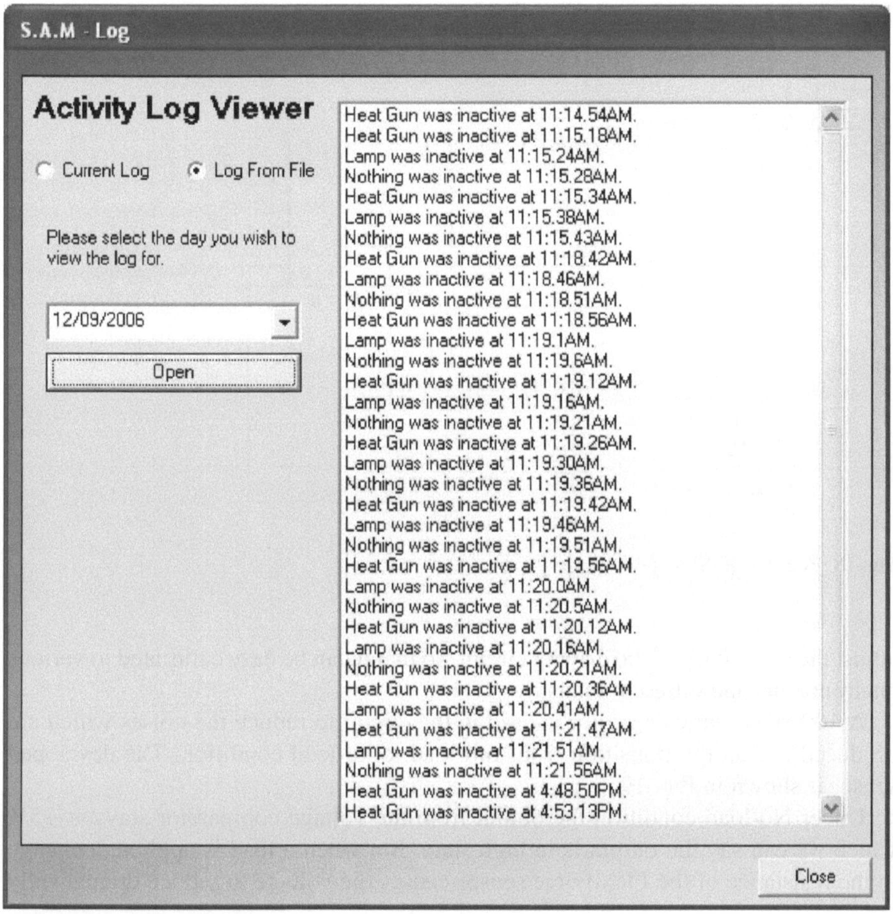

Fig. 16 Activity log viewer

Fig. 17 Flexiforce® sensor used for monitoring bed

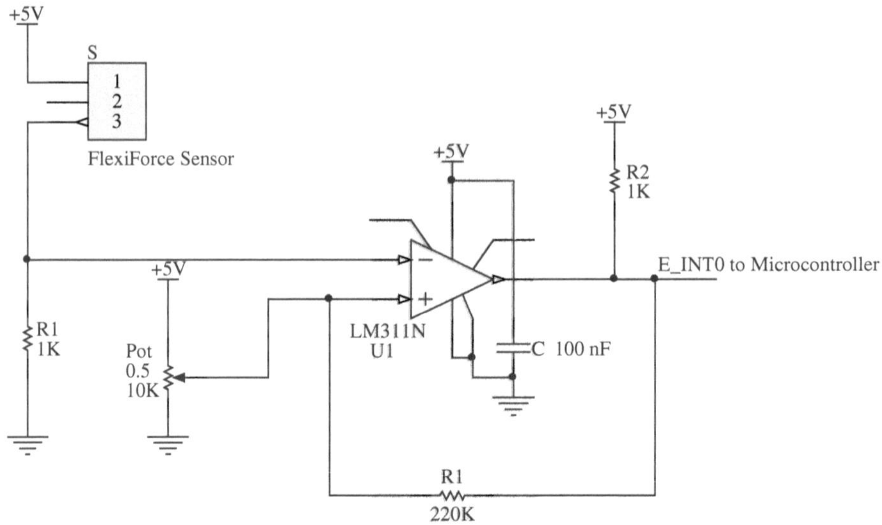

Fig. 18 Bed sensor driving circuit

adjust the sensitivity of the driving circuit, so that it can be easy calibrated to various environments and different loads.

A 100 nF ceramic capacitor is used in the circuit to reduce the noises which are produced during the transition from full load to no-load condition. The developed sensor is shown in Fig. 19.

Under No-load condition the output from the voltage comparator stays at +5V, which we can say the output is in high state. But when a load is applied, a change in the resistance of the FlexiForce sensor causes the voltage to reduce dramatically

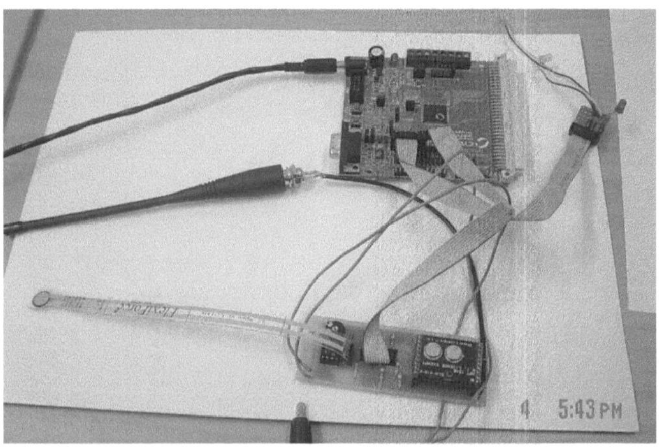

Fig. 19 The developed bed sensor

to the order of milli-volts, in this circuit the voltage drops from +5 to +0.1178 V. So when the load is applied on the sensor the output from the voltage comparator is in low state. The output then is fed as an interrupt signal to a micro controller trough the port pins on the Silabs® C8051F020 microcontroller development board. The port pin which receives the interrupt is always monitored by the hardware program in the microcontroller through a process called "polling". And thus the status of the bed sensor is monitored. By monitoring the persons daily activity, data is collated by Personal Computer which saves all data for processing as well as future use. The habit of the life-style of the person under care is stored in the system. The collected data are compared with the stored pattern and depending on the situation the actions are defined as unusual or abnormal. If the system detects any abnormal activity a warning or alarm message can be transmitted to the care-giver or concern person.

The Bed monitoring sensor system consists of a RF module for wireless communication with the SAM Controller, and a 10 pin port for communication with the microcontroller and a Force sensor driving circuit. The necessary electronic circuit is shown in Fig. 20.

For the Bed Monitoring the sensor unit consists of a Radiometrix BiM-418-40chip for RF communication, a force sensor FlexiForce Sensor fitted to a driving circuit and a Silab C8051F020 Microcontroller development board.

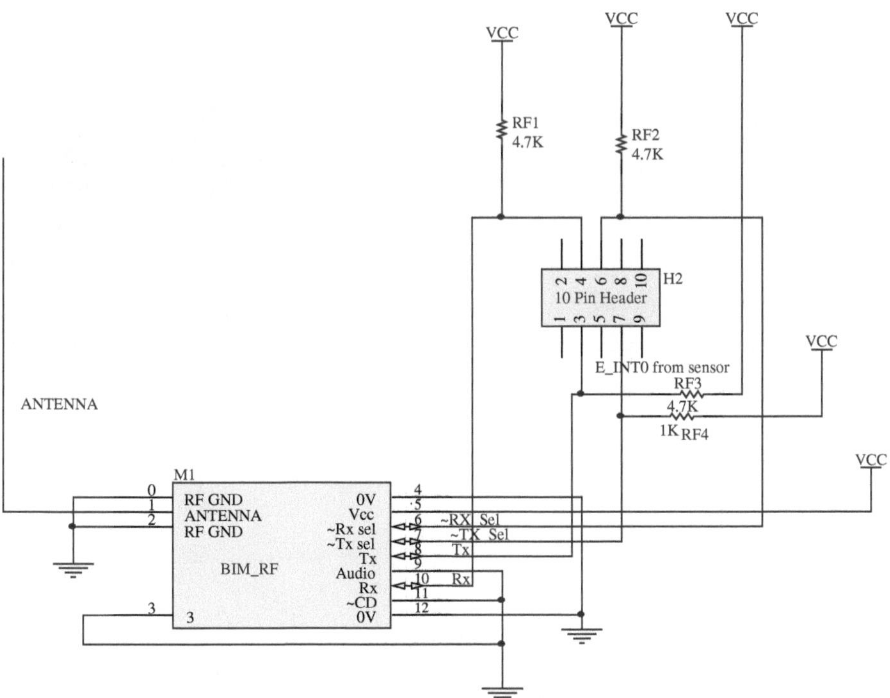

Fig. 20 RF module/ port pins

The Radiometrix BiM-418-40 is a Miniature PCB Mounting module which operated at UHF 418Mhz and capable of half duplex data transmission at speeds up to 40 Kbit/s over distances of 30 m "in-building" and 120 m open ground [9].

7 Conclusions

In this paper we have presented the design of a wireless sensor system based digital home and the applied research into the implementation issues. A proof of concept has been successfully established. The system successfully monitors the appliances in a house. Rules are set for monitoring based on the daily pattern of appliance use of a person. When the rules are violated, an alarm is generated and a SMS text message is sent to a mobile phone. The system is capable of automatically signaling for help, without requiring action from the monitored person. The system is non intrusive and will thus be more acceptable. A suggested design of the future unit, which will be very easy to integrate in the wall just like a normal socket, is shown in Fig. 21.

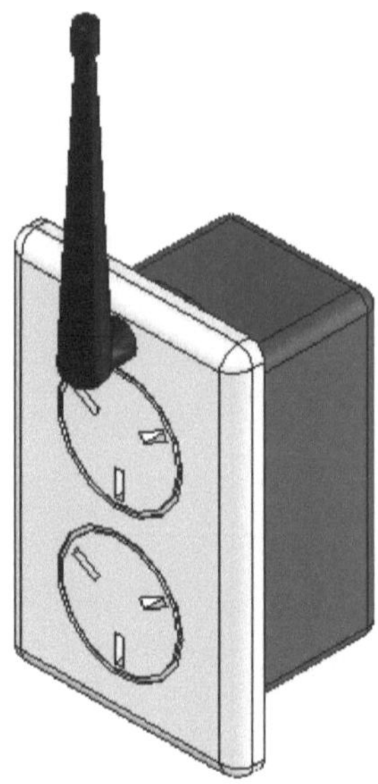

Fig. 21 Future design of the sensor unit (SU)

References

1. A. Tapus, M. J. Mataric and B. Scassellati, "Socially Assistive Robotics", IEEE Robotics & Automation Magazine, March 2007, pp 35–42.
2. S. Ohta, H. Nakamoto, Y. Shinagawa, T. Tanikawa, "A health monitoring system for elderly people living alone", PMID: 12097176 [PubMed – indexed for MEDLINE].

3. A. Dittmar, F. Axisa, G. Delhomme, C. Gehin, "New concepts and technologies in home care and ambulatory monitoring", Stud Health Technol Inform, 2004, pp 9–35.
4. E. Jovanov, D. Raskovic, J. Price, J. Chapman, A. Moore, A. Krishnamurthy, "Patient Monitoring Using Personal Area Networks of Wireless Intelligent Sensors", Biomedical Sciences Instrumentation, 2001, pp 373–378
5. H. Maki, Y. Yonczawa, H. Ogawa, H. Sato, A. W. Hahn, W. M. Caldwell, "A welfare facility resident care support system", Biomedical Sciences Instrumentation, 2004, pp 480–483.
6. G. Sen Gupta, S.C. Mukhopadhyay, M. Sutherland and S. Demidenko, Wireless Sensor Network for Selective Activity Monitoring in a home for the Elderly, Proceedings of 2007 IEEE IMTC conference, Warsaw, Poland, (6 pages), May 1–3, 2007.
7. Flexiforce Sensor User Manual, ver 09/23/05.
8. Philips Semiconductors Linear Products, LM111/211/311/LM311B Voltage comparator.
9. Radiometrix, Low Power UHF Data Transceiver Module, Issue 3, 01 November 2004.

Optimisation of the Nelder-Mead Simplex Method for Its Implementation in A Self-Mixing Laser Displacement Sensor

Usman Zabit, Francis Bony and Thierry Bosch

Abstract A laser diode self-mixing sensor enabling us to perform displacement measurements under moderate feedback has been designed with a resolution up to 32 nm. In this paper, we focus on the signal processing required to reach such a resolution and notably on the Nelder-Mead simplex method we used for the joint estimation of parameters like the linewidth enhancement factor and feedback level described by the coupling factor. Further more, influence of initial step size λ of Nelder Mead method on the reduction in the convergence time as well as on the possibilities of introducing parallelism in the algorithm have been studied for implementation in a future integrated sensor. This analysis leads to optimize the number of optical power samples necessary for joint estimation, the adaptations of Nelder-Mead method and its parameter λ on our application, all of which result in the consequent reduction in computing time necessary for convergence to the optimum values.

Keywords Optical sensors · optical feedback · laser interferometry · signal processing · displacement measurement · Nelder-Mead method

1 Introduction

Optical feedback interferometry (OFI), the so-called self-mixing effect, has been widely studied for the last two decades. As a matter of fact, spectral characteristics of a laser diode (LD) are strongly affected by optical feedback, leading to the degradation of the LD performances for applications like DVD readers. However, this optical feedback can also be of practical use for sensing applications like

Usman Zabit
Université de Toulouse, LOSE, INPT, ENSEEIHT, Toulouse, France

Francis Bony
Université de Toulouse, LOSE, INPT, ENSEEIHT, Toulouse, France

Thierry Bosch
Université de Toulouse, LOSE, INPT, ENSEEIHT, Toulouse, France,
e-mail: thierry.bosch@enseeiht.fr

S.C. Mukhopadhyay, G.S. Gupta (eds.), *Smart Sensors and Sensing Technology*,

displacement, vibration, distance, velocity (of both solid targets and fluids) and correlated measurements [1].

The physical principle of our displacement sensor is thus based on moderate feedback occurring when the optical beam reflected by a remote target at a distance $D(t)$ is coupled back into the LD active cavity. The re-injected laser beam is then self-mixed with the light-wave inside the LD cavity, causing strong variations of the Optical Output Power (OOP) when the optical path length varies, i.e. when the target is moving and/or the injection current of the LD is modulated. The OOP is then monitored by means of a photodiode enclosed in the typical LD package, an OOP variation occurring for each half-wavelength ($\lambda_0/2$) displacement of the target, where λ_0 represents the laser wavelength under free running conditions. The feasibility of this OFI sensor has previously been reported, displacements being measured with a resolution up to $\lambda_0/20$, i.e. around 40 nm [2].

In this paper, the operating principle is introduced. The signal processing developed for a self-mixing signal under moderate feedback is then detailed. In particular, the joint estimation of parameters like the coupling coefficient, characterizing the feedback regime, and the linewidth enhancement factor of the laser diode is presented.

Finally, an analysis of the used Nelder-Mead simplex method is presented followed by its application to moderate feedback interferometric sensor in order to optimize some characteristics of our sensor like the number of OOP samples needed and the computing time of the signal processing.

2 The Self-Mixing Effect

2.1 Basic Theory

Self-mixing effect occurs in a laser when a part of the beam back-scattered or back-reflected from a remote target in front of the laser, is added coherently ("self-mixed") with the standing wave inside the active cavity of the LD, modifying the spectral properties of the laser.

The corresponding wavelength fluctuations can then be found by solving the phase equation [3]:

$$x_0(t) - x_F(t) - C\sin[x_F(t) + \arctan(\alpha)] = 0 \tag{1}$$

where α is the linewidth enhancement factor and C, the coupling coefficient. $x_0(t)$ and $x_F(t)$ are expressed as follows:

$$x_0(t) = 2\pi\frac{D(t)}{\lambda_0/2} \text{ and } x_F(t) = 2\pi\frac{D(t)}{\lambda_F(t)/2} \tag{2}$$

λ_0 and $\lambda_F(t)$ are respectively the wavelength of the LD under free running conditions and subject to optical feedback.

Moreover, variations of the OOP due to the feedback can be expressed as follows:

$$P_F(t) = P_0(1 + m\cos(x_F(t))) \quad (3)$$

where P_0 is the emitted optical power on the free running state and m is the constant modulation index.

2.2 Influence of the Feedback Level

The coupling factor C is highly important for OFI applications as it discriminates the shape of the self-mixing signal between different feedback levels:

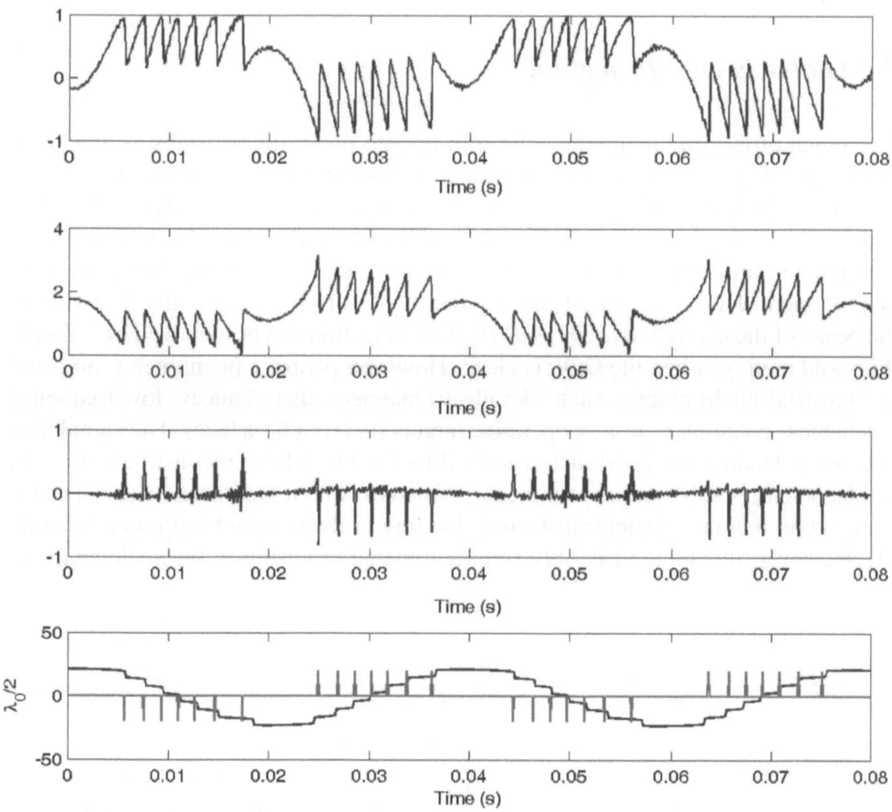

Fig. 1 (**a**) Normalized optical output power P(t) presenting hysteresis for the moderate feedback regime, (**b**) arccos(P(t)), (**c**) derivative of arcos(P(t)), (**d**) transitions and rough reconstruction of $x_F(t)$

$$C = \frac{\varepsilon}{\sqrt{A}}\left(\frac{1-r_2^2}{r_2}\right)r_3\frac{\tau_D\sqrt{1+\alpha^2}}{\tau_l} \qquad (4)$$

τ_D is the time-of-flight between the laser diode and the target (i.e. $2D/c$ where c is the speed of light), A stands for the total power attenuation, τ_l for the diode cavity round trip time delay and ε for the mismatch coefficient.

For $0 < C < 0.1$, the feedback regime is very weak and the OOP variations present a cosine shape. For $0.1 < C < 1$, the feedback regime is weak and the OOP presents a sawtooth-like shape. For larger values of C, the variations of the OOP begin to present hysteresis (Fig. 1-a), typical of the moderate feedback regime, leading to a total disappearance of the sawtooth-like signal in the strong feedback regime.

3 Displacement Measurement

3.1 Global Signal Processing

The signal processing method can be split up into two principal steps as illustrated in Fig. 2 [2]. First step consists in phase unwrapping the cosine function of the normalized OOP P(t) (Fig. 1-b), thresholding the derivative of arcos [p(t)] (Fig. 1-c) and finally converting it in normalized transitions (see Fig. 1-d). Each transition is computed in order to add or substract 2π to $x_F(t)$ depending on the direction of the displacement. However, one of the key points of our algorithm is the detection of the peaks of the derivative of *arcos(P(t))*. This detection can be achieved by a simple threshold method when the OOP is clean. However, parasitic phenomena can occur in industrial environments such like electromagnetic disturbances, low-frequency mechanical coupling or non-cooperative targets (i.e. rough surface). The signal then gets noisy leading the previously reported method to a false reconstruction of the displacement. A solution to optimize the robustness of our algorithm has been tested using a continuous wavelet transform belonging to the reverse biorthogonal family [4]. However, this method is only semi-automated reporting to the scale analysis.

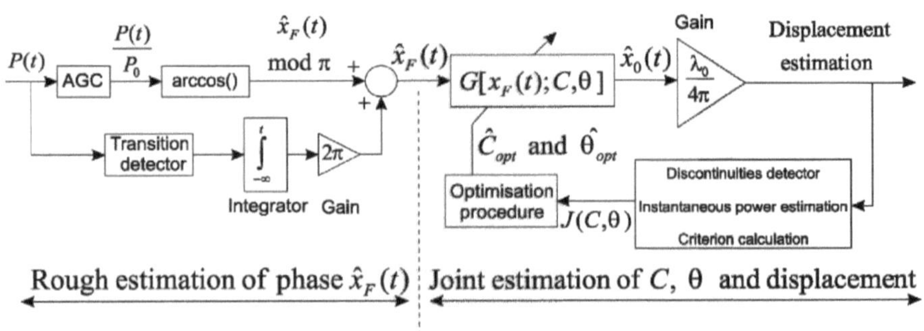

Fig. 2 Principle of the signal processing for the estimation of the displacement

Moreover, this algorithm is still sensitive to noise and consequently a transition may appear due to a false alarm. In order to still improve the robustness of this algorithm, we propose to characterize these OOP singularities by Holder exponent computed from maximum modulus of continuous wavelet transform on all scales (i.e. from low to high frequencies) [5].

The second signal processing step deals with the joint estimation of parameters like the coupling coefficient and the line-width enhancement factor of the LD in order to finally recover the displacement measurement. The algorithm is based on the fact that mechanical target displacement discontinuities are far less frequent than discontinuities of $P(t)$ and $x_F(t)$ caused by the non linearities of the LD subject to moderate feedback [3].

3.2 The Nelder-Mead Simplex Method

The Nelder-Mead Simplex Method (NM) [6] is one of the most widely used direct search methods for multidimensional unconstrained minimization and non-linear optimisation. It attempts to minimize a scalar-valued non-linear function of n real variables using only function values, without any derivative information (explicit or implicit). The NM method thus falls in the general class of direct search methods.

A large subclass of direct search methods, including the NM method, maintain at each step a non-degenerate simplex; a geometric figure in n dimensions of nonzero volume that is the convex hull of $n+1$ vertices (i.e. a pyramid in the case of 3D objective function). Each iteration of a simplex-based direct search method begins with a simplex, specified by its $n+1$ vertices and the associated function values. One or more test points are computed, along with their function values, and the iteration terminates with a new (different) simplex such that the function values at its vertices satisfy some form of descent condition compared to the previous simplex. Among such algorithms, the NM algorithm is particularly parsimonious in function evaluations per iteration, since in practice it typically requires only one or two function evaluations to construct a new simplex. Several popular direct search methods use n or more function evaluations to obtain a new simplex [7].

The NM method is proposed as an algorithm for minimizing a real-valued function $f(\mathrm{x})$ for $x \in \Re^n$. Four scalar parameters must be specified to define a complete NM method: coefficients of reflection (ρ), expansion (χ), contraction (γ), and shrinkage (σ). According to the original NM paper, these parameters should satisfy:

$$\rho > 0,\ \chi > 1,\ \chi > \rho,\ 0 < \gamma < 1,\ 0 < \sigma < 1 \tag{5}$$

The nearly universal choices used in the standard NM algorithm are $\rho = 1, \chi = 2$, $\gamma = 1/2$ and $\sigma = 1/2$ which are also used in our method.

At the beginning of the kth iteration, $k \geq 0$, a non-degenerate simplex Δ_k is given, along with its $n+1$ vertices, each of which is a point in $\Re^n$. It is always assumed that iteration k begins by ordering and labelling these vertices as $x_1^{(k)}, \ldots, x_{n+1}^{(k)}$, such that

$$f_1^{(k)} \leq f_2^{(k)} \leq \ldots \leq f_{n+1}^{(k)} \tag{6}$$

where $f_i^{(k)}$ denotes $f(x_i^{(k)})$. The kth iteration generates a set of $n+1$ vertices that define a different simplex for the next iteration, so that $\Delta_{k+1} \neq \Delta_k$. Because we seek to minimize f, we refer to $x_1^{(k)}$ as the *best* point or vertex, to $x_{n+1}^{(k)}$ as the *worst* point, and to $x_n^{(k)}$ as the *next-worst* point. Similarly, we refer to $f_{n+1}^{(k)}$ as the worst function value, and so on.

A single generic iteration is specified, omitting the superscript k to avoid clutter. The result of each iteration is either (1) a single new vertex- the accepted point - which replaces x_{n+1} in the set of vertices for the next iteration, or (2) if a shrink is performed, a set of n new points that, together with x_1, form the simplex at the next iteration.

3.3 An Iteration of NM Algorithm

Order. Order the $n+1$ vertices to satisfy:

$$f(x_1) \leq f(x_2) \leq \ldots \leq f(x_{n+1}).$$

Reflect. Compute the *reflection point* x_r from

$$x_r = \bar{x} + \rho(\bar{x} - x_{n+1}) \tag{7}$$

where $\bar{x} = \sum_{i=1}^{n} x_i/n$ is the centroid of the n best points (all vertices except for x_{n+1}). We evaluate $f_r = f(x_r)$. For $f_1 \leq f_r < f_n$, we accept the reflected point x_r and terminate the iteration.

Expand. If $f_r < f_1$, the *expansion point* x_e is calculated:

$$x_e = \bar{x} + \chi(x_r - \bar{x}) \tag{8}$$

and $f_e = f(x_e)$ evaluated. For $f_e < f_r$, we accept x_e and terminate the iteration; otherwise (if $f_e \geq f_r$), x_r is accepted and we terminate the iteration.

Contract. If $f_r \geq f_n$, perform a *contraction* between $\bar{x}$ and the better of x_{n+1} and x_r.

- Outside. For $f_n \leq f_r < f_{n+1}$ (i.e., x_r is strictly better than x_{n+1}), we do an outside contraction:

$$x_c = \bar{x} + \gamma(x_r - \bar{x}) \tag{9}$$

and evaluate $f_c = f(x_c)$. For $f_c \leq f_r$, x_c is accepted and we terminate the iteration; otherwise, we perform a shrink step.

- Inside. If $f_r \geq f_{n+1}$, perform an inside contraction:

$$x_{cc} = \bar{x} - \gamma(\bar{x} - x_{n+1}) \tag{10}$$

and evaluate $f_{cc} = f(x_{cc})$. For $f_{cc} \leq f_{n+1}$, we accept x_{cc} and terminate the iteration; otherwise, we do a shrink.

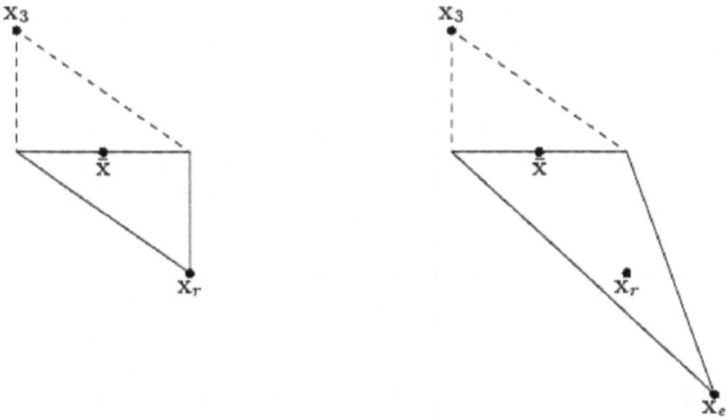

Fig. 3 Nelder-Mead simplexes after a reflection and an expansion step. The original simplex is shown with a dashed line

Perform a shrink step. Evaluate f at the n points $v_i = x_1 + \sigma(x_i - x_1)$, where $i = 2,\ldots,n+1$. The (unordered) vertices of the simplex at the next iteration consist of $x_1, v_2, \ldots, v_{n+1}$.

Figures 3 and 4 show the effects of reflection, expansion, contraction, and shrinkage for a simplex in two dimensions (a triangle), using the standard coefficients [7].

The algorithm is run until the optimised point is reached which may be specified as the obeying the condition, $abs(f_1 - f_{n+1}) \leq tol$ where tol is an infinitesimally small value, approaching machine-level precision for the hardware implementations of the algorithm.

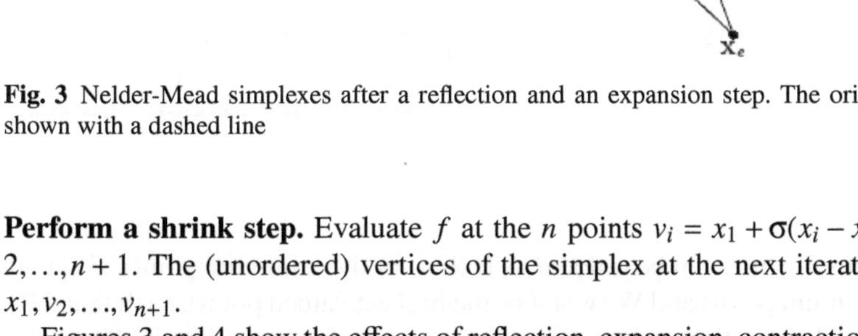

Fig. 4 Nelder-Mead simplexes after an outside contraction, an inside contraction, and a shrink. The original simplex is shown with a dashed line

3.4 Application of the NM Algorithm

Restricting the NM method to two dimensions (as only the coupling coefficient factor and the line-width enhancement factor are being optimised), the number of

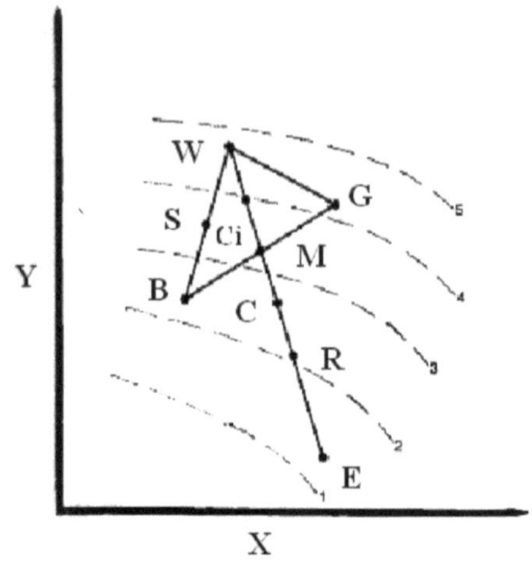

Fig. 5 A two dimensional NM simplex (triangle) on the XY surface with possible subsequent points (see Table 1), where curves 5 to 1 represent descending values of the objective function [8]

vertices becomes 3 which can be ordered as B (best or the minimum point), G (good or the next minimum point) and W (worst or the highest valued point), such that f(B) ≤ f(G) ≤ f(W). The iterations of the algorithm generate further points in 2D space, such as R (reflect), E (expand), C (outside contract), Ci (inside contract), S (shrink) or M (mid-point of B and G).

We observe that the NM method has an *adaptive* feature which enables the simplex (see Fig. 5 for an example in two dimensions) to reflect, expand, contract or shrink so as to conform to the characteristics of the response surface.

These actions are based on the conditions summarised in the Table 1 and are applied repeatedly until a termination criterion is reached. For the particular case illustrated in Fig. 5, reflection of point W (corresponding to the highest result) through the centriod M of the opposite side locates point R. In this instance, an expansion would then follow making the next simplex GBE.

In contrast to other minimization procedures, the simplex procedure approaches the minimum by moving away from high values of the objective function rather than by trying to move in a line toward the minimum [8].

Table 1 Conditions governing the formation of subsequent simplexes (where f(x) means the value of the objective function at point x) [8]

Condition	Action	New simplex
f(B) ≤ f(R) ≤ f(G)	Reflect	G B R
f(R) < f(B)	Extend	G B E
f(W) < f(R)	Inside contract	G B Ci
f(G) < f(R) ≤ f(W)	Outside contract	G B C
f(W) ≤ f(Ci) or f(R) ≤ f(C)	Shrink	S M B

3.5 Application to the Self-Mixing Effect

In this section, to start with, the two lower blocks of the Joint Estimation section of Fig. 2 are explained. The first block provides the criterion $J(C,\theta)$ and afterwards, the NM method is executed on this same parameter in the 'optimization procedure' block.

In order to choose the optimized set of parameters, equation (1) must be rewritten as:

$$x_0(t) = x_F(t) + C\sin\lfloor x_F(t) + \theta\rfloor, \forall t > 0 \tag{11}$$

where $\theta = \arctan(\alpha) + x_F(0)$. Even if $x_F(0)$ is unknown, depending on initial conditions, its influence on the shape of the self-mixing signal remains negligible [3]. During the joint estimation of C and θ, the criterion $J(C,\theta)$ is evaluated , given by:

$$J(C,\theta) = \sum_{k=0}^{K} [\hat{x}_0(k) - \hat{x}_0(k-1)]^2 \tag{12}$$

where $\hat{x}_0(k)$, $k \in \mathrm{N}$, is the discrete form of the phase $x_0(k)$. The criterion is evaluated over a temporal window containing K points of the time derivative of $\hat{x}_0(k)$. Hence the optimal couple $\hat{C}_{opt}$ and $\hat{\theta}_{opt}$ is the one which minimises $J(C,\theta)$ in order to allow the reconstruction of optimal displacement:

$$\underset{C,\theta}{Arg\min}[J(C,\theta)] = Arg\min \sum_{k=0}^{K} [\hat{x}_0(k) - \hat{x}_0(k-1)]^2$$

It has been observed during the optimisation of the demodulation algorithm that the window size of K points (12) in time domain can be drastically reduced to one-third or even one-fourth size with only a cost of less than 5% error in the accuracy of the $\hat{C}_{opt}$ and error of less than 1% in the estimation of $\hat{\theta}_{opt}$. The evolution of %age error in $\hat{C}_{opt}$ and $\hat{\theta}_{opt}$ as against the window size can be seen along the y-axis of Figs. 6 and 7 respectively. Thus, considerable gain in the speed of execution of the algorithm can be achieved without an excessive error in the evaluation of the desired parameters.

Moreover, during the optimisations, the influence of the tolerance criterion was also observed. This parameter actually helps to finally identify the optimised point obtained in the multidimensional simplex space. In fact, it ensures the proximity of $n+1$ vertices as to when they are sufficiently close to terminate the execution of the NM iterations.

Hence, with a very small *"tol"* value, the NM algorithm continues to evaluate the objective function for a long period of time (i-e increased number of iterations and objective function evaluations) as opposed to a small period of time for a relatively large *"tol"* value.

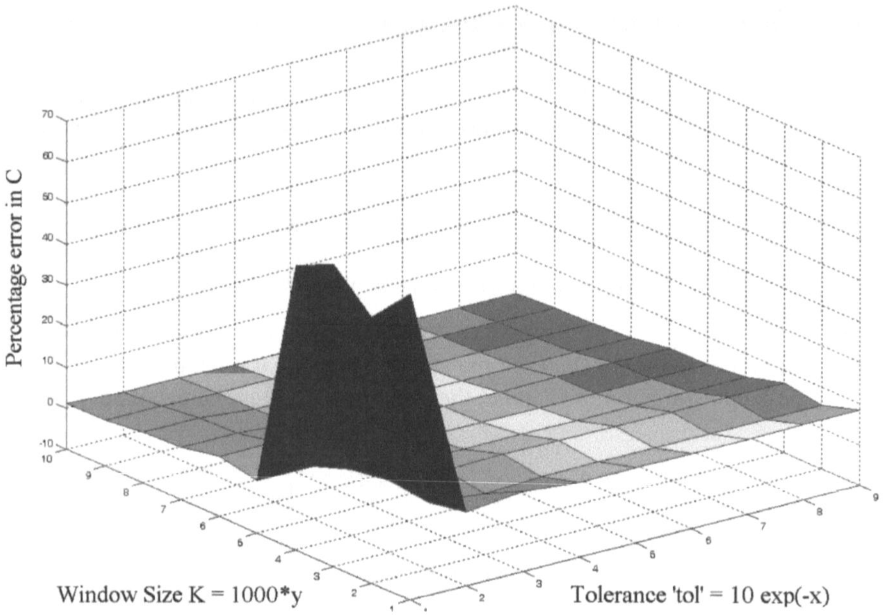

Fig. 6 Percentage error in $\hat{C}_{opt}$ as a function of tolerance "tol" of NM method (x-axis) and Window size K of $J(C,\theta)$(y-axis). The color coding represents the number of objective function utilisations which increases from 8 (*blue*) to 104 (*red*) as a function of x and y

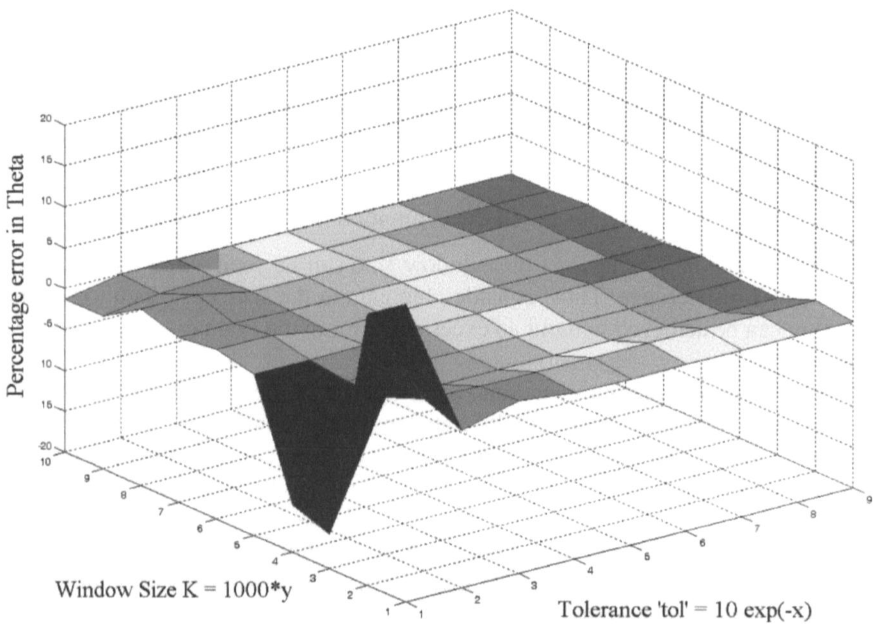

Fig. 7 Percentage error in $\hat{\theta}_{opt}$ as a function of Tolerance "tol" of NM method (x-axis) and Window size K of $J(C,\theta)$(y-axis). The color coding represents the number of objective function utilisations which increases from 8 (*blue*) to 104 (*red*) as a function of x and y

This parameter is theoretically close to zero and is normally set close to machine precision level in the commercial software adaptations of NM method in order to ensure the validity of the 'minimum' reached by the NM method. However, in our case, the influence of 'tol' has been evaluated with a range from 10^{-1} to 10^{-9}. The objective has been to adapt this parameter value in order to decrease the number of function evaluations while maintaining the same levels of accuracy of $\hat{C}_{opt}$ and $\hat{\theta}_{opt}$. It is observed that with a value close to 10^{-3}, the number of objective function evaluations are approx. halved while maintaining the error induced in $\hat{C}_{opt}$ and $\hat{\theta}_{opt}$ to less than 0.5%, as seen along the x-axis of Figs. 6 and 7 respectively.

Lastly, the NM method was adapted to our displacement measurement application with respect to its contraction steps.

As already explained, the NM method utilises two types of contraction, the inside contraction as well as the outside contraction. However, it is observed that the removal of the "inside contraction" step actually helps to decrease the number of objective function utilisations by approx. 25% (Fig. 8) as far as its adaptation to our application is concerned. It is noted that this factor gain is achieved without any error introduced in the joint estimation of $\hat{C}_{opt}$ and $\hat{\theta}_{opt}$.

This could be explained by the fact that our objective function is in two dimensional space and the general shape of the criterion $J(C,\theta)$ is a parabolic valley in 3

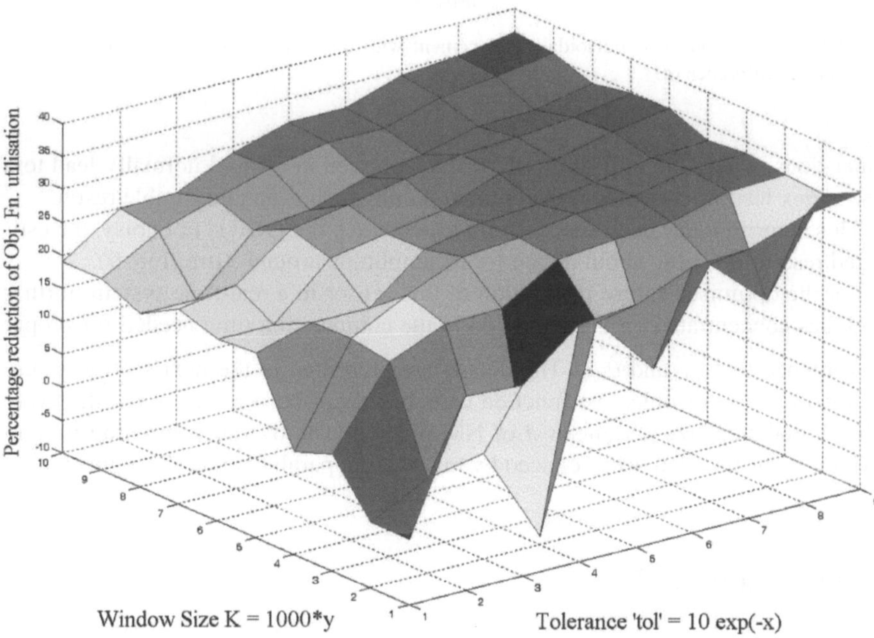

Fig. 8 Percentage reduction in objective function utilisation without using "inside contraction" step of NM method as a function of Tolerance "tol" (x-axis) of NM method and Window size K of $J(C,\theta)$ (y-axis). The color coding represents the same percentage error

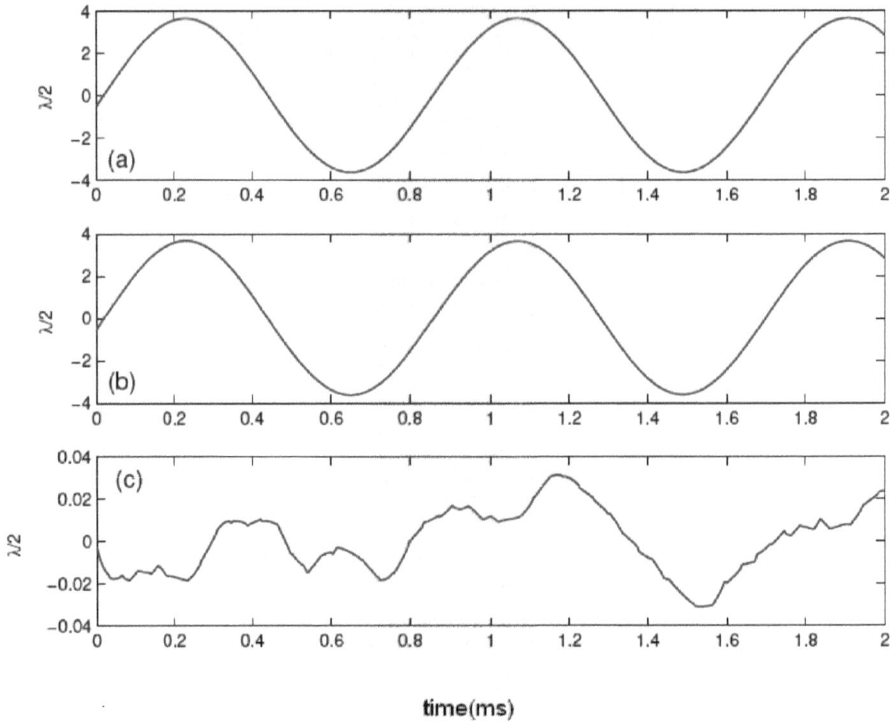

Fig. 9 Reconstruction of a sinusoidal displacement with our sensor (**a**) and a commercial one (**b**) and their difference (**c**)

dimensions, where the utilisation of "inside contraction" would normally lead to a new vertex leading away from the optimal point which is an undesirable result.

This algorithm has been successfully tested when the OOP is not noisy, the estimated resolution being about 32 nm for an amplitude around 3 μm (Fig. 9).

For the optimised future integration of this sensor in a real-time system, further work has concentrated on the reduction in the computation time needed for proper convergence to $\hat{C}_{opt}$ and $\hat{\theta}_{opt}$. The details are presented in the next section where the total number of objective function calls has again been reduced by effectively using initial step size parameter λ of NM method. The overall computing time in NM iterations has also been reduced by introducing parallelism in the algorithm.

4 The Parameter λ

The initial step size parameter λ is used to initialise the starting simplex in the Nelder-Mead Simplex method. This simplex (a triangle in 2D space and a Pyramid in 3D space) is constructed based on an arbitrary initial point in the simplex space.

The required $n+1$ points needed to create a simplex for an n variable function are thus derived based on the value of λ.

Thus, for an initial starting point x_o, the other n points can be found using

$$x_i = x_o + \lambda e_i$$

where the e_i's are N unit vectors in the simplex space and where λ is a constant which is a guess of the objective function's characteristic length scale.

4.1 Influence of λ on Total Number of Objective Function Calls

It has been observed by simulations that this parameter has direct effect on the time needed to find the optimum values (in our case , the values of C and θ that minimize our objective function or the criterion $J(C,\theta)$) through the Nelder-Mead Method.

The simulations have been conducted using three different values of λ, e-g $\lambda = 0.1$, $\lambda = 0.5$ and $\lambda = 0.9$.

In order to study only the influence of λ, the starting values were varied in the objective function space so that its influence on the final convergence can be evened out.

For each starting point e-g C = 4 and theta = 4 etc, 90 simulations were carried out for varying tolerance factor "tol" and total number of Laser Diode Power samples K. The averaged total number of objective function calls for these 90 simulations for a given starting point x_o was used.

Thus, using the same method, averaged number of objective function calls were calculated for 9 different starting points and the resulting curves are plotted in Fig. 10. The same practice has been used for all subsequent data based on varying starting points.

X AXIS	[C Theta]
1	[2 2]
2	[2 4]
3	[2 8]
4	[4 2]
5	[4 4]
6	[4 8]
7	[8 2]
8	[8 4]
9	[8 8]

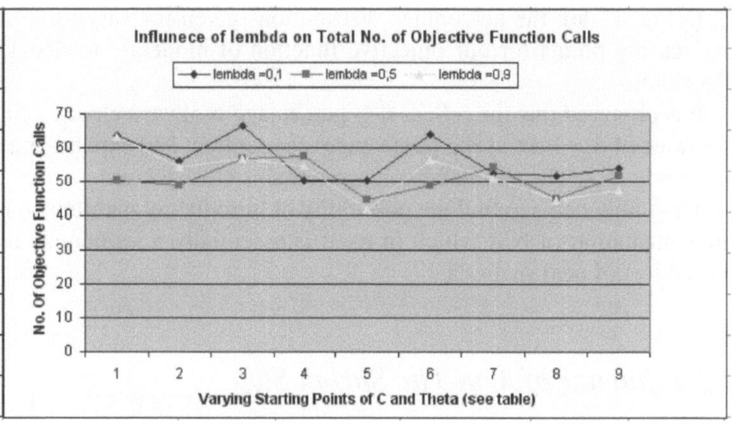

Fig. 10 Influence of λ on the total number of objective function calls needed in order to converge to the optimised values of C and θ for varying starting points of the same two parameters in Nelder-Mead simplex algorithm for minimization of the objective function

Table 2 The mean of total number of objective function call values shown in Fig. 10

$\lambda = 0.1$	$\lambda = 0,5$	$\lambda = 0,9$
56.6	51	52.3

The varying starting points have been selected keeping in mind the usual variation of C and θ for the case of moderate feedback in a self mixing laser diode.

If the mean of these curves is once again taken, it is seen in the Table 2 that for $\lambda = 0,5$, the total number of objective function calls is minimized as compared to other values. This in turn shall lead to a quicker establishment of optimised values of C and θ for our demodulation algorithm of displacement measurement using the Nelder-Mead Method.

4.2 Percentage Distribution of Standard NM Steps for Varying λ

Simulations have been carried out to measure the percentage distribution of the standard NM step related function calls (Reflect, Expand, Contract and Shrink) in the over all number of function calls.

It is seen that a typical NM iteration chooses from the standard four steps; reflect, expand, contract and shrink. Thus every new iteration executes the function call for reflect step, followed by an expansion step or iteration termination in one case or a contraction and/or shrinkage in the other case.

Thus, an averaged percentage distribution can be assumed to contain 50% reflection function calls and 25% each for expansion as well as contraction-shrinkage function calls.

However, these percentages can be altered depending on the function space topography, its initial starting point as well as the parameter λ.

In Fig. 11 are the percentage distributions given for varying λ as well as varying starting points for our objective function of moderate feedback demodulation algorithm.

It is observed that the reflect step percentage is always close to 50 irrespective of the value of λ, where as the percentages for contract and expand or extend step vary according to λ. The shrink percentage remains close to zero.

This study helps to find the possibility of introducing parallelism in the hardware implementation of NM which in itself is essentially a sequential algorithm and is the subject of next sections.

4.3 Influence of λ on The Shrink Step

The number of Shrink Step function calls can also be controlled by varying the parameter λ. This can be seen in the Fig. 11 representing percentage distributions

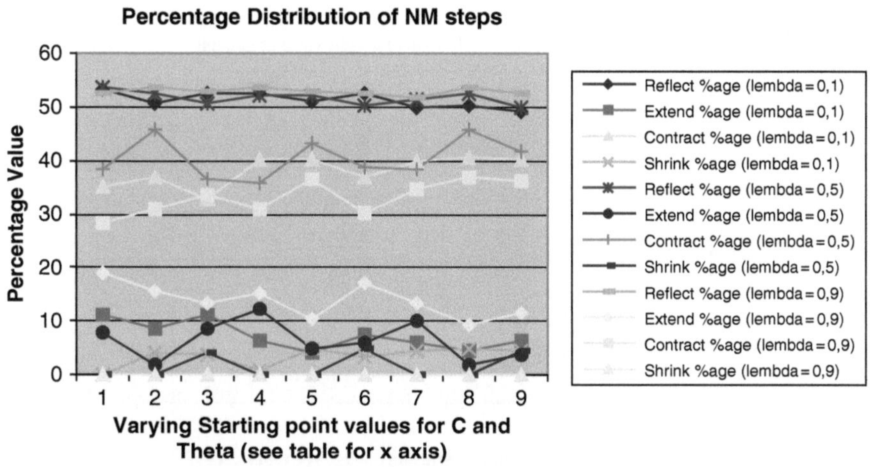

Fig. 11 Percentage distribution of the four basic NM steps in the total number of objective function calls for varying λ and starting points

of the four standard NM steps. For example, for $\lambda = 0.9$, the vertices of the created simplex are fairly close and a shrink step was never needed.

The averaged shrink step function call percentage for different starting points is given below. Once again, a very small value is seen for $\lambda = 0.5$.

The advantage of a very small shrink function call percentage is that its elimination leads to elimination of two function calls per iteration for an objective function in 2D space. Thus for any arbitrary NM iteration, only 3 out of 5 next function calls remain to be possibly executed.

This leads to a direct reduction of total time needed to find optimised values by a factor of 2 by deploying 3 objective function calls in parallel in hardware. The obvious disadvantage is the threefold expense of hardware resources used for implementing the objective function call block for a gain in time of a factor of two.

The mean of shrink percentages of 9 simulations is given in Table 3.

It is evident that the reduction in the shrink percentages results in the increase of the remaining three step percentages and hence allows us to deploy 3 parallel function blocks.

Table 3 The mean of percentage value of Shrink function call curves for varying starting points and λ

$\lambda = 0.1$	$\lambda = 0{,}5$	$\lambda = 0{,}9$
3.4%	1.5%	0%

X AXIS	[C Theta]
1	[2 2]
2	[2 4]
3	[2 8]
4	[4 2]
5	[4 4]
6	[4 8]
7	[8 2]
8	[8 4]
9	[8 8]

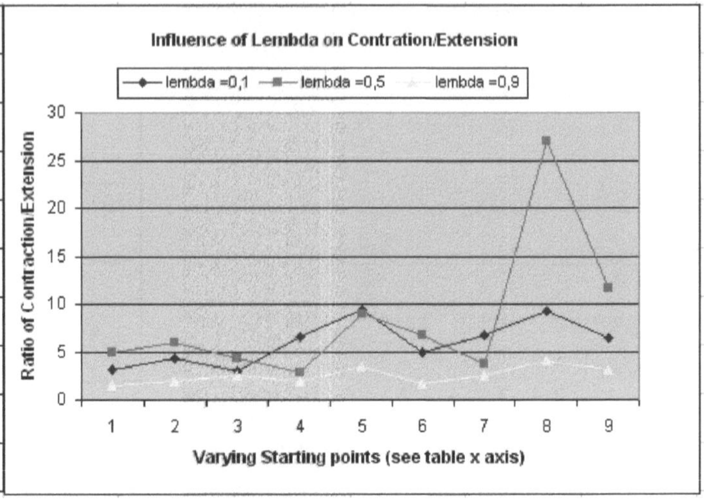

Fig. 12 Influence of λ on the contract –expand ratio for varying starting points

4.4 Influence of λ on the Contraction to Expansion Ratio

It is observed that λ has also an influence on the ratio of contraction steps to expansion steps which are executed in any given NM iteration.

By varying λ, a higher "contraction to expansion" ratio is obtained. Fortunately, as seen in Fig. 12, the best value is again found by $\lambda = 0,5$. Thus the same parameter reduces the number of function calls needed to optimise C and θ as well as to increase this ratio.

This ratio becomes important when parallel objective calls are to be used to reduce the total time needed to find the optimum values. So by using a forward prediction, a contraction step function might be called at the same time as the call for a Reflection step.

With a large probability of a contraction step rather than an expansion step, the parallel call helps to speed up the time needed. The cost of this increase is paid by the duplication of the hardware resources needed to implement the objective function in a real time system.

For example, using $\lambda = 0.5$, and with a quasi elimination of shrink step, the probability of the next function call being a Contraction step after a Reflect step is 8 to 1 as compared to an expansion step, as shown in Table 4. Thus by using double parallel function calls for Reflect and Contract steps, this could lead to a reduction in total time for optimisation by roughly 40%.

Table 4 The mean of contraction/expansion ratios for varying starting points and λ

$\lambda = 0.1$	$\lambda = 0,5$	$\lambda = 0,9$
5.9	8.4	2.5

5 Implementation of Adapted Nelder Mead FSM in Stateflow®

The adapted Nelder Mead Algorithm in the shape of a Finite State Machine (FSM) has been successfully implemented in StateFlow® in MATLAB®. The objective function calls have been executed using estimated phase samples of laser diode power in MATLAB.

The comparison of our FSM with the standard "fminsearch" function of MATLAB is presented in Fig. 13.

As indicated in Table 5, the FSM implementation has provided excellent results while significantly reducing the total number of objective function calls.

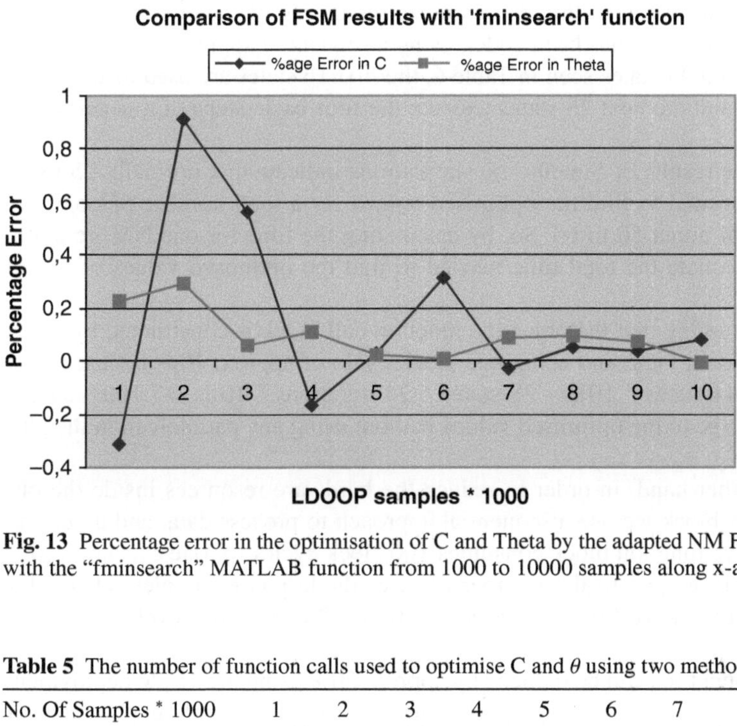

Fig. 13 Percentage error in the optimisation of C and Theta by the adapted NM FSM as compared with the "fminsearch" MATLAB function from 1000 to 10000 samples along x-axis

Table 5 The number of function calls used to optimise C and θ using two methods

No. Of Samples * 1000	1	2	3	4	5	6	7	8	9	10
Fn calls for 'fininsearch'	88	88	93	82	87	83	85	83	87	89
Fn calls for FSM	40	41	52	53	62	56	55	58	55	58

5.1 Implementation of the FSM in VHDL

Based on the FSM tested in StateFlow®, the algorithm has been coded in VHDL. The resulting code has 44 states and assuming a Spartan III Xilinx ® FPGA, maximum frequency of 106.022 MHz has been reported by the simulations.

Table 6 Extract from the Synthesis Report for the VHDL code of the NM FSM

States	44
Transitions	57
Inputs	6
Outputs	57
Clock	clk (rising_edge)
Clock enable	ce (positive)
Power Up State	first
Encoding	automatic
Implementation	LUT

The FSM used for the coding does not use pipelining of the objective function call block and hence this has been coded using a sequential approach.

Out of the total 44 states seen in Table 6, the first 16 states are used to set up the initial simplex and the next 28 states execute the four basic steps of a general NM iteration.

The previous results of Stateflow® simulations indicate that normally 25 to 30 iterations are needed to find the optimised values for a total number of objective function calls of about 50 to 60. So, by calculating the time for one NM iteration, one can approximate the total time needed to find the optimised values of C and theta.

Now, by assuming that the objective function call block is constructed by using only combinational logic and using one clock cycle of approx. 10ns for execution of each state; 44 states * 10 ns + 28 states * 24 iterations * 10 ns = 7.16μs are expected to converge to the optimised values without using any parallelism in the NM algorithm.

If, on the other hand, in order to reduce the hardware resources inside the objective function block we use a sequential approach to process data, and assuming that the objective function block consumes 100 clock cycles for one execution (this value is in itself subject to the number of laser diode phase samples taken); for 50 function calls, the additional time amounts to 50 calls * 100 cycles * 10 ns = 5 μs.

This additional time can be reduced by approx. 50% using the Triple deployment of objective function block or by approx. 40% by using a Double deployment of objective function block. However, it is noted that for about 50% gain in time as compared to a sequential implementation of NM FSM, the hardware resources consumed to implement the objective function block are tripled and are doubled for a gain of about 40%.

So, by a triple deployment of objective function block, we could gain only about 10% in time at an increase in hardware cost of 33% as compared to a double deployment.

6 Conclusion

The signal processing required for a self-mixing laser diode displacement sensor under moderate feedback conditions has been detailed using the Nelder-Mead simplex method, which has resulted in the optimised joint estimation of the coupling coefficient characterizing the feedback regime and the line-width enhancement factor of the laser diode.

The optimization has led to decrease the number of LD optical power samples needed as well as in the reduction of the computing time for the signal processing of the said sensor.

Further more, the influence of the λ on the implementation of the Nelder Mead Method for the optimisation of C and θ in a moderate feedback self mixing laser diode sensor has been detailed.

It has an influence on the number of objective function calls needed to optimise the parameters and is also important in order to obtain a high contraction to expansion ratio as well as on the reduction of the percentage of the Shrink step in a general Nelder Mead iteration.

Thus, by choosing a favourable value of λ, considerable acceleration in the total needed to converge to the optimised parameters can be achieved.

The overall adapted NM algorithm has been tested using StateFlow® and has been coded in VHDL to ascertain the order of delay to be expected for the time duration needed to converge.

Methods for parallelism needed at the objective function level have also been outlined to reduce the delay at the cost of additional hardware resources.

References

1. T. Bosch, C. Bes, L. Scalise and G. Plantier, "Optical feedback interferometry", *the Encyclopedia of Sensors*, American Scientific Publisher, USA, vol. 7, pp 107–127 (2006).
2. C. Bes, G. Plantier and T. Bosch, "Displacement Measurements Using a Self-mixing Laser Diode Under Moderate Feedback", IEEE Trans. on Instrum. and Meas., vol. 55 (4), pp 1101–1105 (2006).
3. G. Plantier, C. Bes and T. Bosch, "Behavioral model of a self-mixing laser diode sensor", IEEE J. Quant. Electron., vol. 41 (9), pp 1157–1167 (2005).
4. C. Bes, T. Bosch, G. Plantier and F. Bony, "Characterisation of a self-mixing displacement sensor under moderate feedback", Opt. Eng., vol. 45 (8), pp 084402-1–084402-6 (2006).
5. A. Doncescu, C. Bes and T. Bosch, "Displacement Estimation with an Optical Feedback Interferometer Using an Evolutionary Algorithm". Proceedings IEEE Sensors, Atlanta, to be published (2007).
6. J.A. Nelder and R. Mead, "A simplex method for function minimization.", Comput. J. Vol. 7, 308–313, (1965).
7. JC Lagarias, JA Reeds, MH Wright, PE Wright, "Convergence properties of the Nelder-Mead simplex algorithm in low dimensions." SIAM Journal on Optimization, Vol. 9, No. 1, pp. 112–147, (1998).
8. DM Olsson and LS Nelson, "The Nelder-Mead Simplex Procedure for Function Minimization.", Technometrics, Vol. 17(1), 45–51. February, (1975).

Index

MIX
Papier aus verantwortungsvollen Quellen
Paper from responsible sources
FSC® C105338

If you have any concerns about our products,
you can contact us on
ProductSafety@springernature.com

In case Publisher is established outside the EU,
the EU authorized representative is:
Springer Nature Customer Service Center GmbH
Europaplatz 3, 69115 Heidelberg, Germany

Printed by Libri Plureos GmbH
in Hamburg, Germany